Communications in Computer and Information Science 652

Commenced Publication in 2007
Founding and Former Series Editors:
Alfredo Cuzzocrea, Dominik Ślęzak, and Xiaokang Yang

More information about this series at http://www.springer.com/series/7899

Michael W. Berry · Azlinah Hj. Mohamed
Bee Wah Yap (Eds.)

Soft Computing in Data Science

Second International Conference, SCDS 2016
Kuala Lumpur, Malaysia, September 21–22, 2016
Proceedings

 Springer

Editors
Michael W. Berry
University of Tennessee
Knoxville, TN
USA

Azlinah Hj. Mohamed
Universiti Teknologi MARA
Shah Alam
Malaysia

Bee Wah Yap
Faculty of Computer and Mathematical
  Sciences
Universiti Teknologi MARA
Shah Alam
Malaysia

ISSN 1865-0929   ISSN 1865-0937 (electronic)
Communications in Computer and Information Science
ISBN 978-981-10-2776-5   ISBN 978-981-10-2777-2 (eBook)
DOI 10.1007/978-981-10-2777-2

Library of Congress Control Number: 2016952875

Printed on acid-free paper

This Springer imprint is published by Springer Nature
The registered company is Springer Nature Singapore Pte Ltd.
The registered company address is: 152 Beach Road, #22-06/08 Gateway East, Singapore 189721, Singapore

Preface

We are pleased to present the proceeding of the Second International Conference on Soft Computing in Data Science 2016 (SCDS 2016). SCDS 2016 was held in Pullman Kuala Lumpur Bangsar, Malaysia during September 21–22, 2016. The theme of the conference was "Science in Analytics: Harnessing Data and Simplifying Solutions." Data science can improve corporate decision-making and performance, personalize medicine and health-care services, and improve the efficiency and performance of organizations. Data science and analytics plays an important role in various disciplines including business, medical and health informatics, social sciences, manufacturing, economics, accounting, and finance.

SCDS 2016 provided a platform for discussions on leading-edge methods and also addressed challenges, problems, and issues in machine learning in data science and analytics. The role of machine learning in data science and analytics is significantly increasing in every field from engineering to life sciences, and with advanced computer algorithms, solutions for complex real problems can be simplified. For the advancement of society in the 21st century, there is a need to transfer knowledge and technology to industrial applications so as to solve real-world problems. Research collaborations between academia and industry can lead to the advancement of useful analytics and computing applications to facilitate real-time insights and solutions.

We are delighted that the conference attracted submissions from a diverse group of national and international researchers. We received a total of 66 paper submissions. SCDS 2016 utilized a double-blind review procedure. All accepted submissions were assigned to at least three independent reviewers (at least one international reviewers) in order to have a rigorous and convincing evaluation process. A total of 57 international and 57 local reviewers were involved in the review process. The conference proceeding volume editors and Springer's CCIS Editorial Board made the final decisions on acceptance, with 27 of the 66 submisssions (41 %) published in the conference proceedings.

We would like to thank the authors who submittted manuscripts to SCDS 2016. We thank the reviewers for voluntarily spending time to review the papers. We thank all Conference Committee members for their tremendous time, ideas, and efforts in ensuring the success of SCDS 2016. We also wish to thank the Springer CCIS Editorial Board, organizations, and sponsors for their continuous support. To all our sponsors, we are deeply grateful for your continuous support.

We sincerely hope that SCDS 2016 provided a venue for knowledge sharing, publication of good research findings, and new research collaborations. Last but not least, we hope everyone gained from the keynote, special, and parallel Sessions, and had an enjoyable and memorable experience at SCDS 2016 and in Malaysia.

September 2016

Michael W. Berry
Azlinah Hj. Mohamed
Bee Wah Yap

Conference Organization

SCDS 2016 was organized by the Faculty of Computer and Mathematical Sciences, Universiti Teknologi MARA in collaboration with the Advanced Analytics Engineering Centre, Universiti Teknologi MARA, University of Tennessee, and Data Analytics and Collaborative Computing Group, University of Macau.

Patron

Hassan Said   Vice Chancellor, Universiti Teknologi MARA, Malaysia

Honorary Chairs

Azlinah Mohamed	Universiti Teknologi MARA, Malaysia
Michael W. Berry	University of Tennessee, USA
Yasmin Mahmood	Multimedia Digital Economy Corporation, Malaysia
Lotfi A. Zadeh	University of California, Berkeley, USA
Fazel Famili	University of Ottawa, Canada
Tan Kay Chen	National University of Singapore, Singapore

Conference Chairs

Bee Wah Yap	Universiti Teknologi MARA, Malaysia
Simon Fong	University of Macau, SAR China

International Scientific Committee Chair

Siti Zaleha Zainal Abidin   Universiti Teknologi MARA, Malaysia

Secretary

Nurin Haniah Asmuni   Universiti Teknologi MARA, Malaysia

Secretariat

Norimah Abu Bakar	Universiti Teknologi MARA, Malaysia
Mastura Jaini	Universiti Teknologi MARA, Malaysia

Finance Committee

Sharifah Aliman	Universiti Teknologi MARA, Malaysia
Zalina Zahid	Universiti Teknologi MARA, Malaysia

Azizah Samsudin	Universiti Teknologi MARA, Malaysia
Salina Yusoff	Universiti Teknologi MARA, Malaysia

Technical Program Committee

Shuzlina Abdul Rahman	Universiti Teknologi MARA, Malaysia
Norizan Mat Diah	Universiti Teknologi MARA, Malaysia
Shuhaida Shuhidan	Universiti Teknologi MARA, Malaysia
Haslizatul Fairuz Mohamed	Universiti Teknologi MARA, Malaysia

Sponsorship Committee

Rokiah Embong	Universiti Teknologi MARA, Malaysia
Khairil Iskandar Othman	Universiti Teknologi MARA, Malaysia
Salmah Abdul Aziz	Universiti Teknologi MARA, Malaysia
Stan Lee	The National ICT Association of Malaysia
Ong Kian Yew	The National ICT Association of Malaysia
Naquib Tajudin	The National ICT Association of Malaysia

Logistics Committee

Azhar Abdul Aziz	Universiti Teknologi MARA, Malaysia
Hamdan Abdul Maad	Universiti Teknologi MARA, Malaysia
Norimah Abu Bakar	Universiti Teknologi MARA, Malaysia
Mudiana Mokhsin@Misron	Universiti Teknologi MARA, Malaysia
Rizauddin Saian	Universiti Teknologi MARA, Malaysia
Muhd Ridwan Mansor	Universiti Teknologi MARA, Malaysia
Abdul Jamal Mat Nasir	Universiti Teknologi MARA, Malaysia

Publication Committee (Program Book)

Nur Atiqah Sia Abdullah	Universiti Teknologi MARA, Malaysia
Anitawati Mohd Lokman	Universiti Teknologi MARA, Malaysia
Zolidah Kasiran	Universiti Teknologi MARA, Malaysia
Syaripah Ruzaini Syed Aris	Universiti Teknologi MARA, Malaysia
Nor Ashikin Mohd Kamal	Universiti Teknologi MARA, Malaysia
Norkhushaini Awang	Universiti Teknologi MARA, Malaysia
Rosanita Adnan	Universiti Teknologi MARA, Malaysia

Website Committee

Mohamad Asyraf Abdul Latif	Universiti Teknologi MARA, Malaysia
Wael Mohamed Shaher Yafooz	Universiti Teknologi MARA, Malaysia

Publicity Committee

Farok Azmat	Universiti Teknologi MARA, Malaysia
Mazani Manaf	Universiti Teknologi MARA, Malaysia
Wael Mohamed Shaher Yafooz	Universiti Teknologi MARA, Malaysia
Muhamad Ridwan Mansor	Universiti Teknologi MARA, Malaysia
Ezzatul Akmal Kamaru Zaman	Universiti Teknologi MARA, Malaysia

Corporate Committee

Nuru'l-'Izzah Othman	Universiti Teknologi MARA, Malaysia
Rosma Mohd Dom	Universiti Teknologi MARA, Malaysia
Jamaliah Taslim	Universiti Teknologi MARA, Malaysia
Nurshahrily Idura Hj Ramli	Universiti Teknologi MARA, Malaysia
Nor Khalidah Mohd Aini	Universiti Teknologi MARA, Malaysia

Booth Exhibition Committee

Mohd Nazrul Mohd-Amin	Universiti Teknologi MARA, Malaysia
Sharifah Nazatul Shima Syed Mohamed Shahruddin	Universiti Teknologi MARA, Malaysia
Arwin Idham Mohamad	Universiti Teknologi MARA, Malaysia
Zahrul Azmir Absl Kamarul Adzhar	Universiti Teknologi MARA, Malaysia
Norkhairunnisa Mohamed Redzwan	Universiti Teknologi MARA, Malaysia

Media/Photography/Montaj Committee

Nazrul Azha Mohamed Shaari	Universiti Teknologi MARA, Malaysia
Kamarul Ariffin Hashim	Universiti Teknologi MARA, Malaysia
Sahifulhamri Sahdi	Universiti Teknologi MARA, Malaysia

Registration Committee

Fuziyah Ishak	Universiti Teknologi MARA, Malaysia
Muhd Azuzuddin Muhran	Universiti Teknologi MARA, Malaysia

Activities and Tour Committee

Saiful Farik Mat Yatin	Universiti Teknologi MARA, Malaysia
Mohd Zairul Masron	Universiti Teknologi MARA, Malaysia

Student Committee

Nur Syahida Zamry	Universiti Teknologi MARA, Malaysia
Hezlin Aryani Abd Rahman	Universiti Teknologi MARA, Malaysia
Ainur Amira Kamaruddin	Universiti Teknologi MARA, Malaysia
Hamzah Abdul Hamid	Universiti Teknologi MARA, Malaysia
Mohd-Yusri Jusoh	Universiti Teknologi MARA, Malaysia
Muhammad Hafidz Bin Anuwar	Universiti Teknologi MARA, Malaysia
Nur'aina Binti Daud	Universiti Teknologi MARA, Malaysia

Events Committee

Noor Latifah Adam	Universiti Teknologi MARA, Malaysia
Nurshahrily Idura Hj Ramli	Universiti Teknologi MARA, Malaysia
Ruhaila Maskat	Universiti Teknologi MARA, Malaysia
Sofianita Mutalib	Universiti Teknologi MARA, Malaysia
Faridah Abdul Halim	Universiti Teknologi MARA, Malaysia
Noor Asiah Ramli	Universiti Teknologi MARA, Malaysia
Sayang Mohd Deni	Universiti Teknologi MARA, Malaysia
Norshahida Shaadan	Universiti Teknologi MARA, Malaysia

International Scientific Committee

Mohammed Bennamoun	University of Western Australia, Australia
Yasue Mitsukura	Keio University, Japan
Dhiya Al-Jumeily	Liverpool John Moores University, UK
Do van Thanh	NTNU — Norwegian University of Science and Technology, Norway
Sridhar (Sri) Krishnan	Ryerson University, Toronto, Canada
Dariusz Krol	University of Wroclaw, Poland
Richard Weber	University of Chile, Santiago, Chile
Jose Maria Pena	Technical University of Madrid, Spain
Adel Al-Jumaily	University of Technology, Sydney, Australia
Ali Selamat	Universiti Teknologi, Malaysia
Daud Mohamed	Universiti Teknologi MARA, Malaysia
Yusuke Nojima	Osaka Perfecture University, Japan
Norita Md Nawawi	Universiti Sains Islam, Malaysia
Siddhivinayak Kulkarni	Universiti of Ballarat, Australia
Tahir Ahmad	Universiti Teknologi, Malaysia
Mohamad Shanudin Zakaria	Universiti Pertahanan, Malaysia
Ku Ruhana Ku Mahamud	Universiti Utara, Malaysia
Azuraliza Abu Bakar	Universiti Kebangsaan, Malaysia
Jasni Bt. Mohamad Zain	Universiti Malaysia Pahang, Malaysia
Ong Seng Huat	Universiti Malaya, Malaysia
Sumanta Guha	Asian University of Technology, Thailand

International Reviewers

Afshin Shaabany	University of Fasa, Iran
Akhil Jabbar Meerja	Vardhaman College of Engineering, India
Alaa Aljanaby	University of Nizwa, Oman
Albert Guvenis	Bogazici University, Turkey
Ali Hakam	UAE University, UAE
Amitkumar Manekar	KL University, India
Anant Baijal	Samsung Electronics, Korea
Anas AL-Badareen	Jerash University, Jordan
Ayan Mondal	Indian Institute of Technology, Kharagpur, India
Bashar Al-Shboul	The University of Jordan, Jordan
Bidyut Mahato	ISM Dhanbad, India
Byeong-jun Han	Korea University, Korea
Chih-Ming Kung	Shih Chien University, Taiwan
Chikouche Djamel	University of M'sila, Algeria
Deepti Theng	G. H. Raisoni College of Engineering, India
Dhananjay Singh	Hankuk University of Foreign Studies, Korea
Dhiya Al-Jumeily	Liverpool John Moores University, UK
Durga Prasad Sharma	AMUIT, MOEFDRE & External Consultant (IT) & Technology Transfer, USA
Ensar Gul	Marmara University, Turkey
Gang Liu	Fortemedia, USA
Gerasimos Spanakis	Maastricht University, The Netherlands
Gnaneswar Nadh Satapathi	National Institute of Technology, Karnataka, India
Habib Rasi	Shiraz University of Technology, Iran
Hanaa Ali	Zagazig University, Egypt
Harihar Kalia	Seemanta Engineering College, India
J. Vimala Jayakumar	Alagappa University, India
Jeong Mook Lim	Electronics and Telco. Research Institute, Korea
JIanglei Yu	Google, USA
July Díaz	Universidad Distrital Francisco José de Caldas, Colombia
K.V. Krishna Kishore	Vignan University, India
Karim Al-Saedi	University of Mustansiriyah, Iraq
Ke Xu	Dell Inc., USA
Komal Batool	National University of Science and Technology, Pakistan
Michael Berry	University of Tennessee, USA
Mohamed Geaiger	Sudan University of Science and Technology, Sudan
Muhammad Malik	National Textile University Pakistan, Pakistan
Neelanjan Bhowmik	University of Paris-Est, France
Nikisha Jariwala	VNSGU, India
Noriko Etani	Kyoto University, Japan
Prashant Verma	Amazon, USA
Raghvendra Kumar	LNCT Group of Colleges, India

Ranjan Dash	College of Engineering and Technology, India
Raunaq Vohra	BITS Pilani KK Birla Goa Campus, India
Rodrigo Campos Bortoletto	São Paulo Federal Institute of Education, Science and Technology, Brazil
Rohit Gupta	Thapar University, India
Rommie Hardy	Army Research Lab, USA
S. Kannadhasan	Raja College of Engineering and Technology, India
S. Sridevi	Dhirajlal Gandhi College of Technology, India
Shadi Khalifa	Queen's University, Canada
Siripurapu Sridhar	Lendi Institute of Engineering and Technology, India
Tousif Khan Nizami	Indian Institute of Technology Guwahati, India
Vijendra Singh	Mody Institute of Technology and Science, India
Vinay Sardana	Abu Dhabi Transmission and Despatch Company, UAE
Vyza Yashwanth Reddy	National Institute of Technology, Rourkela, India
Winai Jaikla	King Mongkut's Institute of Tech. Ladkrabang, Thailand
Xiao Liang	Shanghai Fulxin Intell. Transp. Sol. Co. Ltd., P.R. China
Yashank Tiwari	JSS Academy for Technical Education, India

Local Reviewers

Ahmad Shahrizan Abdul Ghani	TATI University College, Malaysia
Ajab Bai Akbarally	Universiti Teknologi MARA, Malaysia
Asmala Ahmad	Universiti Teknikal Malaysia Melaka, Malaysia
Azilawati Azizan	UiTM Perak, Malaysia
Azlan Iqbal	Universiti Tenaga Nasional, Malaysia
Azlinah Mohamed	Universiti Teknologi MARA, Malaysia
Bahbibi Rahmatullah	Universiti Pendidikan Sultan Idris, Malaysia
Bhagwan Das	Universiti Tun Hussein Onn Malaysia, Malaysia
Bong Chih How	Universiti Malaysia Sarawak, Malaysia
Chin Kim On	Universiti Malaysia Sabah, Malaysia
Daud Mohamad	Universiti Teknologi MARA, Malaysia
Hamidah Jantan	UiTM Terengganu, Malaysia
Haslizatul Mohamed Hanum	Universiti Teknologi MARA, Malaysia
Kamarularifin Abd Jalil	Universiti Teknologi MARA, Malaysia
Leong Siow Hoo	Universiti Teknologi MARA, Malaysia
Marina Yusoff	Universiti Teknologi MARA, Malaysia
Mas Rina Mustaffa	Universiti Putra Malaysia Malaysia
Mazani Manaf	Universiti Teknologi MARA, Malaysia
Mazidah Puteh	UiTM Terengganu, Malaysia
Mohd Izani Mohamed Rawi	Universiti Teknologi Malaysia, Malaysia
Mohd-Shahizan Othman	Universiti Utara Malaysia, Malaysia

Mohd-Shamrie Sainin	Universiti Sains Malaysia, Malaysia
Mohd-Tahir Ismail	Universiti Teknologi MARA, Malaysia
Muthukkaruppan Annamalai	Universiti Teknologi MARA, Malaysia
Nasiroh Omar	Universiti Teknologi MARA, Malaysia
Natrah Abdullah	Universiti Teknologi MARA, Malaysia
Noor Elaiza Abd Khalid	Universiti Teknologi MARA, Malaysia
Nor Ashikin Mohamad Kamal	Universiti Teknologi Malaysia, Malaysia
Nor Azizah Ali	Universiti Teknologi MARA, Malaysia
Nor Hashimah Sulaiman	Universiti Teknologi MARA, Malaysia
Noraini Seman	Universiti Sains Islam Malaysia, Malaysia
Norita Md Norwawi	Universiti Teknologi MARA, Malaysia
Norizan Mat Diah	Universiti Teknologi MARA, Malaysia
Nur Atiqah Sia Abdullah	Universiti Teknologi MARA, Malaysia
Nurin Haniah Asmuni	Universiti Teknologi MARA, Malaysia
Nuru'l-'Izzah Othman	Universiti Malaysia Perlis, Malaysia
Rafikha Aliana A Raof	Universiti Malaysia Sabah, Malaysia
Rayner Alfred	Universiti Tenaga Nasional, Malaysia
Ridha Omar	Universiti Teknologi MARA, Malaysia
Rizauddin Saian	UiTM Perlis, Malaysia
Rogayah Abd. Majid	Universiti Teknologi Malaysia, Malaysia
Roselina Sallehuddin	Universiti Teknologi MARA, Malaysia
Ruhaila Maskat	Universiti Kebangsaan Malaysia, Malaysia
Saidah Saad	Universiti Teknologi MARA, Malaysia
Shaharuddin Cik Soh	Universiti Teknologi MARA, Malaysia
Sharifah Aliman	Universiti Teknologi MARA, Malaysia
Shuhaida Shuhidan	Universiti Teknologi MARA, Malaysia
Shuzlina Abdul-Rahman	Universiti Teknologi MARA, Malaysia
Siti Sakira Kamaruddin	Universiti Utara Malaysia, Malaysia
Sofianita Mutalib	Universiti Teknologi MARA, Malaysia
Sulfeeza Mohd Drus	Universiti Tenaga Nasional, Malaysia
Sumarni Abu Bakar	Universiti Teknologi MARA, Malaysia
Waidah Ismail	Universiti Sains Islam Malaysia, Malaysia
Bee Wah Yap	Universiti Teknologi MARA, Malaysia
Yuhanis Yusof	Universiti Utara Malaysia, Malaysia
Yun Huoy Choo	Universiti Teknikal Malaysia Melaka, Malaysia
Zaidah Ibrahim	Universiti Teknologi MARA, Malaysia

Organizers

Organized by

UNIVERSITI
TEKNOLOGI
MARA

Hosted by

Faculty of Computer and Mathematical Sciences, UiTM Selangor

In Co-operation with

THE UNIVERSITY OF
TENNESSEE
KNOXVILLE

Data Analytics and Collaborative Computing Group

University of Macau

PIKOM

Springer

Sponsoring Institutions

Gold Sponsor

FUSIONEX®
experience.excellence

Silver Sponsors

BANK ISLAM

MDEC°
Driving Transformation

Microsoft

Bronze Sponsors

Contents

Fuzzy Logic

Information and Sentiment Analytics

Artificial Neural Networks

Shallow Network Performance in an Increasing Image Dimension

Mohd Razif Shamsuddin[(✉)], Shuzlina Abdul-Rahman, and Azlinah Mohamed

Faculty of Computer Sciences and Mathematics, Universiti Teknologi MARA, Shah Alam, Selangor, Malaysia
{razif,shuzlina,azlinah}@fskm.uitm.edu.my

Abstract. This paper describes the performance of a shallow network towards increasing complexity of dimension in an image input representation. This paper will highlight the generalization problem in Shallow Neural Network despite its extensive usage. In this experiment, a backpropagation algorithm is chosen to test the network as it is widely used in many classification problems. A set of three different size of binary images are used in this experiment. The idea is to assess how the network performs as the scale of the input dimension increases. In addition, a benchmark MNIST handwritten digit sampling is also used to test the performance of the shallow network. The result of the experiment shows the network performance as the scale of input increases. The result is then discussed and explained. From the conducted experiments it is believed that the complexity of the input size and breadth of the network affects the performance of the Neural Network. Such results can be as a reference and guidance to people that is interested in doing research using backpropagation algorithm.

Keywords: Neural Network · Shallow network · Backpropagation · Image recognition

1 Introduction

The complexity of data in the forthcoming age is increasing rapidly, consistent with the advances of new technology. The trends of user behaviour intently show the rapid growth of information in this new era. This information may consist of high dimensional imagery, GPS location, textual data and etc. While Neural Network (NN) is becoming a popular trend, it may be a necessity in the near future. Many significant corporations and big social media have already made their move towards incorporating Artificial Intelligence (AI) in their product. These entities are either focused on creating their research teams or buying AI assets of their own. Most of the research and development that were done by these entities would involve a learning system. Thus, an experiment is done in order to address the relation of high complex data to NN performance.

This experiment is done to analyse the effects of high complexity data to NN learning algorithm performance. Backpropagation Neural Network (BPNN) is chosen as the learning algorithm as it is currently applied in numerous applications ranging from imagery recognition [1], acoustic classification [2] to financial data analysis [3].

© Springer Nature Singapore Pte Ltd. 2016
M.W. Berry et al. (Eds.): SCDS 2016, CCIS 652, pp. 3–12, 2016.
DOI: 10.1007/978-981-10-2777-2_1

Although BPNN has been traced to be introduced long ago in the early 19th Century, it became more relevant and critical with the evolution of the current technology. BPNN has been known to work well to handle pattern recognition and classification, however there are several issues needs to be attended in order to work well with the current technological standards. This experiments aims to analyse the effects of increasing length of the dimension of imagery data in effect to the BPNN learning performance. Further details will be discussed in the next subsection.

1.1   Overview of Neural Network

Inspired from a biologically analytical model, NN has the aptitude to learn linear and complex non-linear relationships. Similar to the function of an animal biological neuron, theoretically, a Multi Layered NN is considered as general approximators [4, 5]. Due to its capability as approximators it is used in numerous applications, such as perception [6, 7], reasoning [8], intelligent control [9, 10], digital forensics [11] and similar intelligent behaviour. In addition, according to [5], NN is most widely and still applied for problem solving in pattern recognition, data analysis, control and clustering.

A single layer NN is proven to be able to classify a simple linear problem. According to [12] a single layer network has one layer of connection weights. It has two type of layer which consist number of nodes which are input layer and output layer. One of the popular architecture of single layer NN is the Perceptron model. Although the perceptron have been proven to emulate a learning capability, it is known that the perceptron could not solve non-linearly separable functions. The perceptron could not cater for the complex real world problems and may not be suitable for large scale data mining. However, this problem can be solved using multi-layer perceptron or multi-layer neural network.

1.2   Shallow Network

Although multilayer Neural Network is a solution to linearly separable function problem, most of the earlier network is considered as shallow. Shallow Neural Network has been introduced for many decades, it all started when Warren McCulloch and Pitts [19] introduces their idea of calculus on nervous activity which uses the McCulloch and Pitts model. Some of the notable successful NN models were dated back in the 60's. Distinguished successful researchers published several key papers during that era, particularly Rosenblatt (1958), Widrow and Hoff (1962) and Grossberg (1969). At the early 70's the Artificial Intelligence community was introduced with more advanced neural network model.

Many notable NN Scientist such as Kohonen network, focuses on unsupervised networks. Early NN models that was introduced such as established by Palm [13] and Hopfield [14] has a network depth of one. This is due to its early age of research in neural network. Most computers cannot compute a network that has a higher depth than one layer. In the late 90's LeCun el al uses Backpropagation to show the capability of 3 layer BPNN on a digital handwritten number to show the capability of BPNN to categorized a given number sampling by LeCun [15–17]. However, most shallow network cannot cope with the demanding complexity issues in problem solving, [18]. Although shallow network has notably successfully been applied in many areas it needs more depth in order to compute higher abstract functions. This issue however, can probably be solved by using a Deep Networks. Figure 1, shows a typical architecture of a shallow NN.

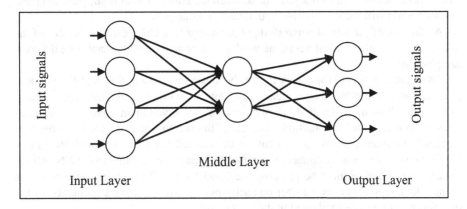

Fig. 1. Typical shallow neural network architecture

2 Challenges in Data Preparation

In order to test the performance of NN learning algorithm, three different data with different dimensional sizes are constructed. Each of the data categories is segmented into ten groups of classifications. The first data category is a set of ten numbers consisting of numbers varying from zero to nine with a dimension size of 5×7 pixels. The second and third data category is the same segmented data with a dimension size of 10×14 pixels and 15×28 pixels respectively. Each of the number images is formatted in binary image representation as shown in Fig. 2(a), (b) and (c).

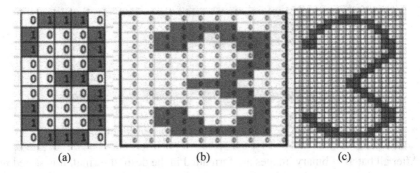

Fig. 2. (a) Binary image representation of 5×7 pixel dimension. (b) 10×14 pixel dimension. (c) 15×28 pixel dimension.

The idea of preparing three sets of input samples is to see the tested performance of the neural network in three different input sizes. This will simulate the increasing dimension size of the input image. The input size of the neural network will increase as it is fed with a higher image size. The computation of the network will also increase. This is because, the sum of dot product between the connection of input parameter and hidden neuron will increase as the input neuron is enlarged.

As the size of the neural network input parameter will change while the size of the image increases, the created neural network program will have to adapt with different sampling sets.

The representation of the binary images becomes smoother as the size of the dimension increases. Smoother images means more detail and less jagged edge. Higher image dimensions allow more detailed information to be preserved in an image. However, as the size increases the computational burden of the network is expected to increase as there will be more processing elements in the created network. Although the binary number images are readily created, it cannot immediately be applied to the NN. All the binary representation must be pre-processed and formatted to a batch of continuous string which represents one number on each rows of string. Further explanation of the data formatting will be explained in the next subsection.

2.1 Formatting the Data Sampling

The binary image must be represented in a continuous string of information in order to enable it to be fed to the neural network. One of the ways to represent the binary image is to concatenate each row of the image binary information to the consecutive rows of binary values. This concatenation process is reiterated until the last row of the binary image. Figure 3, shows how the data sampling are concatenated to a long string.

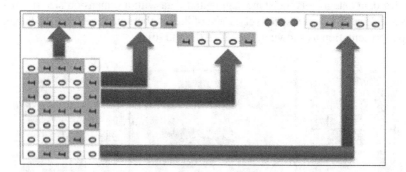

Fig. 3. Formatting the binary image to a continuous long binary string.

After all batch of binary images are formatted in the desired format, it is stored in a database ready to be read by the NN program. Each sets of number are stored in the same group of dimension size which later will be fed to different network architecture as it will have a different input size.

2.2   Designing the Network Architecture

The network architecture will determine the complexity of NN. As the number of neuron increases, so does its computation. This is due to the increased neurons that will need to cater with the increasing dimension size of the NN. In this study, six sets of neural network architecture is created to cater for different performance testing. Figure 4, shows an example of a network architecture which is consisted of 35 input neuron, 10 hidden neuron and 10 output neuron. The complete description of all the created network architecture is shown in Table 1.

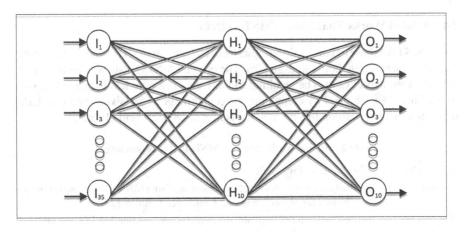

Fig. 4. Architecture of the created network (35 input neuron, 10 hidden neuron and 10 output neuron).

Table 1. List of created network architecture

Architecture	Input neuron	Hidden neuron	Output neuron
1	35	10	10
2	140	10	10
3	420	10	10
4	35	20	10
5	140	20	10
6	420	20	10

From the input datasets as shown in Fig. 2(a), (b) and (c), each datasets is supplied with two different network architectures. This is to test whether the increase of hidden neuron would give better performance in each test datasets. The idea of having an increased hidden neuron is to alter the network without having a deeper layer to the network architecture. According to Jeff [20] there are many ways of choosing the numbers of hidden neuron. In this experiment, the common method optimal size of hidden neuron is used. The optimal number is usually between the size of the input and size of the output layers. In our experiment, the minimum hidden neuron is 10.

2.3   MNIST as Additional Data Sampling

This study will also use benchmark data samples on the shallow NN architecture to test its performance. The learning algorithm will be tested with MNIST benchmark data. It is known that MNIST has been widely used to test NN performances [19] since [16, 17] introduces the handwritten digit image database sampling for Machine Learning. Many researches on Machine Learning and NN use MNIST sampling as benchmark to ease the comparison of the learning performance of their newly created algorithms or experiments.

2.4   Related Works That Uses MNIST as Data

The MNIST database of hand written digits has a training set of sixty thousand examples and testing sets of ten thousand examples. The original MNIST data was normalized to preserve their aspect ratio. The resulting MNIST data is a grey levels image that were centred in a 28 × 28 pixel by computing the center mass of the pixels, and translating the image so as to position this point at the center of 28 × 28 field (Table 2).

Table 2.   Similar works that uses MNIST data as benchmark

Author (Year)	Description of research
Angelo (2014)	Test of his algorithm using Neocognitron. Using MNIST sampling the author showed his algorithm has high tolerance to noise. [27]
Yangcheng (2013)	Test of several different algorithm performance using MNIST and other several benchmark data. [28]
Tapson (2013)	Indicate that there were issues reading high dimensions and big datasets. His research shows lower percentage of error of every digit when the hidden nodes are increased. [29]
Micheal (2013)	Tested an enhanced Spiking NN(SNN), and achieved 92 % of correct classification. [30]
Jung (2011)	Found out that CMOL Crossnet Classifier could not handle MNIST dataset because of its higher dimensions. [31]
Vladmir (2009)	Discusses a new learning paradigm over the classical ones. Added advanced learning algorithms must have some sort of supervised training with extra information. [32]

3   Results

The result of the conducted experiments is explained. The next subsection is divided to three experimental interpretations and conclusions. The first experiment is done with an initial minimal 10 hidden neuron. This method is used because the number of created output is 10 classifications. The next experiment is intended to show how a slight increase in the size of the hidden neuron will affect the performance of the network. Lastly MNIST sampling is used upon the last network to show how it works on the created network.

3.1   Result of First Experiments

The results of the experiments for 10 hidden neuron architecture shows a steep learning for both 5 × 7 image dimension and 10 × 14 image dimension. The learning progress is converging rapidly until the 2000 epoch and shows signs of stabilization. Meanwhile, for 15 × 28 image training shows a very slow progress of learning. This was indicated by the slow convergence of Sum of Squared Errors in the learning curve. Figure 5, shows the test results of the first three datasets. The specified graph shows that, as the complexity of the input increases the network will be affected in terms of its performance. A large gap of difference in terms of errors showed that the highest image pixels learning curve converges early. This is also an indication of overfitting. As indicated by Hinton [18] Neural Network is supposed to work as a generalization function. Overfitting will limit the generalization capability of the neural network.

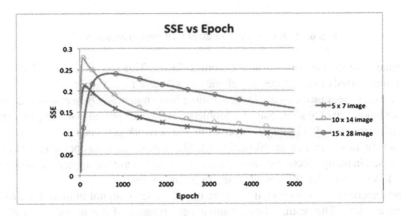

Fig. 5. SSE vs. Epoch for 10 hidden neuron architecture

From this experiment, it is shown that a NN with a shallow architecture will encounter generalization problem.

3.2   Result of the Second Shallow Network

The second experiments show similar results as the input size is increased. This is shown in Fig. 6. The second sets of experimental results for 20 hidden neuron architecture shows a steep learning for both 5 × 7 image dimension and 10 × 14 image dimension. Again, it shows that as the complexity of the input increases, the performance of the network decreases.

The learning progress is converging rapidly until the 1000 epoch and shows signs of stabilization. This time, both datasets converges and stabilizes with minimal differences. Meanwhile, for 15 × 28 image training again shows a very slow progress of learning. This was indicated by the slow convergence of Sum of Squared Errors in the learning curve as shown in Fig. 6. It is shown as the dimension sizes increases the

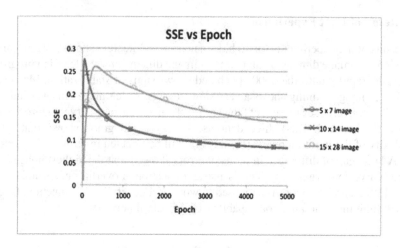

Fig. 6. SSE vs. Epoch for 20 hidden neuron architecture.

performance of the network learning becomes slower. This proves that a higher dimensional image needs more processing elements to cater for its complexity.

The experiment proves that NN needs more hidden neuron connections and maybe a deeper hidden layer to solve complex problems. It is also reported that backpropagation will suffer from training its earlier layer while the network gets deeper. This will be the drive for the next experiment. Nevertheless, Shallow network has been reported to be widely used in many successful cases. However, in most case the implementation of the network is not complex that requires deeper architectures.

The experiments is continued with the use of MNIST digital number handwriting benchmark data. The results shows similar performance of the neural network. The performance of the NN using MNIST datasets is explained next.

3.3 MNIST Experimental Results

Experiments on the MNIST data were prepared. The MNIST samples were trained using the minimum output method to determine the number of the hidden neuron.

The result of the experiment shows similar results from the previous experiments as the accuracy of the validation and testing of the NN shows a large gap from the accuracy of the training samples. This is an indication of overfitting. The results show that the NN could not recognize and classify some of the test and validation data. This may due to some of the test data have similar shapes but it is differentiated with a different classification. According to [21] the existence of many irrelevant and redundant features may lead to overfitting and less cost-effective models. Similarly looking shapes such as the number 3 and 8 datasets may be difficult for the network to learn. This is due to the nature of the shapes these numbers were written. There are too many styles of writing these two numbers and may make it looked too similar whereas it is a completely different number.

The solution to address the complex input pattern and similarity of datasets from different classification may need a more robust NN architecture to tackle this issue (Fig. 7).

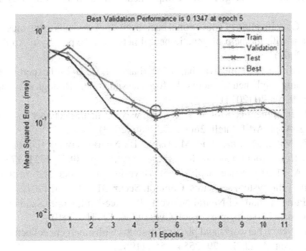

Fig. 7. MSE vs. Epoch for 20 hidden neuron architecture using MNIST benchmark dataset.

4   Conclusion

From the results of the experiments, it is shown that the complexity of the input greatly affects the performance of the network. Although the breadth of the architecture of the second network are wider, there is no significant improvement of performance. This concludes that a shallow network will have a decrease in performance as the input complexity increases.

This paper has contributed on how the complexity of input neuron affects the performance of a shallow network. Although it is shown that the performance is dropping as the complexity increases. It is hoped, this setback can be avoided by using deep network in future research.

References

1. Mona, M.M., Amr, B., Abdelhalim, M.B.: Image classification and retrieval using optimized pulse-coupled neural network. Expert Syst. Appl. **42**(11), 4927–4936 (2015)
2. Tara, N.S., Brian, K., George, S., Hagen, S., Abdel-rahman, M., George, D., Bhuvana, R.: Deep convolutional neural networks for large-scale speech tasks. Neural Netw. **64**, 39–48 (2014)
3. Iturriaga, F.J.L., Sanz, I.P.: Bankruptcy visualization and prediction using neural networks: a study of U.S. commercial banks. Expert Syst. Appl. **42**(6), 2857–2869 (2015)
4. Khosravi, A., Nahavandi, S., Creighton, D., Atiya, A.F.: Comprehensive review of neural network-based prediction intervals and new advances. IEEE Trans. Neural Netw. **22**(9), 1341–1356 (2011)

5. Kumar, K., Thakur, G.S.M.: Advanced applications of neural networks and artificial intelligence: a review. Int. J. Inf. Technol. Comput. Sci. **6**, 57–68 (2012)
6. Rolls, E.T., Deco, G.: Networks for memory, perception, and decision-making, and beyond to how the syntax for language might be implemented in the brain. Brain Res. **1621**, 316–334 (2014)
7. Hardalaç, F.: Classification of educational backgrounds of students using musical intelligence and perception with the help of genetic neural networks. Expert Syst. Appl. **36**(3, Part 2), 6708–6713 (2009)
8. Biswas, S.K., Sinha, N., Purakayastha, B., Marbaniang, L.: Hybrid expert system using case based reasoning and neural network for classification. Biologically Inspired Cogn. Architectures **9**, 57–70 (2014)
9. Farahani, M.: Intelligent control of SVC using wavelet neural network to enhance transient stability. Eng. Appl. Artif. Intell. **26**(1), 273–280 (2013)
10. Miljković, Z., Mitić, M., Lazarević, M., Babić, B.: Neural network reinforcement Learning for visual control of robot manipulators. Expert Syst. Appl. **40**(5), 1721–1736 (2013)
11. Khan, M.N.A.: Performance analysis of Bayesian networks and neural networks in classification of file system activities. Comput. Secur. **31**, 391–401 (2012)
12. Fausett, L.: Fundamentals of Neural Network. Prentice-Hall, Upper Saddle River (1994)
13. Palm, G.: On associative memory. Biol. Cybern. **36**, 19–31 (1980)
14. Hopfield, J.J.: Neural networks and physical systems with emergent collective computational abilities. Proc. Nat. Acad. Sci. **79**, 2554–2558 (1982)
15. LeCun, Y., Boser, B., Denker, J.S., Henderson, D., Howard, R.E., Hubbard, W., Jackel, L.D.: Back-propagation applied to handwritten zip code recognition. Neural Comput. **1**(4), 541–551 (1989)
16. LeCun, Y., Boser, B., Denker, J.S., Henderson, D., Howard, R.E., Hubbard, W., Jackel, L.D.: Handwritten digit recognition with a back-propagation network. In: Touretzky, D.S. (ed.) Advances in Neural Information Processing Systems, vol. 2, pp. 396–404 (1990)
17. LeCun, Y., Bottou, L., Bengio, Y., Haffner, P.: Gradient-based learning applied to document recognition. Proc. IEEE **86**(11), 2278–2324 (2006)
18. Hinton, G.E., Osindero, S., Teh, Y.-W.: A fast learning algorithm for deep belief nets. Neural Comput. **18**(7), 1527–1554 (2006)
19. Warren, S., Mcculloch, W.P.: A logical calculus of the ideas immanent in nervous activity **5**, 115–133 (1943)
20. Jeff, H.: Introduction to Neural Networks for Java, 2nd edn., pp. 1–440 (2008)
21. Abdul-Rahman, S., Bakar, A.A., Mohamed-Hussein, Z.-A.: An intelligent data pre-processing of complex datasets. Intell. Data Anal. **16**, 305–325 (2012)

Applied Neural Network Model to Search for Target Credit Card Customers

Jong-Peir Li[✉]

Department of Management Information Systems, National Chengchi University,
Taipei, Taiwan (R.O.C.)
johnli8139@gmail.com

Abstract. Many credit card businesses are no longer profitable due to antiquated and increasingly obsolete methods of acquiring customers, and as importantly, they followed suit when identifying ideal customers. The objective of this study is to identify the high spending and revolving customers through the development of proper parameters. We combined the back propagation neural network, decision tree and logistic methods as a way to overcome each method's deficiency. Two sets of data were used to develop key eigenvalues that more accurately predict ideal customers. Eventually, after many rounds of testing, we settled on 14 eigenvalues with the lowest error rates when acquiring credit card customers with a significantly improved level of accuracy. It is our hope that data mining and big data can successfully utilize these advantages in data classification and prediction.

Keywords: Credit card · Target customer · Data mining · Neural network

1 Introduction

More than a century has elapsed since banks introduced credit cards. In Taiwan, credit card development has undergone four crucial stages of development. Stage 1 marked a time when 3 banks dominated the market. Most cardholders were required to pay an annual fee. Such a fee subsequently became one of the primary sources of income for issuing banks. Stage 2 began in the 1990s, in which the CTBC Bank became the primary credit card issuer; its success in credit card and consumer financing-related businesses became the learning model for China's China Merchants Bank, which ultimately became one of the early issuers of credit cards in China.

Following the huge hype created by the media, credit cards entered into their third stage of development. During Stage 3, Taiwanese and foreign banks viewed credit cards and cash cards as highly profitable businesses, prompting them to introduce numerous marketing schemes (e.g., gifts and bonuses to customers for their first credit card purchase, airport pickups/drop offs, cashbacks, department store anniversary deals, co-branded cards, and balance transfers) to boost credit card issuance. The competition resulted in the waiver of annual fees, decline in credit quality, surge in revolving balance, and yearly drop in credit card interest, leading to the Credit Card and Cash Card Crisis in Taiwan, the fourth stage of credit card development.

© Springer Nature Singapore Pte Ltd. 2016
M.W. Berry et al. (Eds.): SCDS 2016, CCIS 652, pp. 13–24, 2016.
DOI: 10.1007/978-981-10-2777-2_2

Although the credit card business has grown rapidly, most of the new customers were not the ideal customers, which means they didn't spend or use the revolving credit line. Therefore per card revenue was much lower than per card cost resulting in a low breakeven percentage. This is a very crucial issue for banks because the incremental and incidental costs exceed the revenue resulting a non-profitable business. Traditionally, eigenvalues related to revolving lines of credit currently and commonly used included age, gender, education level, marital status (in which singles account for more than 50 %), and salary. However, banks that use only these variables and bank supervisors' judgments are inadequate to handle the influx of new customers in the Internet era. Banks possess adequate customer data, however they cannot accurately define and predict ideal customers [1]. This study aimed to quickly identify customers with revolving credit demands and issue them credit cards based on data mining techniques.

2 Literature Review

2.1 Credit Card Business in Taiwan

By approximately 2005, many credit card cardholders are in possession of multiple credit cards. Their credit card balance far exceeded their ability to repay. When these cardholders overspent, they simply applied for another credit card and used the new credit card to pay off old credit card balances. This ultimately engendered the Credit Card and Cash Card Crisis, in which credit card and cash card issuing banks experienced bad debts. A number of medium-sized banks were even forced to be sold to private funds, resulting in significant financial losses. This led to a change in credit card applications, in which a stringent credit checking process became the norm. Issuing banks avoided bad debts at all costs, even if it meant foregoing potential profits. As a result, most data mining-related literature after 2005 covered the topics of how to avoid bad debts as well as the early detection of bad debts [2, 3]. So very little literature concerning our topic of finding new customers using the 3 methods has been published after 2005.

However, according to statistics compiled by the Financial Supervisory Commission, in 2015, 36 banks still issued credit cards in Taiwan and the number of cards in circulation was 38 million. Questions thus arose concerning why banks continued to invest in the market as well as how many issuing banks actually made a profit in their credit card business. In reality, profits generated by credit card companies markedly declined since 2005. It is shown in Table 1 that, in 2005, the number of credit cards in circulation was 45 million and the revolving balance was NT$494.7 billion. In 2015, revolving balance diminished to $108.0 billion. This is a significant decline in the net interest income in the whole market and the possible negative trends for the credit card market. In addition, service charges from cash advances decreased by NT$6.7 billion during this period.

The good news is that credit card transaction volume increased from NT$1.4 trillion in 2005 to NT$2.2 trillion in 2015 according to the Financial Supervisory Commission, resulting in higher fee income also shown in Table 1.

The increase of fee income cannot off-set the decrease in the interest income. Further, banks invest considerable amounts of money in hardware, marketing, credit card

Table 1. Credit card interest, credit card transaction services charges, and cash advances in Taiwan. Source: FSC Banking Bureau.

	Revolving balance	Credit card transactions	Cash advance balance
2005	NT$494.7 billion	NT$1.4 trillion	NT$215 billion
2015	NT$108.0 billion	NT$2.2 trillion	NT$27.0 billion
Increase/Decrease	−NT$398.7 billion	+NT$800 billion	−NT$188.0 billion

promotion, bonuses, while customer rewards continue to rise, which significantly cut into the banks' pre-tax income.

Therefore we need a way to identify high spending and revolving customers as a way to maximize the profits of credit card business.

To answer increasingly fierce competition, substantial adjustments were made to the previous customer development model, which involved setting up booths in crowded locations and asking passers-by to apply for a credit card. In addition to challenges from banking industry competitors, credit card issuing banks face competition from online and microfinancing companies. The emergence of online banking is rapidly eroding profits that can made from the credit card and personal credit/loan markets. Because online companies enjoy an enormous customer base (e.g., Tencent has more than 600 million WeChat customers), revenue that can be generated from the introduction of payment and loan services by online companies is unmatched by traditional banks, which possess a limited customer base and relatively small business models. When faced with such tests, the banking industry must find effective contingency plans to elevate the profitability of credit cards. Banks should avoid issuing credit cards to customers who might never use them and only increase the banks' costs.

2.2 A Brief Introduction of the Neural Network

In 1943, Warren McCulloch and Walter Pitts proposed a mathematical model that imitated neurons which facilitated the birth of artificial intelligence-related studies. However, the model itself possessed no computational capability. In 1949, neurologists argued that the human brain learning process begins with neural synapses. In 1957, Rosenblatt combined the two aforementioned theories to create the first-generation perceptron, which could be used to identify text. However, in 1969, artificial intelligence scientist Marvin Minsky reported that perceptron was unable to solve the exclusive or gate (XOR) problem, driving the neural network to its "dark age" that lasted more than a decade. In 1982, John Hopfield invented the back propagation method. This method trains a neutral network to automatically correct its weights when processing data; the neural network then reprocesses the data until an optimal weight is found. Such a break-through re-popularized neural network research and resulted in the usage of neural networks in various domains (e.g., image recognition, stock market forecasts, and industrial engineering) over the last decade [4].

The steps of back propagation network are as follows:

(a) Set up the back propagation architecture by setting the parameters, such as number of layers, number of nodes, learning rates, and maximum number of learning;
(b) Use a computer to randomly select the first weight, while the user selects the activation function (e.g., hard limit, sign, linear S-type, and hyperbolic tangent) that best represents the revolver sign function;
(c) Calculate the forward propagation output transmitted to the hidden layer; enter the eigenvalue (i.e., X) of various data [5]. The eigenvalues are used to determine whether a customer will become a revolver. X is multiplied with the weight automatically and randomly selected by a computer before being added to form the hidden layer output (i.e., Z). Next, calculate the hidden and output layers. Here, the Z value becomes the input value of the next layer and is multiplied with the weight randomly selected by the computer before being added. Using the sign function, calculate the Y value. Compare the Y value with the target value d (where a d value of 1 signifies a revolver). However, if the differences are extensive, the back propagation process is initiated;
(d) Identify ways for the error function E [3] to diminish error(s) between target and output values (i.e., a process that minimizes the error function), which is referred to as the gradient descent learning method (1);

$$E = \frac{1}{2} \sum_{k=1}^{q} (d_k - y_k)^2 \tag{1}$$

(e) Calculate intervals; compute the partial derivative of the weight W using E to produce the interval δ. Subsequently, readjust the weight;
(f) Calculate the correction; back propagation returns to the area between the original hidden layer and output layer. Modify the weight before entering it back to the hidden layer. Revise the weight until the optimal weight combination is found. Changes in weight Δw = learning rate η * error value δ * input value.
(g) Update weights: Return to Steps c) to g) until convergence or preset maximum learning number/time is reached.

2.3   Related Works

This section will detail two critical data processing procedures, which are data organization and data classification. Concerning data organization, it is asserted that because bank data are multidimensional, data preprocessing must be made [6]. The data preprocessing quality has a significant effect on study results. For example, ATM data provides information regarding customers' cash usage habits, whereas online bill payment (OBP) offers insight into customers' payment behavior. However, if all the data were used as variables, the operation would become markedly tedious; the transaction data comprised 13 data source-related categories, which are automated clearing house (ACH), adjustments to account (ADJ), automated teller machine (ATM), bill payment (BPY), credit card (CC), check (CHK), debit card (DC), account fees (FEE), interest on account (INT), wire transfer (WIR), internal transfer (XFR), miscellaneous deposit or withdrawal

(MSC), and unidentifiable channel (BAD) [6]. These 13 categories can be further divided into 138 subcategories. Therefore, identifying variables that are useful to a study during the preprocessing stage becomes critical.

Regarding data classification, it is indicated that most neural network and credit card-related studies focus on credit analysis scoring and seldom discuss existing clients' consumer behavior [7, 8]. Therefore, the clients were classified into three categories, namely, convenience users, transactors, and revolvers. In addition, a neural-based behavioral scoring model was devised to improve the management strategies of customer types [7]. It is recommended that customer data be divided into account information and transaction information in this study [7]. The clients were divided into three categories (i.e., inactive clients, interest-paying clients, and noninterest-paying clients) and employed the linear discriminant analysis and logistic regression model to identify whether clients were revolvers [9]. However, because variables in this analysis and model were demographic data-type variables, identifying customers thus became exceedingly difficult.

Numerous studies published around the golden age of credit card development of 2005 investigated the use of data mining techniques in the credit card market [10, 11]. However, the credit card market underwent considerable changes between the years 2005–2015: (a) the Credit Card and Cash Card Crisis: many banks that issued credit card and cash card suffered considerable losses. Credit information centers began to build databases for personal credit information, marking the end of an era in which people regularly owned dozens of bank cards and used new credit cards to pay off old credit card balances. Balance transfers were also highly regulated, in which a person's total credit card limit could not exceed 22 times their salary; (b) the subprime mortgage crisis in 2008: following the global economic crisis, banks stopped adopting lenient credit policies for individual customers. This resulted in revolvers' inability to borrow funds from banks, which forced them to borrow money from underground institutions; and (c) the Internet era: online companies employed advanced scoring systems and customer analyses, such as the know your customer (KYC) technique to acquire information needed by the companies to offer their core products and/or services to customers, whereas banks still use traditional and time consuming methods to gather customers' information. Customers subsequently compared the two service providers and ultimately chose online companies because of the level of convenience they provided. Instead of generating new business and acquiring customers, Taiwan data mining literature in the past 10 years primarily focused on bad debt avoidance.

3 Methodology

3.1 Defining Target Customers

In general, credit card customers can be divided into the following categories: (1) customers who normally do not use credit cards and only use them when banks are offering promotions and/or when customers can receive superior discounts or bonuses; (2) customers who use credit cards and who pay the balance at the end of the month so they do not use revolving credit; (3) credit card users who use revolving credit. These users can afford to pay the credit card debt, but need time to pay it back; and (4) credit card users who become bad debt customers. These customers cannot afford to pay back the credit card debts and credit card interest. The third and fourth types of customers look very similar on the surface. However, the revolvers do have the ability to pay the money back. It is important to differentiate the revolvers from the bad debt customers.

3.2 Collecting and Classifying Eigenvalues of Data

Once the target customer was determined, the second step was to collect the eigenvalues of the data. The most important step before analysis was to organize the data [12]. Banks currently have many different types of data that are similar to those described in a previous study [6]. First, we selected revolvers and non-revolvers. We then made the personal data untraceable and organized data and put into two tables. Table 2 shows the users' personal data, which are organized together with their credit status. Table 3 shows customers' purchasing behavior and credit card usage conditions. Data eigenvalues were then examined individually. For example, spending NT$10,000 using the credit card at one instance and slowly accumulating revolving credit is different from spending NT $100,000 once a year. These two items have different eigenvalues. However, accumulating more than NT$5,000 in credit card debt is not much different than accumulating more than NT$10,000 in credit debt, so these two do not require differentiated eigenvalues. Credit cards used in supermarkets to buy daily supplies is different from credit cards used in department stores to buy clothes and cosmetics. Data after initial classification are shown below: personal data had 11 eigenvalues and transaction data had 48 eigenvalues.

Table 2. Demographic data-based eigenvalues.

Data type	Data eigenvalues
Age	
Sex	Male Female
Marital status	Married/not married/divorced/other/ Number of dependents
Education level	
Annual income	
Real estate	House with no mortgage/house with mortgage/individual rental/rental with other/house guest
Profession	Agriculture/manufacturing/transportation and logistics/retail/service industry/financial industry/medical industry/accounting industry/lawyer/military, public servant, or teacher/student/retired/housekeeper/freelancer/other
Title	Chairman of the board/general manager/vice president/manager/other
Total asset	
Savings	Checking account/savings account/fixed rate deposit
Loans	Personal credit and loan/mortgage/credit cards/private loans

Table 3. Eigenvalue of transaction information.

Did the customer use cash advance? (3 items)
Did the customer pay in installments? (3 items)
Did the customer use preauthorized payment?
Credit card limit
Number of credit card transactions
Credit card transactions (in NT$) (7 items)
Number of times that customer applied for credit limit increase
Number of defaults
Number of credit cards that the customer has in his/her possession
Location where credit card is used (17 items)
Number of credit cards held by the customer that were issued by other banks (3 items)
Customer's spending habits using credit card(s) issued by other banks (4 items)
Customer's payment habits with other banks (5 items)

Note: After combing and filtering, we arrived at 13 categories with 48 items.

3.3 Data Preprocessing

The third step involved an essential data preprocessing process. This was performed because the data had different scales and some could not be calculated, which would make the research results biased. For the preprocessing process, we first converted the nominal scales and proportion scales to neural network codes (e.g., the northern area was {1,0,0,0,0} and the central area was {0,1,0,0,0}). Next, we integrated the variables to create new variables (e.g., variables "transaction amount (in NT$)" and "income from transactions" were merged to produce a variable with higher discriminability than the two variables by themselves).

3.4 Selecting Neural Model and Parameters

The fourth step was to select the neural model and set the parameters in SAS sytem. This study used the back propagation model because this study covered the topic of supervised learning, and the model is relatively more suitable for processing nonlinear problems. This study set its target values as Y1 and Y2, which represented the revolving credit usage frequency and revolving balance (in NT$), respectively. The two target values were set to enable us to gain insight into the effect that changes to the definition of revolvers have on revolving credit-related results. Because no standard parameters are available for the selection of neural network parameters, this study chose the following parameters and performed the first round of experiments.

Hidden layer: begins from the first layer; the number of layers is gradually increased
Number of units in the hidden layers: 5
Learning rate: 0.2
Maximum learning number/time: 1,000 or 1 h
Initial variables were randomly selected by the neural network
Data distribution: 60 % training, 30 % verification, and 10 % test

Standardization: 0.2–0.8 interval (self - minimum value)/(maximum value - minimum value)

The reason for disregarding the mean divided by standardization is due to possible distortion because of large variable differences

Function selected: logistic function

Momentum: 0.5

3.5　Choosing Data Eigenvalues with High Discriminability

Our sample size included 10,000 revolvers, which were credit card users who used revolving more than 6 times in the past 12 months, and 10,000 non-revolvers, which were credit card users who never used revolving in the past 1 year. We chose these combined 20,000 samples from an existing credit card user portfolio.

Among the 20,000 samples, the customer dimensions were classified by demographic, transactional behavior and life-style. For example, in our samples, we used marital status, age, mortgage status, education, profession and other demographic variables. Likewise, transactional behavior included total spending, cash advances, purchases and installment amount type of data for our samples. Life-style examples included poor money management skills, impulse buying, just married, first mortgage and first job. Please refer to Tables 2 and 3 for demographic and transactional dimensions. Currently, we do have life-style data such as first mortgage and first job, however, we do not have data concerning impulse buying and money management skills. Further study for developing a meaningful life-style database is needed.

The organization stage of data eigenvalues entailed the longest operation time. This stage involved determining the data to be retained, the data to be omitted, and the fields to be merged. For the first phase, all available variables were narrowed down to selective variables, which were subsequently divided into "demographic information" and "transaction information" groups. We hypothesized that demographic information features low discriminability and can be immediately obtained when customers apply for a credit card, and that transaction information features high discriminability but can only be obtained over time after applicants become formal customers. However, the most useful life-style information includes personality traits and behavioral data (e.g., impulse buyers, uninsured individuals with illnesses, and people with poor financial management skills) is difficult to acquire. Due to limited data sources, this study will only focus on finding valuable transactional eigenvalues.

The first round of computation used the SAS neural network function only. We used two sets of data, the first set was comprised of 10,000 revolvers. The second set was made up of 10,000 non-revolvers. The first round result using the neural network and two sets of data yielded eigenvalues with low discriminability and an error rate of 0.49, which proved the limitations of the neural network. The first round computation result from neural network showed that the ranking of eigenvalues by weight was not significant. Therefore, to supplement and overcome the neural network limitations, we also used the decision tree and logistic regression models to obtain and test data eigenvalues with higher discriminability.

Using the SAS decision tree function with the same two sets of data, we came up with four classification layers and five critical eigenvalues with high discriminability. Applying the SAS logistic function to the same two sets of data, we developed 4 eigenvalues with high discriminability.

Finally, as the way to retain the super calculability characteristic of the neural network, we picked the top 5 eigenvalues from the first round computation results, plus 5 from the decision tree and 4 from the logistic regression model results. For the second computation round, the 14 eigenvalues were again put back into the neural network system yielding a new classification error rate of 0.128. This suggested that the 14 eigenvalues possess very high discriminability features.

4   Result

The chart information was sourced from SAS system. The Blue line represents the decision tree method. The Red line represents the logistic regression method and the Dark Brown represents the neural network method (Fig. 1).

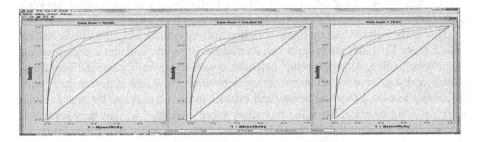

Fig. 1. Training and testing data result (Color figure online)

The first round computation in the chart on the left-hand side, shows that 8,000 training data were used to test 3 models. The result was that the decision tree was the most effective predictive data mining method with 16 % misclassification rate, followed by the logistic method with 20 % and neural network as the least effective.

The chart in the middle used 2,000 testing data with the same probability distribution to validate the training result stability shown in the left-hand chart. The similarity of the two charts proves the training result stability of the models.

The chart on the right-hand side used 10,000 data with different probability distribution. The result was consistent with the first two charts.

In the second round computation, we used 14 eigenvalues with high discriminability and put them back into the SAS neural network model to verify the model accuracy shown in Fig. 2. The misclassification rate was further reduced from 16 % to 11 %.

Results from the two computation rounds supported our hypothesis (i.e., demographic information with low discriminability): none of the 14 eigenvalues were from the demographic category. The results also revealed that demographic information filled out by new customers on banks' credit card applications can be used to determine

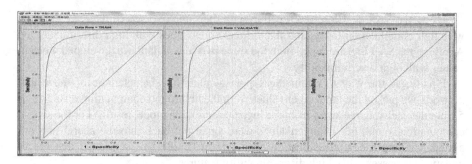

Fig. 2. 14 eigenvalues with the highest discriminability (Color figure online)

whether the applicants have good credit, but not whether they will become major credit card users or revolvers.

To verify whether demographic information truly features low discriminability, we entered 11 demographic eigenvalues into the back propagation model for further tests. The test produced a classification error rate that was approximately 0.5, confirming that eigenvalues of demographic information are ineffective in discriminating between customers.

This study featured a study framework that differed considerably from other studies in four ways: First, this study separated data eigenvalues into different categories right from the beginning. Although neural networks possesses powerful computational capabilities and can process all eigenvalues, separating data into different categories (according to their attributes) beforehand enables future users to use the study results more easily in practice. To make this study more complete, we not only separated customer data into demographic and transaction information, but also divided all original eigenvalues into three groups of eigenvalues to be entered into the system, which facilitated result comparisons. Second, this study identified data eigenvalues with high discriminability by employing back propagation, decision trees, and logistic functions, which enabled subsequent comparisons. The results were highly similar: all three methods produced similar eigenvalues with high discriminability. However, all three methods also had their own limitations: for example, neural networks have the worst explanatory power, whereas decision trees (despite featuring rapid data classification speed) are unable to guarantee that all omitted eigenvalues are irrelevant. Third, this study constantly changed eigenvalues, in which eigenvalues that were easily obtained but featured low discriminability were replaced with those that were not easily obtained but featured high discriminability. Fourth, this study enabled general rules derived from eigenvalue combination searches to be applied to finance-related operations (e.g., identifying ways to find active financial management clients and safe investment-type insurance clients) and even non finance-related operations (e.g., tourism industry and department stores looking for ways to locate their target customer base, and prosecutors finding the physical features of criminals), illustrating the high practical values of such general rules in information management.

5 Conclusions

When searching for new customers, the method introduced in this study enables users to easily identify valuable customers, eliminates instances in which credit cards are issued to customers with no credit card needs, and allows more meaningful questions to be provided on credit card applications to obtain eigenvalues with high discriminability that allows you to better identify ideal customers. Another method for improving credit card-related business performance is to combine several transaction-based eigenvalues into life-style eigenvalues that feature higher discriminability.

Nevertheless, the present study is merely a beginning. Traditional banking has used traditional labels to identify customers, i.e. age, occupation and income. These labels are no longer adequate. With the arrival of the online economy, everyone has been given numerous personality traits and behavioral labels with various eigenvalues. Therefore, businesses that have a full understanding of consumers' personality traits and behavioral labels will be able to create new value for themselves and their customers.

The neural network model has its own advantages and disadvantages. Therefore, combining the decision tree and the logistic models work to overcome the neural network model disadvantages and help to better improve the search for target credit card customers. This will increase the effectiveness of each model used in data classification and prediction as well as enhance their practical values in regards to financial industries.

References

1. Malhotra, R., Malhotra, D.K.: Evaluating consumer loans using neural networks. Omega **31**(2), 83–96 (2003)
2. Khandani, A.E., Kim, A.J., Lo, A.W.: Consumer credit-risk models via machine-learning algorithms. J. Bank. Finance **34**(11), 2767–2787 (2010)
3. Zakaryazad, A., Duman, E., Kibekbaev, A.: Profit-based artificial neural network (ANN) trained by migrating birds optimization: a case study in credit card fraud detection. Department of Industrial Engineering, Ozyegin University, Istanbul, Turkey. A Kibekbaev Proceedings of European Conference on Data Mining, pp. 28–36 (2015)
4. Ye, Y.C.: Application and Implementation of Neural Network Models. Rulin Publishing House, Taipei (2004)
5. Chen, C.L., Kaber, D.B., Dempsey, P.G.: A new approach to applying feedforward neural networks to the prediction of musculoskeletal disorder. Appl. Ergon. **31**(3), 269–282 (2000)
6. He, C.Z., Zhu, B., Zhang, M.Z., Zhuang, Y.Y., He, X.L., Du, D.Y.: Customers' risk type prediction based on analog complexing. Procedia Computer Science, Business School, Sichuan University, Chengdu, China **55**, 939–943 (2105)
7. Harrison, P., Gray, C.T.: Profiling for profit: a report on target marketing and profiling practices in the credit industry. A Joint Research Project by Deakin University and Consumer Action Law Centre (2012)
8. Rhomas, L.C.: A survey of credit and behavioral scoring: forecasting financial risk of lending to consumers. Int. J. Forecast. **16**(2), 149–172 (2000)
9. Hamilton, R., Khan, M.: Revolving credit card holders: who are they and how can they be identified ? Serv. Ind. J. **21**(3), 37–48 (2001)

10. Chuang, C.C.: Applying neural networks to factors analysis of credit card loan decisions. Master's thesis, Department of Information Management, National Taiwan University of Science and Technology, Taipei (2005)
11. Hsieh, N.C.: An integrated data mining and behavioral scoring model of analyzing bank customers. Expert Syst. Appl. **27**(4), 623–633 (2004)
12. Oreski, S., Oreski, D., Oreski, G.: Hybrid system with genetic algorithm and artificial neural networks and its application to retail credit risk assessment. Expert Syst. Appl. **39**(16), 12605–12617 (2012)

Selection Probe of EEG Using Dynamic Graph of Autocatalytic Set (ACS)

Azmirul Ashaari[1], Tahir Ahmad[2(✉)], Suzelawati Zenian[1,3], and Noorsufia Abdul Shukor[4]

[1] Department of Mathematical Science, Faculty of Science,
Universiti Teknologi Malaysia, UTM, 81310 Skudai, Johor, Malaysia
[2] Centre for Sustainable Nanomaterials, Ibnu Sina Institute for Scientific
and Industrial Research, Universiti Teknologi Malaysia,
UTM, 81310 Skudai, Johor, Malaysia
tahir@ibnusina.utm.my
[3] Department of Mathematics with Computer Graphics, Faculty of Science
and Natural Resources, Universiti Malaysia Sabah,
Jalan UMS, 88400 Kota Kinabalu, Sabah, Malaysia
[4] Fakulti Sains Komputer dan Matematik, Universiti Teknologi Mara,
Cawangan Negeri Sembilan, Kampus Seremban 3,
70300 Seremban, Negeri Sembilan, Malaysia

Abstract. Electroencephalography (EEG) machine is a medical equipment which is used to diagnose seizure. EEG signal records data in the form of graph which consist of abnormal patterns such as spikes, sharp waves and also spikes and wave complexes. This pattern also come in multiple line series which then give some difficulties to analyze. This paper introduce the implementation of dynamic graph of Autocatalytic Set (ACS) for EEG signal during seizure. The result is then compared with other publish method namely Principal Component Analysis (PCA) of same EEG data.

Keywords: Dynamic graph · Electroencephalography (EEG) · Epilepsy · Autocatalytic Set (ACS)

1 Introduction

Electroencephalography (EEG) system (see Fig. 1) is used to record brain electrical activity from the scalp. EEG reads voltage differences on the head, relative to a given point. There are various kinds of electrical brain activities recordings such as scalp EEG, depth EEG and many more. Each of them use different types of electrodes.

Scalp EEG refers to the recording of the brain activity from the outer surface of the head and is non-invasive [2]. The recommended standard method for the recording scalp EEG known as the International Ten-Twenty System of electrode placement (see Fig. 1). Epilepsy is a common symptom which define as multiple seizures cause by malfunction of a brain. Approximate one per cent of the population will encounter these epilepsy symptom. Most of the epilepsy attack occurs during the period of young children and elderly rather than young and middle age adults [3]. It describes the

© Springer Nature Singapore Pte Ltd. 2016
M.W. Berry et al. (Eds.): SCDS 2016, CCIS 652, pp. 25–36, 2016.
DOI: 10.1007/978-981-10-2777-2_3

 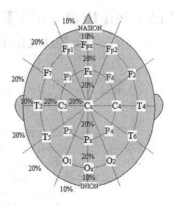

Fig. 1. International Ten-Twenty System [1].

condition of a patient having recurring spontaneous seizures due to the sudden development of synchronous firing in the cerebral cortex caused by lasting cerebral abnormality. During epilepsy there is a miniature brainstorm of certain groups of brain cells that may lead to convulsions. Seizure can be controlled by antiepileptic drug or epilepsy surgery [4]. Surgical treatment may be an option in patients having epileptic seizures refractory to medication [5].

Seizures are the results of sudden excessive electrical discharges in a group of neurons. It is divided into two major groups namely partial and generalized. Partial (local or focal) seizures are those which have a localized onset in the brain [6]. It involves only a part of the cerebral hemisphere at seizure onset and produces symptoms in corresponding parts of the body or disturbances in some related mental functions. Generalized seizures are marked with a lack of evidence for a localized seizure onset. Typically it involves the entire cerebrum at the onset and produce bilateral motor symptoms. Seizure cause patients to experience loss of consciousness at the onset of the seizure. Patient usually is in a state of confusion, loss awareness, uncontrollable shaking and hallucinations.

2 Graph and Autocatalytic Set

Graph is a one of mathematics methods developed from the connection of linked points. Graph was first introduced by Swiss mathematician Leonhard Euler in order to solve the Seven Bridges of Königsberg problem [7]. The problem was to find a walkthrough of the city, where islands can only be reached via bridges. However, the walk path must not cross each bridge more than once, and every bridge must be completely crossed every time. According to Balakrishnan and Ranganathan [8] the development of the graph has grown into a significant area of mathematical researches and has been used in other disciplines such as physics, chemistry, psychology, sociology and computer sciences. Graph can be modelled by interconnection between elements of natural and manmade [8–10]. The definition of directed graph is given as follows:

Definition 1.1 [7]: A directed graph $G = G(V, E)$ is defined by a set V of "vertices" and a set E of "edge" where each link is an ordered pair of vertices.

In short, the set of vertices and edges can be represented as $V = \{v_1, v_2, v_3 \ldots, v_n\}$ and $E = \{e_1, e_2, e_3, \ldots, e_n\}$. The concept of an autocatalytic set was introduced in the context of interaction between compounds as follows:

Definition 1.2 [11]: An autocatalytic set (ACS) is a sub graph, each of whose nodes have at least one incoming link from a node belonging to the same sub graph (see Fig. 2).

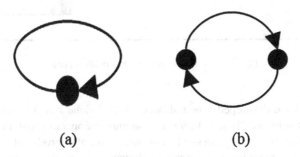

(a)   **(b)**

Fig. 2. (a) A 1-cycle, the simplest ACS. (b) A 2- cycle.

Tahir et al. [12] presented some properties of ACS matrix. Theorem 1.1 indicates that a finite set of vertices requires a finite set of edges.

Theorem 1.1 [12]: If $G(V, E)$ is an autocatalytic set, and $|V|=n$ then $|E| \leq n^2$.

Tahir et al. [12] define that for any autocatalytic set $G(V, E)$ with a finite n vertices the possible maximum number of edges is at most n^2. It can transformed into an adjacency matrix of $n \times n$ matrix with finite edges less or equal to n^2. This statement is justify by the following theorem.

Theorem 1.2 [12]: Any autocatalytic set $G(V, E)$ with n vertices has edges at most n^2. i.e.: $|E(n)| \leq n^2$ where $E(n)$ is edges with n vertices.

3 Brain Neuron

Human brain is composed about ten billion to hundred billions of cells. Based on biological perspective, a brain cell is called a neurons. Each of these neuron is always connected between each other. The networking between each neuron, is operated by the process of tiny burst, which released electrical activities. Figure 3 shows the link or communication between each neuron occurred inside of brain.

The purpose of sharing by communication between neurons occurs is in order to analyze the information. It is due to each neuron has different function and information. In 1943, Warren McCulloch the fact thata neurophysiologist, and Walter Pitts a

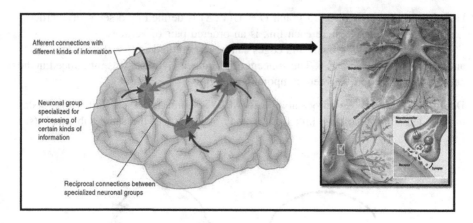

Fig. 3. The link between each neuron [13].

mathematician have developed a neural network technique based on communication system between neurons. These concepts of communication show that, there exist links between each neuron. Each process of communication also produced an electric discharge at different magnitude of strength. However, problem occurs in identifying which neurons transmit electrical signals and which neurons that receive them. The strength of electrical discharge by neurons communication can be measure using probe of EEG. The time and location receive by EEG probes are recorded, which show the communication between neurons. These electrical signal recorded by EEG probes are transformed into graph and matrix at the end. The matrix is then assumed to be the Magnetic Contour (MC) plane given in [14] (see Fig. 4).

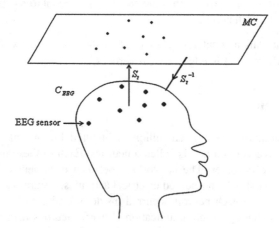

Fig. 4. MC plane [14].

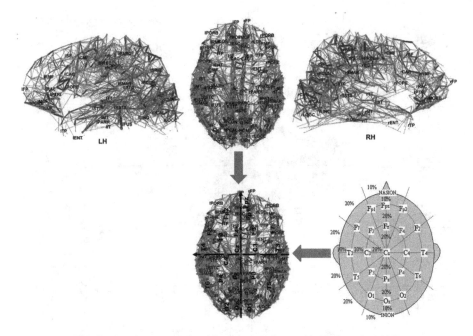

Fig. 5. Graph network for 66 cortical regions with EEG probes [15].

Patric et al. [15] modelled the brain interconnection regions as graph network for 66 cortical regions (see Fig. 5). Whereby each region consists of one link coming from other region.

Each EEG probe capture the signal at every seconds. Therefore, each probe receives at least 66 information coming from other regions (see Fig. 6). Hence, this phenomena can be consider as an ACS if one views it as a graph.

In [1], there were twenty one probes used to capture the signal. Hence a square matrix with dimension 5×5 is produced. Four new probes namely, A_1, A_2, A_3 and A_4 are added to produce the square matrix. Their average potential values for these four 'ghost' probes are zero.

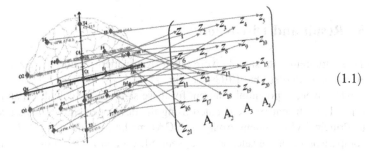

(1.1)

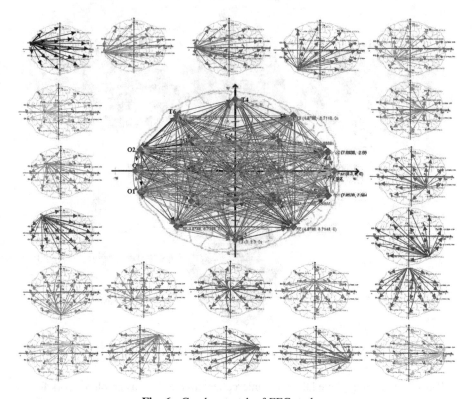

Fig. 6. Graph network of EEG probes.

4   System Dynamic Variable Selection (SDVS)

System Dynamic Variable Selection (SDVS) has received copyright © *2015 Universiti Teknologi Malaysia- All Right Reserved* is a developed software that helps the user to evaluate and analyze the dynamicity of a given system. It is used to select variables through time via ACS algorithm.

5   Result and Discussion

Three different data sample were taken from patient A during seizures for 10, 20 and 30 s using Nicolet One EEG (see Fig. 7).

The average potential different for these samples are calculated and substituted into Eq. 1.1. Each matrix is evaluated using the developed SDVS. This result is compared to Principal Component Analysis (PCA) method obtain in [16]. Figure 8 shows the comparison of probe selections between SDVS and PCA. These two methods produce almost similar pattern.

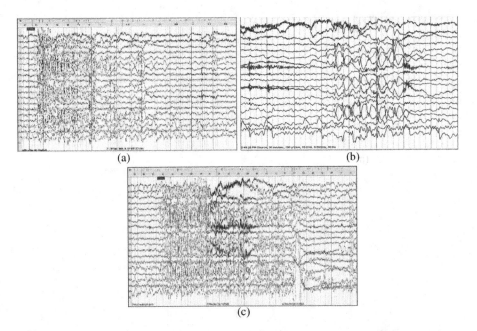

Fig. 7. Sample data of EEG signal from epilepsy seizure (a) 10 s epileptic data (b) 20 s epileptic data (c) 30 s epileptic data

Figure 9 shows the location of sensors on the surface of the patient's head (in red). The locations of clusters center using PCA [16] are in blue and by SDVS in purple. The orange color indicates the identical cluster centers obtained by both methods.

At $t = 1$ and $t = 2$, are the pre-ictal state. However, during $t = 3$, these pattern are demised to the right side hemisphere of the brain. This shows the seizure attack is started for the patient. At $t = 4$ until $t = 7$, signals are scattered in both side hemispheres of brain which indicated the ictal state. At $t = 8$, the signals starts to demise to the right side of brain. Its shows the seizure attack is over (postictal state) at $t = 9$ until $t = 10$. It suggested a generalized type of seizures.

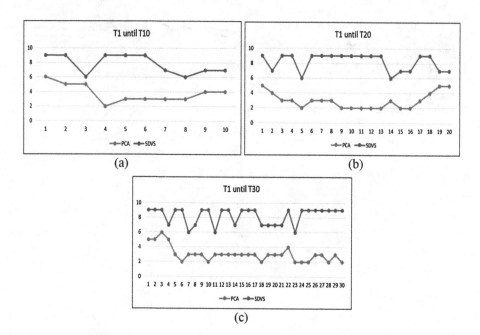

Fig. 8. Comparison of Probe selection between SDVS and PCA from: (a) 10 s epileptic data (b) 20 s epileptic data (c) 30 s epileptic data

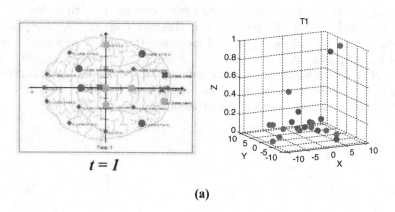

Fig. 9. (a) Processed signal for T1. (b) Processed signal from T2 to T5. (c) Processed signal from T6 to T9. (d) Processed signal for T10 (Color figure online)

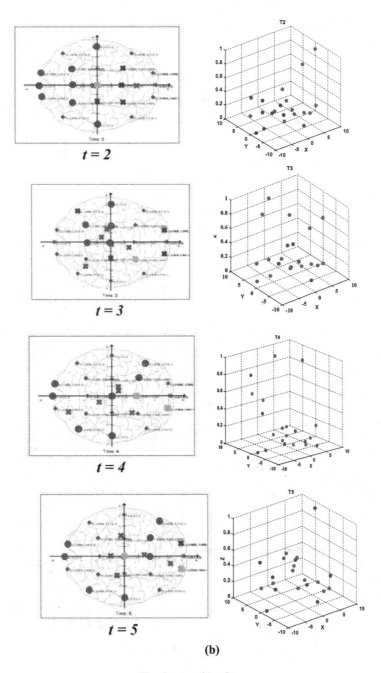

(b)

Fig. 9. (continued)

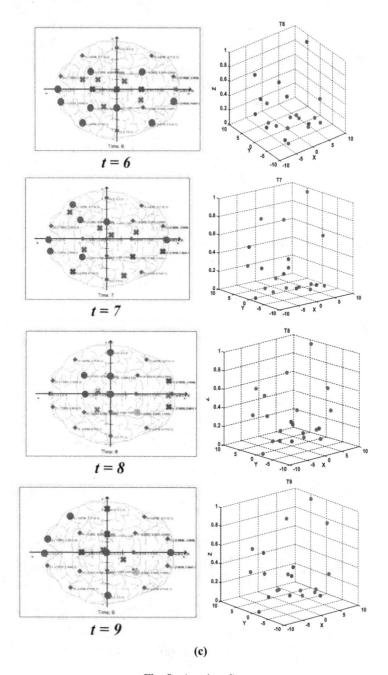

Fig. 9. (continued)

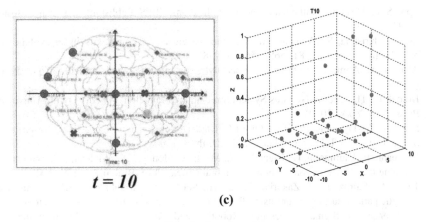

$$t = 10$$

(c)

Fig. 9. (continued)

6 Conclusion

The dynamic graph of ACS for EEG signal during seizure using SDVS has success-fully implemented. Three different samples of data from patient A has successfully applied into SDVS and compared with PCA. It shows that Patient A was having a generalized type of seizure.

Acknowledgment. This work has been supported by Ibnu Sina Institute, MyBrain15 from Ministry of High Education Malaysia and University Teknologi Malaysia.

References

1. Teplan, M.: Fundamentals of EEG measurement. Measur. Sci. Rev. **2**(2), 1–11 (2002)
2. Rowan, J.A., Tolunsky, E.: A Primer of EEG: with a Mini Atlas. Elsevier's Health Sciences, USA (2003)
3. Weaver, Donald: A Guide for Parents, Families and Sufferers Daily Telegraph. Robinson, London (2004)
4. Lasemidis, L.D., Pardalos, P., Sackellares, J.C., Shiau, D.S.: Quadratic binary programming and dynamical system approach to determine the predictability of epileptic seizures. J. Comb. Optim. **5**, 9–26 (2001)
5. Engel, J.R., Van Ness, P.C., Rasmussen, T.B., Ojemann, L.M.: Outcome with respect to epileptic seizures. In: Surgical Treatment of the Epilepsies, pp. 609–622. Raven, New York (1993)
6. Gastaut, H.: Clinical and electroencephalographical classification of epileptic seizures. Epilepsia **11**, 102–113 (1970)
7. Carlson, S.C.: Graph Theory. Encyclopedia Britannica, Chicago (2010)
8. Balakrishnan, R., Ranganathan, K.: A Textbook of Graph Theory. Springer, New York (2012)
9. Harary, F.: Graph Theory. Addison Wesley Publishing Company, California (1969)

10. Epp, S.S.: Discrete Mathematics with Applications. PWS Publishing Company, Boston (1993)
11. Jain, S., Krishna, S.: Autocatalytic sets and the growth of complexity in an evolutionary model. Phys. Rev. Lett. **81**, 5684–5687 (1998)
12. Ashaari, A., Ahmad, T.: On fuzzy autocatalytic set. Int. J. Pure Appl. Math. **107**(1), 59–68 (2016)
13. Brodal, P.: The Central Nervous System Structure and Function, 4th edn. Oxford University Press, New York (2010)
14. Fauziah, Z.: Dynamic Profiling Of EEG Data During Seizure Using Fuzzy Information Space. Universiti Teknologi Malaysia: Ph.D. thesis (2008)
15. Hagmann, P., Cammoun, L., Gigandet, X., Meuli, R., Honey, C.J., Wedeen, V.J., Sporns, O.: Mapping the structural core of human cerebral cortex. PLoS Biol. **6**(7), e159 (2008)
16. Tahir, A., Fairuz, R.A., Zakaria, F., Isa, H.: Selection of a subset of EEG channels of epileptic patient during seizure using PCA. In: Proceedings of the 7th WSEAS International Conference on Signal Processing, Robotics and Automation. World Scientific and Engineering Academy and Society (WSEAS) (2008)

A Comparison of BPNN, RBF, and ENN in Number Plate Recognition

Chin Kim On[1(✉)], Teo Kein Yao[1], Rayner Alfred[1],
Ag Asri Ag Ibrahim[1], Wang Cheng[3], and Tan Tse Guan[2]

[1] Faculty of Computing and Informatics, Universiti Malaysia Sabah,
Kota Kinabalu, Sabah, Malaysia
kimonchin@gmail.com, keinyau@gmail.com,
aaai500@gmail.com, ralfred@ums.edu.my
[2] Faculty of Creative Technology and Heritage,
Universiti Malaysia Kelantan, Kota Bharu, Kelantan, Malaysia
tseguantan@gmail.com
[3] School of Economics and Management, Qiqihar University,
Qiqihar, Heilongjiang Province, China
wang-cheng@sohu.com

Abstract. In this paper, we discuss a research project that related to autonomous recognition of Malaysia car plates using neural network approaches. This research aims to compare the proposed conventional Backpropagation Feed Forward Neural Network (BPNN), Radial Basis Function Network (RBF), and Ensemble Neural Network (ENN). There are numerous research articles discussed the performances of BPNN and RFB in various applications. Interestingly, there is lack of discussion and application of ENN approach as the idea of ENN is still very young. Furthermore, this paper also discusses a novel technique used to localize car plate automatically without labelling them or matching their positions with template. The proposed method could solve most of the localization challenges. The experimental results show the proposed technique could automatically localize most of the car plate. The testing results show that the proposed ENN performed better than the BPNN and RBF. Furthermore, the proposed RBF performed better than BPNN.

Keywords: Backpropagation Feed Forward Neural Network (BPNN) · Radial Basis Function Network (RBF) · Ensemble Neural Network (ENN) · Car plate recognition · Image processing

1 Introduction

Artificial Neural Network (ANN) is a powerful machine learning method used for image recognition [1], image compression [2], signal and pattern recognition [3], weather prediction [4], stock market forecasting [5], solving travelling salesman's problem [6], developing intelligent gaming non-player character [7], loan application [8], medicine research [9], security [10], as named a few. The advantage of ANN is their ability to be used as an arbitrary function approximation mechanism that can learn from very huge data in minimal cost. It involves three learning paradigms; supervised

© Springer Nature Singapore Pte Ltd. 2016
M.W. Berry et al. (Eds.): SCDS 2016, CCIS 652, pp. 37–47, 2016.
DOI: 10.1007/978-981-10-2777-2_4

learning, unsupervised learning and reinforcement learning. Generally, the application that falls within supervised learning are image recognition, signal and pattern recognition, etc., a set of complete data is required in the training phase. Tasks that fall within unsupervised learning involve statistical estimation and distribution such as solving travelling salesman's problem and developing gaming AI, some data is given in the training phase. Reinforcement learning used to solve control problems, games, and sequential decision making tasks. In most cases, data is not given but generated through the learning process when interact with the environment.

Backpropagation is a supervised learning algorithm that is commonly used in training ANN. The backpropagation training algorithms are usually classified into steepest descent, quasi-Newton and Levenberg-Marquardt and conjugate gradient. They are different due to learning rate variable, momentum factor, and the activation functions used [11]. In fact, the training mechanism is an inverse problem in achieving the estimated parameters or goals of the NN model. There are numerous articles discussed the performance and application of Backpropagation Neural Network (BPNN) [5]. Some compared the performance of NN based on different NN topology such as BPNN, Radial Basis Function network (RBF), Recurrent Network, Probabilities Neural Network, etc. Interestingly, there is very lack of Ensemble Neural Network (ENN) discussion particularly in the image processing perspective. ENN has been proposed and proven to be superior in gaming AI design [12], rainfall forecasting [13], and numbering classification [14]. However, its performance in car plate recognition is still remain unknown.

In this research paper, we discuss the performance of the BPNN, RBF, and ENN in solving number plate recognition problems. Number plate recognition system plays an important role in numerous real-life application such as toll collection, traffic control, and road traffic monitoring, as name a few [15–17]. The research is challenging as the image captured could be vary in the perspective of location, quantity, size, color, font type and font size, occlusion, inclination, illumination, etc. As a result, some countries have to develop their own automatic recognition system as the existing car plate recognition systems do not suit to their case. As such, many approaches have been proposed since last decade in order to localize and recognize car plate such as edge detection [18], morphology based method [19], wavelet transform [20], The Omhining rough set theory [15], self-organizing feature map (SOFM) [15], modified Hausdorff distance template matching [21], as named a few. However, none of the algorithm have been tested for Malaysia license plates thus far, except [22, 23]. [22, 23] proposed stroke extraction and analysis, pixel compactness, angles, and projection histogram to localize and recognize car plates. Although most proposed methods could achieve 96 % or more success rate, there are still challenges to be solved; (1) car plate localization - most of the proposed localization methods have to manually localize and detect the characters on the car plate, labelling them or matching their positions with template, and (2) number/ alphabet recognition - in some cases, the proposed method failed to deal with the characters that are similar to their type, such as (B and 8), (O and 0), (A and 4), (C and G), (D and O), and (K and X). In the project, we proposed image deblurring technique to automatically locate the alphanumeric characters from the number plate instead of manually searching the region of the plate. Furthermore, the experimental results show the proposed ENN could solve the characters similarity problem.

The remainder of this paper is organized as follows. Section 2 discusses the methodologies used in the research. This section also describes the pre-processing steps in localizing the license plate. The experimental setup is presented in Sect. 3, results and discussions are provided in Sect. 4 and finally the conclusion and future work are described in Sect. 5.

2 Methodology

In general, car plate recognition system can be divided into four processing phases: (1) image acquisition or data collection, (2) license plate localization, (3) character segmentation, and (4) recognition phase.

2.1 Image Acquisition

The hardware used during data collection plays an important role. A high spec camera or video recorder should be considered as the camera resolution and shutter speed could determine the image quality. However, these hardware could be costly and the process could be time consuming when conveying and processing high quality images.

In our case, due to very limited resources, a digital camera with 5 megapixels resolution was used during data collection. There were 500 images had been collected during daytime in a shopping mall's car park. The images were saved in 1920 × 1080 resolution. The images were taken within 5 feet distance from the car plate.

2.2 Car Plate Localization

The car plate localization process could be challenging. There are many factors could influence the effectiveness and efficiency for correctly localizing the car plate. In our case, the factors are summarized as below.

- Location - car plate exist in different locations of an image. Car plates also exist in different locations of a car. It is not always positioned on the center of the car body. Hence, it is impossible to label and match the car plates' positions with template.
- Size – car plates have different sizes. Some car plates exist in a rectangle shape and others are designed in square form. The sizes are also vary due to camera distance.
- Color – car plate have similar color with car body. There are few cases showed that the car plate and car body colors are same. Moreover, the car plates do not contain a clear invest color of border.
- Illumination – the images were captured and saved with different type of illumination. Mainly due to car park lighting and light reflection on the license plate surface.
- Objects - car logo, car name, and other objects have greatly increased the difficulty in localizing the car plate.

We proposed a novel technique to solve the problems as mentioned above. The proposed method first binarize the images and then remove the unwanted objects.

The binarization threshold is 0.7, slightly higher than the default threshold used by most researchers. Furthermore, any objects with smaller than 30 pixels coverage and more than 500 pixels coverage are removed automatically. The process is mainly to remove dirt, leaves, small objects, and unwanted large objects such as sticker, logo, side mirror, etc. from the images. The images are "deblurred" after the filtering process. The objective is to create a bounding box for the characters/letters as one object. As the characters/letters on car plate are always individual and next to each other, it created large pixel coverage area after the deblurring process. Then, the filtering process is implemented again for removing unwanted object that is sizing smaller than the car plate.

2.3   Character Detection and Segmentation

It is highly possible for every bounded object be the car plate. In our case, image histogram method is used to differentiate the bounded objects. The properties of the bounded objects and the generated histograms are compared with other bounded objects. This is mainly to check whether the bounded object consists of numbers or alphabets or other unwanted objects. This can be validated by comparing the properties of characters and the properties of the bounded objects. If there is a match, then the bounded object is assumed to be a license plate. Otherwise, it could be the car's logo or other unwanted objects. The detected alphanumeric are then bounded into one object and save separately as a car plate image. Furthermore, the thinning algorithm is then applied to the car plate image. The thinning algorithm is important to remove unwanted foreground pixels whilst remain its topology from the binary images. The thinned images are then used for the recognition purpose.

2.4   Recognition Algorithms

Backpropagation Neural Network (BPNN). BPNN is commonly used in image processing, signal processing, robotics, etc. [6]. It is one of the supervised learning algorithms that is powerful, fast, and reliable. The BPNN algorithm calculates the gradient of a loss function in the network. The gradient are saved as weight in the network. The gradient is then fed to the neural network which is then used to update the BPNN weights. The updated process is to minimize the loss function.

BPNN is divided into two phases: (1) the phase refers to propagation of the neurons, and (2) the phase involves weight update. In the first phase, the training input data is first forward propagate through the NN in order to generate output activations. Next, the output activation is backward propagate through the NN using the generated training pattern target. The deltas of all output and hidden neurons are generated during the backward propagation. The second phase of the training involves multiplying the generated output delta and input activation to obtain the gradient of the weight. Then, the process continues with a subtraction of a ratio of the gradient from the weight. The ratio is also named as learning rate and it influences the quality and speed of the network. The greater the learning rate, the faster the network trains; the lower the learning rate, the more accurate the training is (Heaton, 2008). Phases 1 and 2 are repeated until the performance of the network is satisfied.

Radial Basis Function Network (RBF). The RBF is almost similar to BPNN. It is also one of the supervised learning algorithm that is commonly used in function approximation, time series prediction, classification, etc. Initially, the RBF is different to BPNN due to the function used during training. In RBF, the output units form a linear combination of the basis functions computed by the hidden units. There are three types of activation functions can be used in the RBF, namely Gaussian, Multi-quadratics, and Invest Multiquadratics. The Gaussian function is used in this study as it is commonly used in the research. The learning topology of RBF is divided into two phases: (1) unsupervised learning in the hidden layer, and (2) supervised learning in the output layer.

Ensemble Neural Network. Training a number of NNs and then combining their results is known as ENN. Average or weighted average method is most frequently used in the ENN. In this work, the proposed ENN is the averaged of one BPNN and one RBF network. Figure 1 shows the proposed ENN architecture and the formula used is shown in (1) below.

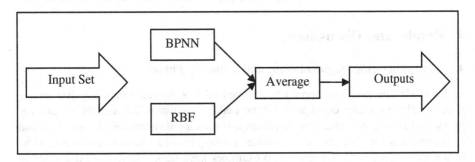

Fig. 1. The proposed ENN

$$Y_{net\,ENN} = (Y_{net\,BPNN} + Y_{net\,RBP})/2 \qquad (1)$$

3 Experimental Setup

The proposed BPNN and RBF used one input layer, one hidden layer, and one output layer. The input layer consists of 35 neurons, the hidden layer has 52 neurons + 1 bias, and the output layer consists of 43 neurons + 1 bias. The preliminary experiments showed optimal results could be generated with learning rate 0.001 and none of the preliminary experiment needed more than 20000 epochs. We set the goal of the training as 0.01. Biases of 1.0 were used in both of NN's input and hidden layers. There were 300 training samples and 200 testing samples used in the recognition phase. The ENN

Table 1. Experimental setting used for BPNN, RBF, and ENN

	BPNN	RBF	ENN
Input Neurons:		35	
Hidden Neurons:		52 + 1 bias	
Output Neurons:		43 + 1 bias	
Learning rate:		0.001	
Goal of Mean Square error:		0.01	
Function:	*Binary Sigmoid*	*Gaussian Function*	-
Training Samples:		300	
Testing Samples:		200	
Maximum Number of Epoch:		20000	
Bias:		1.0,1.0	
Number of NN		One FFNN	Two FFNN
ENN Method	-	-	Averaged output

used is the averaged output of a combination from one BPNN and one RBF architectures. Table 1 shows the summary of the experimental setting used in this research.

4   Results and Discussions

4.1   Pre-processing, Segmentation, and Thinning Phases

In this study, the proposed approach achieved 99.5 % success rate using 200 testing samples. These samples consists of 170 normal license plates, 20 of Malaysia taxi car plates, and 10 of commemorates car plates. It is difficult in collecting Malaysia taxi and commemorates car plates as these car plates are very limited. The failure was caused by an unclear sample used. There are two examples have been selected and discussed in the next section. First example shows the car plate color is similar as car body color and there is no border to differentiate the car plate and car body. Hence, it is difficult to localize the car plate. But, the car plate has been successfully localized and thinned with the proposed deblur technique. The second example shows the result obtained for commemorative car plate case.

Figure 2 shows the experimental results for car plate color that is similar as car body. Figure 2(a) and (b) show that the images have been captured and then binarized. Figure 2(c) shows the binarized image is then applied for noise removal. The large objects and unwanted objects are then selected as shown in Fig. 2(d). Figure 2(e) shows the large objects or unwanted objects are removed. Then, the remaining objects are bounded and deblurred as shown in Fig. 2(f). Figure 2(g) and (h) showed the extracted, segmented and thinned images.

In Malaysia, there are a number of memorial plates, or plates with distinctive prefixes that are made available by the Malaysian Road Transport Department (JPJ) (in Malay language: Jabatan Pengangkutan Jalan Malaysia) [24, 25].These special plates are used to denote the manufacturer of the car or a special event such as "iM4U" stands

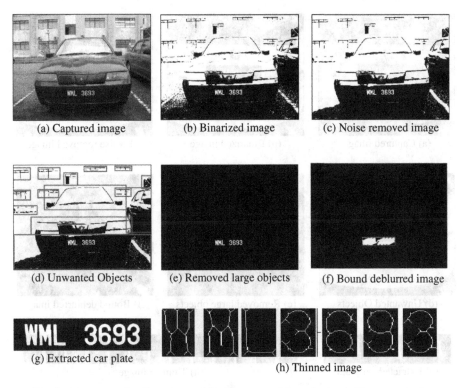

(a) Captured image   (b) Binarized image   (c) Noise removed image

(d) Unwanted Objects   (e) Removed large objects   (f) Bound deblurred image

(g) Extracted car plate

(h) Thinned image

Fig. 2. Pre-processing Experimental Results – car plate color is similar as car body.

for "1 Malaysia for youth". Hence, designing an accurate ANPR system for such license plates is challenging as the commemorative car plates consist of different type of font used, different size of font, and some of the fonts are italic or bolded. Figure 3 shows the experimental results for commemorative car plate. Figure 3(a) shows the image captured consists of car plate "*Putrajaya* 1166". The image is then binarized as shows in Fig. 3(b). Figure 3(c) shows small objects/noise have been removed from the binarized image. Figure 3(d) shows the algorithm used bounded the large unwanted objects and large objects are then removed as shown in Fig. 3(e). Then, the remaining objects are then deblurred and bounded as shown in Fig. 3(f). Figure 3(g) shows the car plate is then extracted and saved separately after the properties are compared among all objects. Finally, Fig. 3(h) shows the image is segmented and thinned with the thinning algorithm proposed by [26].

4.2 Recognition Phase

The experimental results showed the proposed BPNN was capable to recognize most of the characters extracted from the car plate. However, the BPNN model performed slightly worse in recognizing the characters J, M, 7 and 8. The recognition success rates were 90.91 % for character "J", 85.71 % for character "M", 97.22 % for character "7", and 97.14 % for character "8".

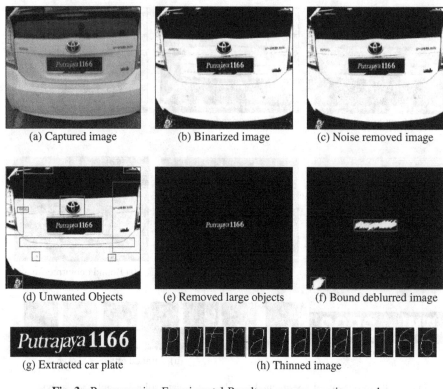

(a) Captured image   (b) Binarized image   (c) Noise removed image

(d) Unwanted Objects   (e) Removed large objects   (f) Bound deblurred image

(g) Extracted car plate   (h) Thinned image

Fig. 3. Pre-processing Experimental Results – commemorative car plate.

Table 2. Character distributions, BPNN, RBF, and ENN comparison results.

Characters	Number of characters involved			Success rate in %		
	500 samples	300 training	200 testing	BPNN	RBF	ENN
A	278	178	100	100 %	100 %	100 %
B	129	93	36	100 %	97.22 %	100 %
C	23	18	5	100 %	100 %	100 %
D	19	12	7	100 %	100 %	100 %
E	27	10	17	100 %	100 %	100 %
F	18	10	8	100 %	100 %	100 %
G	12	4	8	100 %	100 %	100 %
H	23	13	10	100 %	100 %	100 %
J	22	11	11	90.91 %	100 %	100 %
K	25	14	11	100 %	100 %	100 %
L	32	25	7	100 %	100 %	100 %
M	25	18	7	85.71 %	100 %	100 %
N	12	10	2	100 %	100 %	100 %

(*Continued*)

Table 2. *(Continued)*

Characters	Number of characters involved			Success rate in %		
	500 samples	300 training	200 testing	BPNN	RBF	ENN
P	24	14	10	100 %	100 %	100 %
Q	17	13	4	100 %	100 %	100 %
R	17	12	5	100 %	100 %	100 %
S	239	173	66	100 %	96.97 %	98.48 %
T	25	16	9	100 %	100 %	100 %
U	20	14	6	100 %	100 %	100 %
V	22	17	5	100 %	100 %	100 %
W	36	24	12	100 %	100 %	100 %
X	12	6	6	100 %	100 %	100 %
Y	9	4	5	100 %	100 %	100 %
1	113	77	36	100 %	100 %	100 %
2	125	84	41	100 %	100 %	100 %
3	97	59	38	100 %	100 %	100 %
4	102	62	40	100 %	100 %	100 %
5	109	77	32	100 %	100 %	100 %
6	126	83	43	100 %	100 %	100 %
7	116	80	36	97.22 %	100 %	100 %
8	138	103	35	97.14 %	100 %	100 %
9	149	97	52	100 %	100 %	100 %
0	78	48	30	100 %	86.67 %	100 %
e	4	2	2	100 %	100 %	100 %
b	3	2	1	100 %	100 %	100 %
p	3	2	1	100 %	100 %	100 %
u	3	2	1	100 %	100 %	100 %
t	3	2	1	100 %	100 %	100 %
r	3	2	1	100 %	100 %	100 %
a	5	2	3	100 %	100 %	100 %
j	3	2	1	100 %	100 %	100 %
y	3	2	1	100 %	100 %	100 %
PERODUA	3	2	1	100 %	100 %	100 %
Total	2249	1217	1032			

The RBF results showed the NN used performed slightly weak in recognizing the character B, with 97.22 % of success rate, character S with 96.97 % success rate, and slightly low performance for recognizing character 0/zero with 86.67 % of success rate. The experimental results showed the ENN performance is better. The proposed ENN was successfully recognized all characters except character S, with 98.48 % of success rate. Table 2 below summarized the experimental results.

5   Conclusions and Future Works

In this research paper, we discussed the deblur technique used for localizing car plate and compared the performance of BPNN, RBF, and ENN in the recognizing phase. The proposed localization technique is simple and does not required predefined template. However, there was failure as unclear image had been accidently included to the dataset without prior filtered. The proposed technique involves binarization, small and large objects filtering, bounding deblurred objects, car plate extraction, and thinning. The proposed technique solved the five challenges mentioned in Sect. 2.2.

The performance of the proposed BPNN, RBF, and ENN had been compared. The ENN testing result performed better compared to both BPNN and RBF. However, the ENN model was not able to achieve a 100 % success rate in recognizing all characters. Probably it happened due to small sample size used for certain characters. Hence, it is difficult for the ENN to learn the S pattern. The RBF performed slightly better than BPNN as RBF performance weak in recognizing the B, S and zero characters but RBF performed good in recognizing the K, M, 7, and 8 characters.

There are still rooms for improvement although the proposed algorithms could solve the challenges mentioned in Sect. 2.2. Larger data set should be considered in order to prepare fairly characters distribution. Secondly, video source should be considered as well instead of static images. We will also perform further experimentation on images that are captured from various angles as well as images that are taken when it is raining or foggy.

References

1. Sharma, A., Chaudhary, D.R.: Character recognition using neural network. Int. J. Eng. Trends Technol. (IJETT) **4**, 662–667 (2013)
2. Durai, S.A., Saro, E.A.: Image compression with back-propagation neural network using cumulative distribution function. World Acad. Sci. Eng. Technol. **17**, 60–64 (2006)
3. On, C.K., Pandiyan, P.M., Yaacob, S., Saudi, A.: Mel-frequency cepstral coefficient analysis in speech recognition. In: 2006 International Conference on Computing & Informatics, pp. 1–5. IEEE (2006)
4. Taylor, J.W., Buizza, R.: Neural network load forecasting with weather ensemble predictions. Trans. Power Syst. **17**(3), 626–632 (2002)
5. Vui, C.S., Soon, G.K., On, C.K., Alfred, R., Anthony, P.: A review of stock market prediction with Artificial neural network (ANN). In: 2013 IEEE International Conference on Control System, Computing and Engineering (ICCSCE), pp. 477–482. IEEE (2013)
6. Pham, D., Karaboga, D.: Intelligent optimisation techniques: genetic algorithms, tabu search, simulated annealing and neural networks. Springer Science & Business Media, London (2012)
7. Tong, C.K., On, C.K., Teo, J., Kiring, A.M.J.: Evolving neural controllers using GA for Warcraft 3-real time strategy game. In: 2011 Sixth International Conference on Bio-Inspired Computing: Theories and Applications (BIC-TA), pp. 15–20. IEEE (2011)
8. Oreski, S., Oreski, D., Oreski, G.: Hybrid system with genetic algorithm and artificial neural networks and its application to retail credit risk assessment. Expert Syst. Appl. **39**(16), 12605–12617 (2012)

9. Amato, F., López, A., Peña-Méndez, E.M., Vaňhara, P., Hampl, A., Havel, J.: Artificial neural networks in medical diagnosis. J. Appl. Biomed. **11**(2), 47–58 (2013)
10. On, C.K., Pandiyan, P.M., Yaacob, S., Saudi, A.: Fingerprint feature extraction based discrete cosine transformation (DCT). In: 2006 International Conference on Computing & Informatics, pp. 1–5. IEEE (2006)
11. Rumelhart, D.E., Hinton, G.E., Williams, R.J.: Learning representations by back-propagating errors. Cogn. Model. **5**(3), 1 (1988)
12. Shi, J.L., Tan, T.G., Teo, J., Chin, K.O., Alfred, R., Anthony, P.: Evolving Controllers For Simulated Car Racing Using Differential Evolution. Asia-Pacific J. Inf. Technol. Multimedia **2**(1), 57–68 (2013)
13. Nagahamulla, H.R., Ratnayake, U.R., Ratnaweera, A.: An ensemble of artificial neural networks in rainfall forecasting. In: 2012 International Conference on Advances in ICT for Emerging Regions (ICTer), pp. 176–181. IEEE (2012)
14. Wang, J., Yang, J., Li, S., Dai, Q., Xie, J.: Number image recognition based on neural network ensemble. In: Third International Conference on Natural Computation (ICNC 2007), vol. 1, pp. 237–240. IEEE (2007)
15. Fang, M., Liang, C., Zhao, X.: A method based on rough set and SOFM neural network for the car's plate character recognition. In: Fifth World Congress on Intelligent Control and Automation, WCICA 2004, vol. 5, pp. 4037–4040. IEEE (2004)
16. Jusoh, N.A., Zain, J.M., Kadir, T.A.A.: Enhancing thinning method for malaysian car plates recognition. In: Second International Conference on IEEE Innovative Computing, Information and Control, ICICIC 2007, pp. 378–378. IEEE (2007)
17. Lee, E.R., Kim, P.K., Kim, H.J.: Automatic recognition of a car license plate using color image Processing. In: IEEE International Conference on Image Processing, ICIP 1994, vol. 2, pp. 301–305 (1994)
18. Rattanathammawat, P., Chalidabhongse, T.H.: A car plate detector using edge information. In: International Symposium on IEEE Communications and Information Technologies, ISCIT 2006, pp. 1039–1043. IEEE (2006)
19. Nelson, C., Babu, K., Nallaperumal, K.: A license plate localization using morphology and recognition. In: Annual IEEE India Conference, INDICON, pp. 34–39. IEEE (2010)
20. Wen, W., Huang, X., Yang, L., Yang, Z., Zhang, P.: Vehicle license plate location method based-on wavelet transform. In: International Joint Conference on IEEE Computational Sciences and Optimization, CSO 2009, vol. 2, pp. 381–384. IEEE (2010)
21. You-qing, Z., Cui-hua, L.: A recognition method of Car License Plate Characters based on template matching using modified Hausdorff distance. In: International Conference on IEEE Computer, Mechatronics, Control and Electronic Engineering (CMCE), vol. 6, pp. 25–28. IEEE (2010)
22. Setumin, S., Sheikh, U.U., Abu-Bakar, S.A.R.: Character-based car plate de tection and localization. In: Computer Vision, Video and Image Processing, Universiti Teknologi Malaysia, pp. 737–740. IEEE (2010)
23. Setumin, S., Sheikh, U.U., Abu-Bakar, S.A.R.: Car plate character extraction and recognition using stroke analysis. In: Sixth International Conference on Signal-Image Technology and Internet-Based Systems (SITIS), pp. 30–34. IEEE (2010)
24. IM4U number plate launched. http://paultan.org/2013/03/04/im4u-registration-plate-launched-on-sale-march-10/
25. Businessman Buys Regal Plate Number. http://www.thestar.com.my/News/Community/2014/09/09/Businessman-buys-regal-plate-number/
26. Certain Improvement in Pixel Based Thinning Algorithm using Neural Network. http://ofuturescholar.com/paperpage?docid=312198

Multi-step Time Series Forecasting Using Ridge Polynomial Neural Network with Error-Output Feedbacks

Waddah Waheeb[1,2(✉)] and Rozaida Ghazali[1]

[1] Faculty of Computer Science and Information Technology, Universiti Tun Hussein Onn Malaysia, Batu Pahat, 86400 Parit Raja, Johor, Malaysia
rozaida@uthm.edu.my
[2] Computer Science Department, Hodeidah University, P.O. Box 3114 Alduraihimi, Hodeidah, Yemen
waddah.waheeb@gmail.com

Abstract. Time series forecasting gets much attention due to its impact on many practical applications. Higher-order neural network with recurrent feedback is a powerful technique which used successfully for forecasting. It maintains fast learning and the ability to learn the dynamics of the series over time. For that, in this paper, we propose a novel model, called Ridge Polynomial Neural Network with Error-Output Feedbacks (RPNN-EOF), which combines three powerful properties: higher order terms, output feedback and error feedback. The well-known Mackey–Glass time series is used to evaluate the forecasting capability of RPNN-EOF. Results show that the proposed RPNN-EOF provides better understanding for the Mackey–Glass time series with root mean square error equal to 0.00416. This error is smaller than other models in the literature. Therefore, we can conclude that the RPNN-EOF can be applied successfully for time series forecasting. Furthermore, the error-output feedbacks can be investigated and applied with different neural network models.

Keywords: Time series forecasting · Ridge polynomial neural network with error-output feedbacks · Higher order neural networks · Recurrent neural networks · Mackey–glass equation

1 Introduction

Time series forecasting approaches have been widely applied to many fields such as financial forecasting, weather forecasting, traffic forecasting, etc. The aim of time series forecasting is building an approach that use past observations to forecast the future. For example, using a series of data $x_{t-n}, \ldots, x_{t-2}, x_{t-1}, x_t$ to forecasts data values $x_{t+1}, \ldots, x_{t+m}$. Generally, time series forecasting approaches can be classified into two approaches; statistical-based and intelligent-based approaches. Due to the nonlinear nature of most of time series signals, intelligent-based approaches have shown better performance than statistical approaches in time series forecasting [1].

Artificial Neural network (ANN), which is inspired by biological nervous systems, is an example of intelligent-based approaches. ANN can learn from historical data and

© Springer Nature Singapore Pte Ltd. 2016
M.W. Berry et al. (Eds.): SCDS 2016, CCIS 652, pp. 48–58, 2016.
DOI: 10.1007/978-981-10-2777-2_5

learn its weight matrices to construct model that can forecast the future. Due to the nonlinear nature and the ability to produce complex nonlinear input-output mapping, ANN have been used successfully for time series forecasting [2].

Generally, ANNs can be grouped into two groups based on network structure; feedforward and recurrent networks [2]. In feedforward networks, the data flows in one direction only from the input nodes to the output nodes through network connections (i.e. weights). On other hand, the connections between the nodes in recurrent networks form a cycle.

Multilayer perceptron (MLP) is one of the most used feedforward ANNs in forecasting tasks [3]. However, due to the multilayered structure of MLP, it needs a large number of units to solve complex nonlinear mapping problems, which results in low learning rate and poor generalization [4]. To overcome these drawbacks, different types of single layer higher order neural networks (HONNs) with product neurons were introduced. Ridge Polynomial Neural Network (RPNN) [5] is a feedforward HONNs that maintain fast learning and powerful mapping properties, and is thus suitable for solving complex problems [3].

Two recurrent versions of RPNNs are existed, namely Dynamic Ridge Polynomial Neural Networks (DRPNN) [6] and Ridge Polynomial Neural Networks with Error Feedback (RPNN-EF) [7]. DRPNN uses the output value from the output layer as a feedback connection to the input layer. On other hand, RPNN-EF uses the network error which is calculated by subtracting the desired value from the forecast value. The idea behind recurrent networks is learning the network the dynamics of the series over time. As a result, the network should use this memory to improve the results of the forecasting [8]. DRPNN and RPNN-EF have been successfully used for time series forecasting [1, 6, 7, 9].

Due to the success of DRPNN and RPNN-EF, in this paper we propose a model that combine the properties of RPNNs and output-error feedbacks recurrent neural networks. This model is called Ridge Polynomial Neural Network with Error-Output Feedbacks (RPNN-EOF). We applied the RPNN-EOF to the chaotic Mackey-Glass differential delay equation series which is recognized as a benchmark problem that has been used and reported by many researchers for comparing the generalization ability of different models [10–18].

This paper consists of six sections. Section 2 introduces the basic concepts of RPNN, DRPNN and RPNN-EF. We describe the proposed model in Sect. 3. Section 4 covers the experimental settings. Section 5 is about results and discussion. And finally, Sect. 6 concludes the paper.

2 The Existing Ridge Polynomial Neural Network Based Models

This section discusses the existing Ridge Polynomial Neural Networks based models, namely Ridge Polynomial Neural Networks (RPNNs), Dynamic Ridge Polynomial Neural Networks (DRPNNs) and Ridge Polynomial Neural Networks with Error Feedback (RPNN-EF).

2.1   Ridge Polynomial Neural Networks (RPNNs)

RPNN [5] is an example of feedforward higher order neural networks (HONNs) that uses one layer of trainable weights. RPNNs maintain powerful mapping capabilities and fast learning properties of single layer HONNs [3]. They are constructed by adding different degrees of Pi-Sigma Neural Networks (PSNNs) blocks [19] until a defined goal is achieved. RPNNs can approximate any continuous function on a compact set in multidimensional input space with arbitrary degree of accuracy [5]. RPNNs utilize univariate polynomials which help to avoid an explosion of free parameters that found in some types of higher order feedforward neural networks [5].

2.2   Dynamic Ridge Polynomial Neural Networks (DRPNNs)

DRPNN [6] is one type of recurrent version of RPNNs. DRPNNs take advantage of the network output value as an additional input to the input layer. They are provided with memories that help to retain information to be used later [6]. DRPNN is trained by the real time recurrent learning algorithm (RTRL) [20].

DRPNNs can be considered to be more suitable than RPNNs for time series forecasting due the fact that the behavior of some time series signals related to some past inputs on which the present inputs depends. Interested readers for the application of DRPNNs for time series forecasting may be referred to [1, 6, 9].

2.3   Ridge Polynomial Neural Network with Error Feedback (RPNN-EF)

Another recurrent type of RPNNs is RPNN-EF [7]. Unlike DRPNN, RPNN-EF takes advantage of the network error value as an additional input to the input layer. This error is calculated by taking the difference between the desired output and network output. Such error feedback is also used in the literature with Functional Link Network (FLN) and Adaptive Neuro-Fuzzy Inference System (ANFIS) model [21, 22]. Like DRPNN, RPNN-EF is trained by RTRL algorithm.

RPNN-EF showed better understanding for multi-step ahead forecasting than RPNN and DRPNN. Furthermore, RPNN-EF was significantly faster than other RPNN-based models for one-step ahead forecasting [7].

3   The Proposed Model: Ridge Polynomial Neural Network with Error-Output Feedbacks (RPNN-EOF)

Due to the success of DRPNN, RPNN-EF for time series forecasting [1, 6, 7, 9], we propose Ridge Polynomial Neural Networks with Error-Output Feedbacks (RPNN-EOF). This model combines the properties of RPNNs, and the powerful of both error and output feedbacks.

The generic network architecture of the RPNN-EOF using Pi-Sigma neural networks as basic building blocks is shown in Fig. 1. Like other RPNN based models, RPNN-EOF uses constructive learning method. That means the network structure

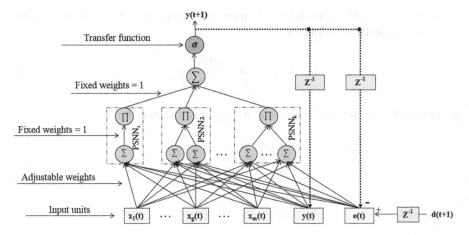

Fig. 1. Ridge Polynomial Neural Networks with Error-Output Feedbacks. *PSNN, d(t)* stands for Pi-Sigma Neural Network and the desired output at time *t*. Bias nodes are not shown here for reason of simplicity.

grows from a small network and the network becomes larger as learning proceeds until the desired level of defined error is reached [5].

RPNN-EOF is trained by RTRL algorithm. The output of RPNN-EOF, which is denoted by $y(t)$, is calculated as follows:

$$y(t) = \sigma\left(\sum_{i=1}^{k} P_i(t)\right),$$

$$P_i(t) = \prod_{j=1}^{i} h_j(t),$$ (1)

$$h_j(t) = \sum_{g=1}^{m+2} (w_{gj} * Z_g(t)) + w_{oj}.$$

where $P_i(t)$ is the output of Pi-Sigma block, σ is the transfer function, $h_j(t)$ is the net sum of the sigma unit j, w_{oj} is the bias, w_{gj} is the weights between input and sigma units, and $Z(t)$ is the inputs which given as follow:

$$Z_g(t) = \begin{cases} x_g(t) & 1 \leq g \leq M \\ e(t-1) = d(t-1) - y(t-1) & g = M+1 \\ y(t-1) & g = M+2 \end{cases}$$ (2)

Network error is calculated using the sum squared error as follows:

$$E(t) = \frac{1}{2}\sum e(t)^2,$$ (3)

$$e(t) = d(t) - y(t)$$ (4)

where d is the desired output and y is the predicted output. At every time t, the weights changes are calculated as follows:

$$\Delta w_{gl} = -\eta * \left(\frac{\partial E(t)}{\partial w_{gl}} \right) \tag{5}$$

where η is the learning rate. The value of $\left(\frac{\partial E(t)}{\partial w_{gl}} \right)$ is determined as:

$$\frac{\partial E(t)}{\partial w_{gl}} = e(t) * \left(\frac{\partial e(t)}{\partial w_{gl}} \right) \tag{6}$$

$$\frac{\partial e(t)}{\partial w_{gl}} = \frac{\partial e(t)}{\partial y(t)} * \frac{\partial y(t)}{\partial w_{gl}} \tag{7}$$

$$\frac{\partial e(t)}{\partial w_{gl}} = -\frac{\partial y(t)}{\partial w_{gl}} \tag{8}$$

$$\frac{\partial y(t)}{\partial w_{gl}} = \frac{\partial y(t)}{\partial P_i(t)} * \frac{\partial P_i(t)}{\partial w_{gl}} \tag{9}$$

From Eq. (1), we have

$$
\begin{aligned}
\frac{\partial y(t)}{\partial w_{gl}} = \frac{\partial y(t)}{\partial P_i(t)} * \frac{\partial P_i(t)}{\partial w_{gl}} &= (y(t))' * \left(\prod_{j=1, j \neq l}^{i} h_j(t) \right) \\
&* \left(Z_g(t) + \left(w_{(M+1)j} * \frac{\partial e(t-1)}{\partial w_{gl}} \right) + \left(w_{(M+2)j} * \frac{\partial y(t-1)}{\partial w_{gl}} \right) \right)
\end{aligned}
\tag{10}
$$

Assume D^Y and D^E as dynamic system variables [2], where D^Y and D^E are:

$$D_{gl}^Y(t) = \frac{\partial y(t)}{\partial w_{gl}} \tag{11}$$

$$D_{gl}^E(t) = \frac{\partial e(t)}{\partial w_{gl}} \tag{12}$$

Substituting Eqs. (11) and (12) into Eq. (10), we have

$$
\begin{aligned}
\frac{\partial y(t)}{\partial w_{gl}} &= (y(t))' * \left(\prod_{j=1, j \neq l}^{i} h_j(t) \right) \\
&* \left(Z_g(t) + \left(w_{(M+1)j} * D_{gl}^E(t-1) \right) + \left(w_{(M+2)j} * D_{gl}^Y(t-1) \right) \right)
\end{aligned}
\tag{13}
$$

Recall Eq. (8), from Eq. (12), we have

$$D_{gl}^E(t) = -D_{gl}^Y(t) \tag{14}$$

Substituting Eq. (14) into Eq. (13), we have

$$D_{gl}^Y(t) = (y(t))' * \left(\prod_{j=1, j\neq l}^{i} h_j(t) \right)$$
$$* \left(Z_g(t) + D_{gl}^Y(t) \left(w_{(M+2)j} - w_{(M+1)j} \right) \right) \tag{15}$$

For simplification, the initial values for $D_{gl}^Y(t) = 0$, and $e(t) = y(t) = 0.5$ to avoid zero value of $D_{gl}^Y(t)$. Then the weights updating rule is derived by substituting Eqs. (15) and (8) into Eq. (6), we have

$$\frac{\partial E(t)}{\partial w_{gl}} = -e(t) * D_{gl}^Y(t) \tag{16}$$

Then, substituting Eq. (16) into Eq. (5), we have

$$\Delta w_{gl} = \eta * e(t) * D_{gl}^Y(t) \tag{17}$$

4 Experimental Design

4.1 Mackey-Glass Differential Delay Equation

In this paper, we used the well-known chaotic Mackey-Glass differential delay equation series. This series is recognized as a benchmark problem that has been used and reported by many researchers for comparing the generalization ability of different models [10–18]. Mackey-Glass time series is given by the following delay differential equation:

$$\frac{dx}{dt} = \beta x(t) + \frac{\alpha x(t - \tau)}{1 + x^{10}(t - \tau)} \tag{18}$$

where τ is the time delay. We chose the following values for the variables $\alpha = 0.2$, $\beta = -0.1$, $x(0) = 1.2$ and $\tau = 17$. With this setting the series produce chaotic behavior. One thousand data point were generated. The first 500 points of the series were used as a training sample, while the remaining 500 points were used as out-of-sample data. We used four input variables $x(t)$, $x(t - 6)$, $x(t - 12)$, $x(t - 18)$, to predict $x(t + 6)$. All these settings were used for fair comparison with other studies in the literature [10–18].

4.2 Data Preprocessing

We scaled the points to the range [0.2, 0.8] to avoid getting network output too close to the two endpoints of sigmoid transfer function [1]. We used the minimum and maximum normalization method which is given by:

$$\hat{x} = (\max{}_2 - \min{}_2) * \left(\frac{x - \min_1}{\max{}_1 - \min_1} \right) + \min{}_2. \tag{19}$$

where x refers to the observed (original) value, $\hat{x}$ is the normalized version of x, min_1 and max_1 are the respective minimum and maximum values of all observations, and min_2 and max_2 refer to the desired minimum and maximum of the new scaled series.

4.3 Network Topology

The topology of the RPNN-EOF that we used is shown in Table 1. Most of the settings are either based on the previous works with RPNN based models that found in the literature [1, 6, 7, 9] or by trial and error.

Table 1. Network topology.

Setting	Value
Activation function	Sigmoid function
Number of Pi-Sigma block (PSNN)	Incrementally grown from 1 to 5
Stopping criteria	Maximum number of epochs = 3000 or after accomplishing the 5th order network learning
Initial weights	[−0.5,0.5]
Momentum	[0.4−0.8][a]
Learning rate (n)	[0.01−1][a]
Decreasing factors for n	0.8
Threshold of successive PSNN addition (r)	[0.00001-0.1][a]
Decreasing factors for r	[0.05, 0.2][a]

[a]This setting is based on trial and error.

4.4 Performance Metrics

Because we aim to compare our proposed model's performance with other models in the literature, we used the Root Mean Squared Error (RMSE) metric. RMSE is the standard metric which used by many researchers with Mackey-Glass series [10–18]. The equation for RMSE is given by:

$$RMSE = \sqrt{\frac{1}{N} \sum_{i=1}^{N} (y_i - \hat{y}_i)^2} \qquad (20)$$

5 Results and Discussion

The forecasting model of Mackey-Glass time series is built via the experimental design settings. The best out-of-sample data forecasting for RPNN-EOF is shown in Fig. 2. It can be seen that the RPNN-EOF can follow the dynamic behavior of the series precisely. To show the difference between the output of the RPNN-EOF and the out-of-sample points, which is called forecasting error, the forecasting error is plotted in Fig. 3.

For fair comparison, we de-normalized the results of RPNN-EOF model. Table 2 lists the generalization capabilities of the compared methods [10–18]. The generalization capabilities were measured by applying each model to forecast the out-of-sample data. The results show that RPNN-EOF offers a smaller RMSE than the other models. Based on these results, we can conclude that the RPNN-EOF alone provides better understanding for the series and smaller error in comparison to other hybrid models.

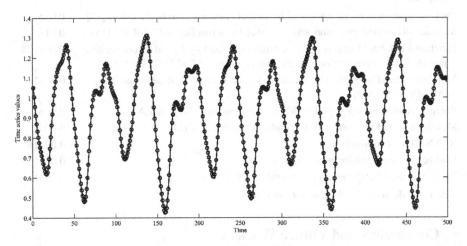

Fig. 2. The best forecasting results for Mackey–Glass time series using RPNN-EOF. *Circle* is the original series while the *solid* line is the forecast series.

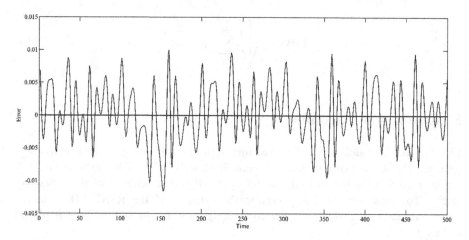

Fig. 3. Forecasting error for Mackey–Glass time series using RPNN-EOF.

Table 2. Comparison of the performance of various existing models.

Model	RMSE
Differential evolution - beta basis function neural networks (DE-BBFNN) [10]	0.030
Dynamic evolving computation system (DECS) [11]	0.0289
Orthogonal function neural network [12]	0.016
Multilayer feedforward neural network - Backpropagation algorithm (MLFBP) [13]	0.0155
Backpropagation network optimized by hybrid K-means-greedy algorithm [14]	0.015
Modified differential evolution and the radial basis function (MDE-RBF) [15]	0.013
Functional-link-based neural fuzzy network optimized by hybrid of cooperative particle swarm optimization and cultural algorithm (FLNFN-CCPSO) [16]	0.008274
Multilayer neural network with the multi-valued neurons- QR decomposition (MLMVN-QR) [13]	0.0065
Wavelet neural network with hybrid learning approach (WNN-HLA) [17]	0.006
Multilayer neural network with the multi-valued neurons (MLMVN) [13]	0.0056
RPNN-EOF (proposed)	**0.00416**[a]
Grid-based fuzzy systems with 192 rules [18]	0.0041
Multigrid-based fuzzy system 3 sub-grids with 120 rules [18]	0.0031

[a]This is the de-normalized value for the RMSE.

6   Conclusions and Future Works

This paper investigated the forecasting capability of the Ridge Polynomial Neural Network with Error-Output Feedbacks (RPNN-EOF) for multi-step time series forecasting. The well-known Mackey–Glass differential delay equation was used to evaluate the forecasting capability of RPNN-EOF. Simulation results showed that the proposed RPNN-EOF provides better understanding for the Mackey–Glass time series and smaller error in comparison to other models in the literature. Therefore, we can conclude that the RPNN-EOF can be applied successfully for time series forecasting. The future works will be using more time series to ensure the good performance of the proposed model. Furthermore, the error-output feedbacks can be investigated and applied with different neural network models.

Acknowledgments. The authors would like to thank Universiti Tun Hussein Onn Malaysia (UTHM) and Ministry of Higher Education (MOHE) Malaysia for financially supporting this research under the Fundamental Research Grant Scheme (FRGS), Vote No. 1235.

References

1. Al-Jumeily, D., Ghazali, R., Hussain, A.: Predicting Physical Time Series Using Dynamic Ridge Polynomial Neural Networks. PLOS ONE (2014)
2. Haykin, S.S.: Neural Networks and Learning Machines. Prentice Hall, New Jersey (2009)
3. Ghazali, R., Hussain, A.J., Liatsis, P., Tawfik, H.: The application of ridge polynomial neural network to multi-step ahead financial time series prediction. Neural Comput. Appl. **17**(3), 311–323 (2008)
4. Yu, X., Tang, L., Chen, Q., Xu, C.: Monotonicity and convergence of asynchronous update gradient method for ridge polynomial neural network. Neurocomputing **129**, 437–444 (2014)
5. Shin, Y., Ghosh, J.: Ridge polynomial networks. IEEE T. Neural Networ. **6**(3), 610–622 (1995)
6. Ghazali, R., Hussain, A.J., Nawi, N.M., Mohamad, B.: Non-stationary and stationary prediction of financial time series using dynamic ridge polynomial neural network. Neurocomputing **72**(10), 2359–2367 (2009)
7. Waheeb, W., Ghazali, R., Herawan, T.: Time series forecasting using ridge polynomial neural network with error feedback. In: Proceedings of the Second International Conference on Soft Computing and Data Mining (SCDM-2016) (in press)
8. Samarasinghe, S.: Neural Networks for Applied Sciences and Engineering: From Fundamentals to Complex Pattern Recognition. CRC Press, New York (2006)
9. Ghazali, R., Hussain, A.J., Liatsis, P.: Dynamic ridge polynomial neural network: forecasting the univariate non-stationary and stationary trading signals. Expert Syst. Appl. **38**(4), 3765–3776 (2011)
10. Dhahri, H., Alimi, A.: Automatic selection for the beta basis function neural networks. In: Krasnogor, N., Nicosia, G., Pavone, M., Pelta, D. (eds.) Nature Inspired Cooperative Strategies for Optimization (NICSO 2007). Studies in Computational Intelligence, vol. 129, pp. 461–474. Springer, Heidelberg (2008)
11. Chen, Y.M., Lin, C.T.: Dynamic parameter optimization of evolutionary computation for on-line prediction of time series with changing dynamics. Appl. Soft Comput. **7**(4), 1170–1176 (2007)
12. Wang, H., Gu, H.: Prediction of chaotic time series based on neural network with legendre polynomials. In: Yu, W., He, H., Zhang, N. (eds.) ISNN 2009, Part I. LNCS, vol. 5551, pp. 836–843. Springer, Heidelberg (2009)
13. Aizenberg, I., Luchetta, A., Manetti, S.: A modified learning algorithm for the multilayer neural network with multi-valued neurons based on the complex QR decomposition. Soft. Comput. **16**(4), 563–575 (2012)

14. Tan, J.Y., Bong, D.B., Rigit, A.R.: Time series prediction using backpropagation network optimized by hybrid K-means-greedy algorithm. Eng. Lett. **20**(3), 203–210 (2012)
15. Dhahri, H., Alimi, A.M.: The modified differential evolution and the RBF (MDE-RBF) neural network for time series prediction. In: The 2006 IEEE International Joint Conference on Neural Network Proceedings, pp. 2938–2943. IEEE (2006)
16. Lin, C.J., Chen, C.H., Lin, C.T.: A hybrid of cooperative particle swarm optimization and cultural algorithm for neural fuzzy networks and its prediction applications. IEEE Trans. Syst. Man Cybernetics Part C (Applications and Reviews) **39**(1), 55–68 (2009)
17. Lin, C.J.: Wavelet neural networks with a hybrid learning approach. J. Inf. Sci. Eng. **22**(6), 1367–1387 (2006)
18. Herrera, L.J., Pomares, H., Rojas, I., Guillén, A., González, J., Awad, M., Herrera, A.: Multigrid-based fuzzy systems for time series prediction: CATS competition. Neurocomputing. **70**(13), 2410–2425 (2007)
19. Shin, Y., Ghosh, J.: The Pi-sigma network: an efficient higher-order neural network for pattern classification and function approximation. In: IJCNN-91-Seattle International Joint Conference Neural Networks, vol. 1, pp. 13–18. IEEE (1991)
20. Williams, R.J., Zipser, D.: A learning algorithm for continually running fully recurrent neural networks. Neural Comput. **1**(2), 270–280 (1989)
21. Dash, P.K., Satpathy, H.P., Liew, A.C., Rahman, S.: A real-time short-term load forecasting system using functional link network. IEEE Trans. Power Syst. **12**(2), 675–680 (1997)
22. Mahmud, M.S., Meesad, P.: An innovative recurrent error-based neuro-fuzzy system with momentum for stock price prediction. Soft. Comput. 1–19 (2015)

Classification/Clustering/Visualization

Comparative Feature Selection of Crime Data in Thailand

Tanavich Sithiprom[✉] and Anongnart Srivihok[✉]

Department of Computer Science, Kasetsart University,
Bangkok, Thailand
{tanavich.s, fsciang}@ku.ac.th

Abstract. The crime is a major problem of community and society which is increasing day by day. Especially in Thailand, crime is a major problem that affects all aspects of the country such as tourism, administration of government and problem in daily life. Therefore, government and private sectors have to understand the several crime patterns for planning, preventing and solving solution of crime correctly. The purposes of this study are to generate a crime model for Thailand using data mining techniques. Data were collected from Dailynews and Thairath online newspapers. The proposed model can be generated by using more feature selection and more classification techniques to different model. Experiments show feature selection with the wrapper of attribute evaluator seems to be an appropriate evaluation algorithm because data set mostly is the best accuracy rate. This improves efficiency in identifying offenders more quickly and accurately. The model can be used for the prevention of crime that will occur in Thailand in the future.

Keywords: Crime · Feature selection · Classification · Data mining

1 Introduction

A crime is an act that violates the law. A criminal is a person who commits a crime [1]. Crime in Thailand is divided to 5 groups according to the police department [2], the Ministry of Interior. So, 5 groups include (1) crimes related to violent and horrible, (2) crimes related to life body and sex, (3) crimes related to asset, (4) crimes related to burglary and (5) crimes related to government are victim.

Nowadays, there are too many crimes reported in online news and the numbers of crime are increased day by day. Criminals included several types such as domestic violence, fraud, drugs and drugging, computer crimes, rape and human trafficking and prostitution [3]. Hence, officers who are responsible for crime cannot recognize data thoroughly. Thus, currently it is necessary to apply information technology to support data processing and analyzing crime for planning to preventing crimes that may occur in the future.

The statistics of reports and arrested for violent crime groups for the whole kingdom of Thailand [4] including murder, gang robbery, robbery, kidnapping and arson, was reported in 2006. Crimes of total 8,738 cases included 4,687 murder cases, 1,244

© Springer Nature Singapore Pte Ltd. 2016
M.W. Berry et al. (Eds.): SCDS 2016, CCIS 652, pp. 61–71, 2016.
DOI: 10.1007/978-981-10-2777-2_6

gang robbery cases, 2,319 robbery cases, 11 kidnapping cases and 477 arson cases, In 2015, they had reported total 3,874 cases included 2,228 murder cases, 364 gang robbery cases, 1,044 robbery cases, 7 kidnapping cases, and 232 arson cases. The arrested crimes in 2006 were total 3,774 cases included 1,888 murder cases, 687 gang robbery cases, 1,068 robbery cases, 5 kidnapping cases, and 126 arson cases but in 2015 they had arrested total 2,774 cases included 1,570 murder cases, 281 gang robbery cases, 756 robbery cases, 6 kidnapping cases and 161 arson cases. Thus the numbers of crimes and the arrested cases was decreased in 2015.

2 Related Study

Data mining can be used to analyze crime data in several ways. For example, in 2006, there was a study about crime pattern detection using data mining [5], by clustering of crime data set from sheriff's office Metairie Louisiana United States. This information is useful for investigators. The data set included 6 features: crime type, suspect race, suspect sex, suspect age, victim age and weapon. Moreover, they used K-Means clustering with weighting features. Therefore, the results are useful for investigators and predicts group of crime.

The collection of crime data which were unstructured for mining by using association rules mining [6]. The data set included 12 features: victim name, offender name, location, organization, crime type, crime subcategory, vehicle, weapon, time, date, event summary, and crime group involved. The performances of two algorithms Naïve Bayes and Support Vector Machine (SVM) were compared. The results show that precision of SVM is better than NB, but NB uses less memory and less time than SVM. The association rules mining, Apriori algorithm was used to find pattern which was happening repeatedly and suggesting locations with risks of crimes. Finally, Decision Tree was used to predict pattern of crimes.

3 Related Theory

3.1 Feature Selection

The feature selection process was using for choosing the best features which were met the criterion. This will reduce dimensional of features and enhance the efficiency of prediction. Thus, the data mining will process faster and the model will learn better. The feature selection included four steps: subset generation, evaluation, stopping criterion and result validation [7]. The process was started by generating subset of all features and evaluating each subset. That was repeated until stop when criteria are defined and then evaluated subset with classification algorithms. (as depicted in Fig. 1).

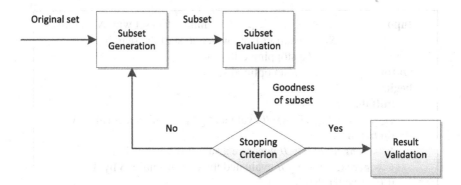

Fig. 1. Four key steps of feature selection [7].

OPEN = [initial state]
CLOSED = []
While OPEN is not empty
Do
    1. Remove the best node from OPEN, call it n, and add it to CLOSED.
    2. If n is the goal state, backtrace path to n (through recorded parents)
and return path.
    3. Create n's successors.
    4. For each successor do:
        a. If it is not in CLOSED and it is not in OPEN: evaluate it, add it
to OPEN, and record its parent.
        b. Otherwise, if this new path is better than previous one, change
its record parent.
            i. If it is not in OPEN add it to OPEN.
            ii. Otherwise, adjust its priority in OPEN using this new evalu-
ation.
Done

Fig. 2. Best-first search algorithm [9].

3.1.1   Search Method

Best-First Search is a search method by combining advantage of Breath-first search and depth-first search with heuristic function to select the best variable by greedy algorithm. The complexity is $O(b^m)$, it follows that b is average of child node of each node and m is height depth of the tree [8] (Figs. 2 and 3)

```
Input:       D(F_0, F_1, ..., F_{n-1}) // a training data set with N features
             S_0  // a subset from which to start the search
             δ   // a stopping criterion
Output:      S_best   // an optimal subset
begin
    initialize: S_best = S_0;
    Y_best = eval(S_0, D, A); // evaluate S_0 by a mining algorithm A
    do begin
    S = generate(D); // generate a subset for evaluation
    γ = eval(S, D, A);   // evaluated the current subset S by A
    if (γ is better than Y_best)
        Y_best = Y;
        S_best = S;
    end until (δ is reached);
    Return S_best;
end;
```

Fig. 3. Wrapper algorithm [7].

Attribute Evaluator.

The feature selection with wrapper [7] method required subset evaluator by selecting a subset of features which has the highest a goodness of fit.

3.2 Classification

The classification algorithm, Naïve Bayes [10] is used to classify data set. The Naïve Bayes is based on Bayes' Theorem which is based on probability theory with the hypothesis of all variables are independent. It predicts analyze.

$$P(C_i|E) = \frac{P(C_i)P(C|C_i)}{P(E)}$$

In equation are defined E as the data for calculated probability of hypothesis C_i as $P(C_i|E)$

$P(C_i)$ as priori that is probability occur hypotheses C_i.

$P(E)$ as evidence that is probability of data set has occur E.

$P(C_i|E)$ as posterior that is probability has occur C_i when occurred E early.

$P(E|C_i)$ as likelihood that is probability has occur E when occurred C_i early.

4 Methodology

4.1 Data Collection

The Criminal data in Thailand are collected from Dailynews and Thairath online news starting from 1 August 2015 to 21 September 2015 for Thairath news and 23 October 2013 to 30 July 2015 for Dailynews. Data were combined to single data set and incomplete and missing data were deleted. Finally, 645 instances were selected. The data set included 15 indexes which are applied in this study. They are crime type, offender, victim, offender size, offender race, victim race, offender age, victim age, offender skin, time, motivation, weapon, vehicle, place, and region then the crime type is index class. (see Table 1)

Table 1. Features of Crime data set

Feature	Description
1. crime type	Type of crimes included domestic violence, fraud, accident, drugs and drugging, human trafficking and prostitution, computer and rape.
2. offender	Sex of offender included male, female and male & female.
3. victim	Sex of victim included male, female and male & female.
4. offender size	Size of offender group included small, medium and large.
5. offender race	Race of offender included Thai, AEC and other.
6. victim race	Race of victim included Thai, AEC and other.
7. offender age	Age of offender divided into 5 groups age included 0-15, 16-30, 31-45, 46-60 and over than 60 years old.
8. victim age	Age of victim divided into 5 groups age included 0-15, 16-30, 31-45, 46-60 and over than 60 years old.
9. offender skin	Skin of offender included dark, yellow and white.
10. time	Time of incident divided into 4 range included 0.00-5.59 AM, 6.00-11.59 AM, 0.00-5.59 PM and 6.00-11.59 PM.
11. motivation	Motivation for incident included angry, money, fun, drugs and negligent.
12. weapon	Weapon for incident included gun, knife, cyber and other.
13. vehicle	Vehicle for incident included car, motorcycle, truck and other.
14. place	Type of place for incident included community, resident and other.
15. region	Region to incident included north, northeastern, east, central, west, Bangkok and south.

Data set in Table 1 was added Bangkok in feature **region** because Bangkok is the capital of Thailand which a lot of crimes occurred. **Races** of offenders and victims in real data included several races such as countries of Asia, countries of Europe, countries of America and countries of Africa etc. Therefore, We combined all of the countries and divided 3 groups including 1. Thai, 2. AEC, and 3. other.

In this study, data set divided to 3 types: data set of **all** records, data set of **foreigners** records which are data set foreigner offenders or victims of Thai race, and data set of **Thai** records which are Thai offenders and Thai victims races. All are data set of all are 645 records, data set of foreigners are 139 records and data set of Thai are 491 records.

4.2   Feature Selection

In this process, we used various search methods including Best-First Search (BFS), Greedy Stepwise Search (GS), Genetic Search (GA) and Exhaustive Search (ES) for generating subset. The attribute evaluators included Classifier Subset Evaluator (CS), Wrapper Subset Evaluator (WP), Consistency Subset Evaluator (CSE), and Cfs Subset Evaluator for evaluating attributes from each of data set.

4.3   Classification

The classification technique used 10-fold cross-validation for dividing data of data set to testing by six classifiers included NB, MLP, JRIP, PART, J48, and REPTree are used to test the accuracy of the prediction model which generated from selected features.

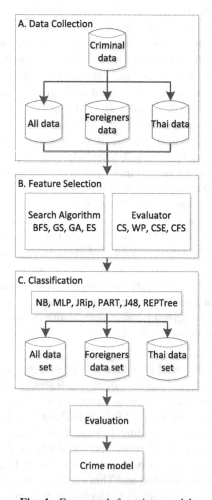

Fig. 4. Framework for crime model.

Figure 4 showed that crime model was divided to 3 steps including Step A, data were collected from two online newspapers sources and combined into a single data set then the records of incomplete and missing data were deleted. Next, the combined data set were divided to 3 data sets: (1) all data set, (2) foreigner data set and (3) Thai data set. Step B, feature selection by using 4 search methods including BFS (Best-First Search), GS (Greedy Stepwise), GA (Genetic Search), and ES (Exhaustive Search) and then evaluating by 4 evaluators including CS (Classifier Subset Evaluator), WP (Wrapper Subset Evaluator), CSE (Consistency Subset Evaluator), and Cfs Subset Evaluator. Step C, data classification for calculating accuracy of each feature selection by six algorithms including Bayesian as NB (Naïve Bayes), Neural Network as MLP (Multilayer Perceptron), 2 Rules as JRip and PART, 2 Trees as J48 and REPTree. Finally evaluate an experiment.

5   Experiment

This experiment used Weka 3.6.13 software [11] for data mining: feature selections were applied and using six classifier algorithms including NB, MLP, JRip, PART, J48, and REPTree to generate prediction models. Table 2 showed accuracy rates of classifiers by using BFS for the Feature Selection techniques. Therefore, the data set of all BFS WP and None has highest accuracy rates as 85.74 % however average accuracy rate was BFS WP with the highest as 84.24 %. Lastly, feature selection by BFS WP provided highest accuracy rate for both foreigners and Thai data sets as 88.49 % and 86.56 % respective.

Table 2. Classification of data set using Best-First Search (BFS) for feature selection

Data set	Evaluator	No. of attr.	Bayesian and neural		Rule based		Decision tree		Avg.	Std.
			NB	MLP	JRip	Part	J48	RepTree		
All	CS	10	85.43	83.72	84.03	84.50	84.81	84.50	84.5	0.60
	WP	10	**85.74**	82.02	83.41	85.43	85.12	83.72	**84.24**	1.44
	CSE	11	85.12	84.96	81.86	83.88	84.50	83.26	83.93	1.23
	CFS	10	84.19	83.88	83.57	85.12	84.96	82.79	84.08	0.88
	None	15	84.81	**85.74**	83.10	83.41	84.81	82.95	**84.13**	1.14
Foreigners	CS	10	81.30	77.70	82.73	83.45	84.17	84.89	82.37	2.60
	WP	5	**88.49**	82.01	82.73	82.73	84.17	84.17	84.05	2.34
	CSE	7	81.30	79.14	81.30	83.45	84.17	84.17	82.25	2.01
	CFS	8	83.45	81.30	81.30	82.73	84.17	83.45	82.73	1.20
	None	15	84.17	79.86	79.14	82.01	84.17	84.17	82.25	2.30
Thai	CS	10	86.35	85.54	82.89	85.54	85.95	83.91	85.03	1.34
	WP	8	**86.56**	83.30	82.69	84.93	85.54	82.08	84.18	1.76
	CSE	10	85.54	84.11	81.67	82.89	82.89	82.48	83.27	1.37
	CFS	10	84.93	82.69	85.54	83.30	85.54	81.87	83.98	1.57
	None	15	85.54	84.73	82.89	83.71	85.54	82.08	84.08	1.43

Table 3 showed the accuracy rate of six classifiers using GS for search method in Feature Selection technique. Therefore, the data set of all, it was no feature selection which has the highest accuracy as 85.74 %, however for the data set of foreigners and Thai, GS WP provided the highest accuracy as 85.61 % and 86.56 % respectively.

Table 3. Classification of data set using Greedy Stepwise Search (GS) for feature selection

Data set	Evaluator	No. of attr.	Bayesian and neural		Rule based		Decision tree		Avg.	Std.
			NB	MLP	JRip	Part	J48	RepTree		
All	CS	4	84.96	84.81	81.24	83.26	83.26	83.26	83.46	1.35
	WP	3	85.27	84.65	80.78	83.26	83.26	83.26	83.41	1.55
	CSE	10	84.50	84.34	83.88	83.57	84.03	83.26	83.93	0.47
	CFS	10	84.19	83.88	83.57	85.12	84.96	82.79	84.08	0.88
	None	15	84.81	**85.74**	83.10	83.41	84.81	82.95	84.13	1.14
Foreigners	CS	4	83.45	84.17	82.01	83.45	84.17	84.17	83.57	0.84
	WP	**3**	**85.61**	83.45	83.45	81.30	84.17	84.90	83.81	1.49
	CSE	7	81.30	79.14	81.30	83.45	84.17	84.17	82.25	2.01
	CFS	8	83.45	81.30	81.30	82.73	84.17	83.45	82.73	1.20
	None	15	84.17	79.86	79.14	82.01	84.17	84.17	82.25	2.30
Thai	CS	10	86.35	85.54	82.89	85.54	85.95	83.91	85.03	1.34
	WP	**8**	**86.56**	83.30	82.69	84.93	85.54	82.08	84.18	1.76
	CSE	9	85.95	83.91	83.30	82.28	82.69	81.87	83.33	1.47
	CFS	10	84.93	82.69	85.54	83.30	85.54	81.87	83.98	1.57
	None	15	85.54	84.73	82.89	83.71	85.54	82.08	84.08	1.43

In feature selection using Genetic Search was showed in Table 4. For data set of all, it was GA CS and None provided the highest accuracy rate as 85.74 % but for average accuracy rate it was None which was the highest as 84.13 %, The data set of foreigners and Thai were GA WP have highest accuracy as 88.49 % and 86.97 % respectively.

Table 4. Classification of data set using Genetic Search (GA) for feature selection

Data set	Evaluator	No. of attr.	Bayesian and neural		Rule based		Decision tree		Avg.	Std.
			NB	MLP	JRip	Part	J48	RepTree		
All	**CS**	**10**	84.96	83.41	81.71	85.12	**85.74**	83.10	**84.01**	1.52
	WP	10	85.43	83.72	84.03	84.50	84.81	84.50	84.50	0.60
	CSE	12	84.65	84.50	82.48	82.95	84.50	83.26	83.72	0.94
	CFS	10	84.19	83.88	83.57	85.12	84.96	82.79	84.08	0.88
	None	15	84.81	**85.74**	83.10	83.41	84.81	82.95	**84.13**	1.14
Foreigners	CS	10	82.01	81.30	82.01	82.73	84.17	84.89	82.85	1.40
	WP	**5**	**88.49**	82.01	82.73	82.73	84.17	84.17	84.05	2.34

(Continued)

Table 4. (*Continued*)

Data set	Evaluator	No. of attr.	Bayesian and neural		Rule based		Decision tree		Avg.	Std.
			NB	MLP	JRip	Part	J48	RepTree		
	CSE	7						51.08		6.75
	CFS	8	83.45	81.30	81.30	82.73	84.17	83.45	82.73	1.20
	None	15	84.17	79.86	79.14	82.01	84.17	84.17	82.25	2.30
Thai	CS	10	86.35	85.54	82.89	85.54	85.95	83.91	85.03	1.34
	WP	**10**	**86.97**	85.13	85.34	84.73	85.95	83.71	85.30	1.10
	CSE	9	85.95	83.71	83.30	82.08	81.06	81.47	82.93	1.80
	CFS	10	84.93	82.69	85.54	83.30	85.54	81.87	83.98	1.57
	None	15	85.54	84.73	82.89	83.71	85.54	82.08	84.08	1.43

The results of feature selection using Exhaustive Search was showed in Table 5. The data set of all was ES WP and None with the highest accuracy as 85.74 % but average accuracy rate was ES WP has the highest as 84.34 %, The data set of foreigners and Thai were ES WP have highest accuracy as 88.49 % and 86.56 % respectively.

Table 5. Classification of data set using Exhaustive Search (ES) for feature selection

Data set	Evaluator	No. of attr.	Bayesian and neural		Rule based		Decision tree		Avg.	Std.
			NB	MLP	JRip	Part	J48	RepTree		
All	CS	10	85.43	83.72	84.03	84.50	84.81	84.50	84.50	0.60
	WP	**8**	**85.74**	83.41	84.65	84.65	84.81	82.79	**84.34**	1.06
	CSE	11	84.19	83.57	82.79	82.64	84.34	82.95	83.41	0.73
	CFS	10	84.19	83.88	83.57	85.12	84.96	82.79	84.08	0.88
	None	**15**	84.81	**85.74**	83.10	83.41	84.81	82.95	**84.13**	1.14
Foreigners	CS	10	82.73	82.01	81.30	82.01	84.17	84.17	82.73	1.20
	WP	**5**	**88.49**	82.01	82.73	82.73	84.17	84.17	84.05	2.34
	CSE	6	84.17	79.14	81.30	83.45	84.17	84.17	82.73	2.08
	CFS	8	83.45	81.30	81.30	82.73	84.17	83.45	82.73	1.20
	None	15	84.17	79.86	79.14	82.01	84.17	84.17	82.25	2.30
Thai	CS	10	86.35	85.54	82.89	85.54	85.95	83.91	85.03	1.34
	WP	**8**	**86.56**	83.30	82.69	84.93	85.54	82.08	84.18	1.76
	CSE	10	83.10	83.71	82.69	82.48	83.91	81.26	82.86	0.96
	CFS	10	84.93	82.69	85.54	83.30	85.54	81.87	83.98	1.57
	None	15	85.54	84.73	82.89	83.71	85.54	82.08	84.08	1.43

Finally, the summary of experiments was presented in Table 6. The data set of all model with the highest accuracy rate were equal as 85.74 % but the average compared were ES WP has the highest as 84.34 %. The data set of foreigners using BFS WP,

Table 6. Summarize of feature selection and classifications of crime data set

Data set	Feature selection									
	None	No. of attr.	BFS WP	No. of attr.	GS WP	No. of attr.	GA WP	No. of attr.	ES WP	No. of attr.
All	85.74 Avg. 84.13	15	85.74 Avg. 84.24	10	–	–	–	–	**85.74 Avg. 84.34**	8
Foreigners	–	–	**88.49 Avg. 84.05**	5	85.61 Avg. 83.81	3	**88.49 Avg. 84.05**	5	**88.49 Avg. 84.05**	5
Thai	–	–	86.56 Avg. 84.18	8	86.56 Avg. 84.18	8	**86.97 Avg. 85.30**	10	86.56 Avg. 84.18	8

GA WP and ES WP accuracy rates equals as 88.49 % and all three data sets have the same average is 84.05 %. On the basic of average accuracy Data set of Thai was GA WP with the highest accuracy as 86.97 %.

6 Conclusion

In this study, there are 3 steps: first, data collection from online news included Dailynews and Thairath then they were combined to a single data set. Next, data were divided into 3 data sets according to nationality included data set of All, data set of foreigners and data set of Thai. Second, Feature Selection was performed by using 4 search methods included BFS (Best-First Search), GS (Greedy Stepwise), GA (Genetic Search) and ES (Exhaustive Search) for generated subset and selected attribute. Then 4 evaluators included CS (Classifier Subset Evaluator), WP (Wrapper Subset Evaluator), CSE (Consistency Subset Evaluator) and CFS Subset Evaluator were used for evaluating subset from each data set. Third, Classification was applied for calculating accuracy rate by 6 classifiers including NB, Multilayer Perceptron (MLP), JRip, PART, J48, and REPTree.

Results of feature selection of **All data set**, ES WP was the best algorithm with accuracy rate 85.74 %. The selected features included eight features *crime type, victim, skin offender, time, motivation, weapon, place, and region* were selected, **foreigners data set** BFS WP, GA WP, and ES WP were the best for feature selection with accuracy rate 88.49 %, then all three feature selections have same features included five features *crime type, offender, motivation, weapon, and place* were selected. For **Thai data set** GA WP was the best with accuracy rate 86.97 % then ten features consist of *crime type, offender, victim, offender age, victim age, time, motivation, weapon, vehicle, and place* were selected. Therefore, the feature selection with GS has selected the smallest number of features, however the accuracy rates of all algorithms were close to each other search methods. WP seems to be an appropriate evaluation algorithm because feature selection in three data sets using WP as an evaluator provide the highest accuracy rates. According to foreigner and Thai data sets, feature selections in both data sets are different. There are more features selected included *victim, offender age, victim age, time* and *vehicle* in Thai. This information will be benefit to the administrators or police for preventing and managing crimes.

References

1. Oxford University Press: Definition of crime in English. In: Book Definition of Crime in English (2016)
2. Matcha, S.: Five Criminal in Thailand, What is Criminal 5 Groups (2009)
3. Royal Thai Police: Crime by type, Crime in Thailand (2014)
4. Royal Thai Police: Statistics of Reported and Arrested for The Violent Crime Group by Type of Reported Cases, Whole Kingdom, The Criminal Case. National Statistical Office (2015)
5. Nath, S.V.: Crime pattern detection using data mining. In: Web Intelligence and Intelligent Agent Technology Workshops, pp. 41–44 (2006)
6. Sathyadevan, S., Sreemandiram, D.M., Gangadharan, S.S.: Crime analysis and prediction using data mining. In: 2014 First International Conference on Networks & Soft Computing (ICNSC), pp. 406–412, 19–20 August 2014
7. Huan, L., Lei, Y.: Toward integrating feature selection algorithms for classification and clustering. IEEE Trans. Knowl. Data Eng. **17**(4), 491–502 (2005)
8. Pearl, J.: Heuristics: Intelligent Search Strategies for Computer Problem Solving, p. 382 (1984)
9. Russell, S.J., Norvig, P.: Artificial Intelligence: A Modern Approach, p. 1132. Pearson Education, Upper Saddle River (2003)
10. Domingos, P., Pazzani, M.: On the optimality of the simple Bayesian classifier under zero-one loss. Mach. Learn. **29**(2), 103–130 (1997)
11. Smith, T.C., Frank, E.: Weka 3: Data Mining Software in Java: Book Weka 3, Machine Learning Group at the University of Waikato (2015)

The Impact of Data Normalization on Stock Market Prediction: Using SVM and Technical Indicators

Jiaqi Pan[(✉)], Yan Zhuang, and Simon Fong

Department of Computer Information Science,
University of Macau, Taipa, Macau SAR
{mb25470, syz, ccfong}@umac.mo

Abstract. Predicting stock index and its movement has never been lack of attention among traders and professional analysts, because of the attractive financial gains. For the last two decades, extensive researches combined technical indicators with machine learning techniques to construct effective prediction models. This study is to investigate the impact of various data normalization methods on using support vector machine (SVM) and technical indicators to predict the price movement of stock index. The experimental results suggested that, the prediction system based on SVM and technical indicators, should carefully choose an appropriate data normalization method so as to avoid its negative influence on prediction accuracy and the processing time on training.

Keywords: Stock market prediction · Technical indicator · Support vector machine (SVM) · Data normalization

1 Introduction

Predicting stock index and its movement has never been lack of attention among traders and professional analysts. Typically, professional traders use technical indicators to analyze stocks and make investment decisions [1]. Nonetheless, it's always to find that two opposite trading signals (buy and sell) are generated among different selected indicators. For that moment, traders have to make a judgement based on personal experience or decision models.

For the last two decades, extensive researches combined technical indicators with machine learning techniques to construct effective prediction models [2]. Some machine learning techniques, like artificial neural network (ANN) and support vector machine (SVM), had proved to be useful tools for prediction and classification in financial market [3, 4]. Compared to the empirical risk minimization principle of ANN, SVM implements the structural risk minimization principle [5], specifically searches to minimize the upper bound of the generalization error instead of minimizing the empirical error. In addition, the solution of SVM may be global optimum while neural network models may tend to fall into a local optimal solution, thus overfitting is unlikely to occur with SVM.

© Springer Nature Singapore Pte Ltd. 2016
M.W. Berry et al. (Eds.): SCDS 2016, CCIS 652, pp. 72–88, 2016.
DOI: 10.1007/978-981-10-2777-2_7

The success of applying SVM on stock market prediction can be affected by many factors, and the representation and quality of instance data is first and foremost [6]. As one of the widely used data preprocessing techniques, data normalization may have great effect on preparing the data to be suitable for training. The purpose of this study is to investigate the impact of various normalization techniques on using SVM and technical indicators to predict the directional movement of stock index. The experimental results suggested that, the prediction system based on SVM and technical indicators, should carefully choose an appropriate data normalization method so as to avoid its negative influence on prediction accuracy and the processing time on training.

The remainder of this paper is organized as follows. Section 2 covers the related work of this study. Section 3 describes different types of data normalization methods. The details of experiment are presented in Sect. 4, and Sect. 5 gives the experimental result and discussion. Finally, Sect. 6 concludes this study.

2 Related Work

Even though the Efficient Markets Hypothesis [7] and the Random Walk Theory [8] had been proven through a number of empirical studies, many academic studies [2] indicate that stock markets do not in fact follow a random walk and price prediction is possible. Especially, support vector machine (SVM) is one of the most popular machine learning techniques used for predicting stock market. In [9], author used SVM to predict the direction of daily stock price change in the Korea composite stock price index (KOSPI). He selected 12 technical indicators to make up the initial attributes, and scaled the value of data into the range of [−1.0, +1.0]. This study compared the prediction performance of SVM with respect to back-propagation neural network (BPN) and case-based reasoning (CBR). The experimental results showed that SVM outperformed BPN and CBR. Similarly, [10] used SVM and random forest to predict the directional movement of daily price of S&P CNX NIFTY Market Index of the National Stock Exchange, compared the results with that of traditional discriminant and logit model and artificial neural network (ANN). Results showed that SVM outperformed random forest, ANN and other traditional models. [11] investigated the predictability of financial movement direction with SVM by forecasting the weekly movement direction of NIKKEI 225 index. To evaluate the forecasting ability of SVM, they compared its performance with those of Linear Discriminant Analysis, Quadratic Discriminant Analysis and Elman Back-propagation Neural Networks. The experiment results indicated that SVM outperformed the other classification methods.

In most of studies, researchers adopted Min-Max normalization as an essential data preprocessing procedure in data mining, which linearly scaled data into a specified range like [−1, +1] or [0,1]. However, there are various data normalization methods for selecting, rather than one single method. [12] tested the effect of three normalization methods on applying induction decision tree (ID3) to classify the HSV data. The experimental results suggested choosing the Min-Max normalization data set as the best design for training data set because of the highest accuracy, least complexity and shortest in learning speed. [13] investigated the use of three normalization techniques in predicting dengue outbreak. They found that LS-SVM and ANN achieved better accuracy and minimized error by using Decimal Scaling Normalization compared to the other two techniques. In [14], various statistical normalization techniques were used for diabetes data classification. The experimental results showed that the performance of the diabetes data classification model using the neural networks was dependent on the normalization methods. [15] evaluated impact of various normalization methods for forecasting the daily closing price of Bombay stock exchange (BSE) indices in Indian stock market. From the experimental results, it was observed that a single normalization method is not superior over all others. [16] investigated impact of feature normalization techniques on classifier's performance in breast tumor classification. Results showed that normalization of features had significant effect on the classification accuracy.

In summary, it had been proved that SVM is a powerful tool for prediction and classification in financial market. However, most of these studies selected Min-Max normalization as an essential data preprocessing procedure. To the author's knowledge, there was little information available in literature about the impact of various data normalization method on using SVM and technical indicators to predict stock price index.

3 Data Normalization

As one of the widely used preprocessing techniques, data normalization is more like a "scaling down" transformation of the attributes [17]. After performing normalization, the value magnitudes are scaled to appreciably low values. It should be noted that, the rule of normalization method applied in testing data must be the same to training data. In this paper, the following normalization methods were discussed.

3.1 Min-Max Normalization

This method of normalization linearly rescales the data from one range of values to a new range of values [17], such as [0, 1] or [−1, 1]. Min-Max method normalizes the values of the attribute X according to its minimum and maximum values, X_{min} and X_{max}. It converted a value v_i ($i = 1, 2, 3, \ldots,$ n) of attribute X to v_i' in the predefined range [new_X_{min}, new_X_{max}] by computing:

$$v_i' = \frac{v_i - X_{min}}{X_{max} - X_{min}}(new\_X_{max} - new\_X_{min}) + new\_X_{min} \tag{1}$$

Although Min-Max normalization has the advantage of preserving the relationships among the original data values, it's highly sensitive to extreme value and outliers, as the result may be dominated by a single large value. It will generate an "out-of-bound" error if a future input case for normalization falls outside of the original data range of attribute X.

Moreover, in order to explore the difference between the normalized results in the range of [0,1] and [−1,1], the Min-Max normalization method is divided into two categories: Min-Max1 and Min-Max2 normalization. The normalized results in the range of [0,1] is represented by Min-Max1 normalization, and the range of [−1,1] for Min-Max2 normalization.

3.2 Z-Score Normalization

This method uses the mean and standard deviation for each feature attribute to rescale the data values [17]. The mean and standard deviation are computed for each feature. In Z-Score normalization, a value v_i of attribute X is normalized to v_i' by computing:

$$v_i' = \frac{v_i - \mu_X}{\sigma_X} \tag{2}$$

Where, μ_X is the mean value and σ_X is the standard deviation of attribute X, respectively.

Z-Score normalization is useful to reduce the effect of outliers that dominates the result in Min-Max normalization, and still deal well when the actual minimum and maximum value of attribute X are unknown. However, this method fails to transform the value of different feature into a common numerical range, and the mean and standard deviation of time series will vary over time.

3.3 Decimal Scaling Normalization

This method normalizes the feature data by moving the decimal point of the value [17]. The number of decimal points moved depends on the maximum absolute value of attribute X, and a value v_i of attribute X is rescaled to v_i' by computing:

$$v_i' = \frac{v_i}{10^j} \tag{3}$$

Where, j is the smallest value such that $\max(|v_i'|) < 1$.

The procedure of Decimal Scaling normalization is similar to Min-Max normalization, which depending on knowing the minimum and/or maximum value of each attribute. And it also has the same problem of Min-Max normalization when applied to time series data.

3.4 Median and Median Absolute Deviation (MMAD) Normalization

The procedure of this normalization method is somewhat alike to Z-Score normalization, but uses the median and the median absolute deviation (MAD) to convert the data values, instead of the mean and standard deviation. X value v_i of attribute X is normalized to v_i' by computing [15]:

$$v'_i = \frac{v_i - X_{median}}{MAD} \qquad (4)$$

Where, X_{median} is the median value of attribute X, and MAD = median $(|v_i - X_{median}|)$.

The median and MAD are insensitive to extreme value and outliers, especially the median is more robust than the mean. Nevertheless, in time series data, these two estimators also vary over the time.

4   Experiments

In order to investigate the impact of data normalization on using SVM and technical indicators to predict the directional movement of stock price index, five popular methods of data normalization were separately applied in different experiments. The work started with collecting the research data from the stock market, and then assigned the class labels and initial features (based on technical indicators) for each research data. The relevant features were filtered out by applying a feature selection method. Data sets were normalized before feeding to the learning algorithm. The SVM with radial basis function (RBF) kernel was implemented to construct the prediction model. The performance of models was evaluated at the end of experiment process.

4.1   Data Collection

The research data used in this study were collected from the historical data of the SSE Composite Index (SSECI) in China's stock market. Historical data contained opening, high, low and closing prices, and volume of each trading day. The entire data set covered a seven-year period, from May 5, 2008 to June 30, 2015, and totally 1741 trading day were made. The entire data set was split into nine subsets by slicing the time period. Each subset contained a five-year period for training and ten-month period for testing, and the number of cases for each data subset is given in Table 1. The period of training and testing are starting from the first trading day of a month and ending at the last trading day of a month (weekend and holiday are excepted).

Table 1. The number of cases for each data subset

Dataset	Training			Testing			Total
	Period	Total	%	Period	Total	%	
1	2008/05/05-2013/04/26	1213	85.79	2013/05/02-2014/02/28	201	14.21	1414
2	2008/07/01-2013/06/28	1212	85.59	2013/07/01-2014/04/30	204	14.41	1416
3	2008/09/01-2013/08/30	1213	85.91	2013/09/02-2014/06/30	199	14.09	1412
4	2008/11/03-2013/10/31	1211	85.46	2013/11/01-2014/08/29	206	14.54	1417
5	2009/01/05-2013/12/31	1211	85.70	2014/01/02-2014/10/31	202	14.30	1413

(Continued)

Table 1. (*Continued*)

Dataset	Training			Testing			Total
	Period	Total	%	Period	Total	%	
6	2009/03/02-2014/02/28	1213	85.36	2014/03/03-2014/12/31	208	14.64	1421
7	2009/05/04-2014/04/30	1212	85.77	2014/05/05-2015/02/27	201	14.23	1413
8	2009/07/01-2014/06/30	1212	85.59	2014/07/01-2015/04/30	204	14.41	1416
9	2009/09/01-2014/08/29	1212	85.77	2014/09/01-2015/06/30	201	14.23	1413

4.2 Class Labeling

This study is to predict the daily price movement of SSECI, which is a two-class problem either increase or decrease. Class labels are assigned to each case according to the forthcoming behavior of the SSECI closing price. The label "Increase" or "1" is assigned to a case if the closing price goes up the next day, and the label "Decrease" or "-1" if the closing price goes down the next day.

4.3 Features Generation

In [4], authors suggested that using different or additional input features can improve the prediction performance of SVM models. Hence, in this study, twenty-seven technical indicators were selected to make up the input features to forecasting the directional movement of stock price index. Previous researches indicated that some of these indicators are relevant for predicting and classifying stock price index [4, 9, 18]. Other technical indicators suggested by the domain experts were adopted. The description and definitions of these technical indicators are presented in Appendix A.

4.4 Feature Selection

For a specified period of data set, maybe only a few of selected technical indicators have the important information to support building an effective prediction model, rather than all of them. Therefore, it's necessary to apply a feature selection procedure before feeding the training data to the learning algorithm.

The Pearson correlation coefficient is a simple measure of correlation between two numeric attributes. Although its calculation requires numeric value, the use of Pearson correlation coefficient can still be extended to the case of two-class classification, for which each class label is mapped to a given value, e.g., ± 1 [19, 20]. Given a set of m examples $\{x_k, y_k\}(k = 1, 2, 3, \ldots, m)$ consisting of n input features $x_{k,i}(i = 1, 2, 3, \ldots, n)$ and the corresponding class $y_k \in \{-1, 1\}$, and the Pearson correlation coefficient $R(i)$ is defined as:

$$R(i) = \frac{\sum_{k=1}^{m} (x_{k,i} - \overline{x_i})(y_k - \overline{y})}{\sqrt{\sum_{k=1}^{m} (x_{k,i} - \overline{x_i})^2} \sqrt{\sum_{i=1}^{n} (y_k - y)^2}} \tag{5}$$

Where, $\bar{x}_i$ and $\bar{y}$ is the mean of the feature x_i and class y respectively.

The larger $R(i)$ is, the more likely this feature is discriminative. This study evaluated the worth of each feature by measuring the Pearson correlation coefficient between it and the class, and made up the relevant feature subsets to train the prediction models by the following procedure:

Step 1. Calculate $R(i)$ for the m features.
Step 2. Sort $R(i)$ in decreasing order, and set possible number of features by $f = 1 + i, i \in \{0, 1, 2, \ldots, m - 1\}$.
Step 3. For each f, do the following

(a) Keep the first f features according to the sorted $R(i)$.
(b) Train the prediction model and measure its prediction performance.

Step 4. Summarize the prediction performance of all prediction models.

4.5   Prediction Model

In this paper, the radial basis function (RBF) kernel is utilized in constructing the SVM models. Unlike the Linear kernel, the RBF kernel can nonlinearly map input data into a higher dimensional space, when the relation between class labels and features is nonlinear. The sigmoid kernel behaves like RBF for certain parameters, and polynomial kernel was excluded because of a comparatively low computational time efficiency [21].

Moreover, the parameter setting is critical for improving the SVM prediction accuracy. There are two parameters for an SVM model with radial basis function (RBF) kernel: the parameter gamma (γ) of the RBF kernel and the regularization parameter (C). It is unknown beforehand which C and γ are best for a given task. The grid search approach is a common method to search for the best C and γ. As suggested in [21], the grid search method uses an exponentially growing sequences to identify the best parameter combination of C and γ, i.e., $C = 2^{-5}, 2^{-3}, 2^{-1} \ldots, 2^{15}, \gamma = 2^{-15}, 2^{-13}, 2^{-11} \ldots, 2^3$. The number of parameter setting yielded a total of $11 \times 10 = 110$ treatments for RBF-based SVM, and each model was evaluated.

4.6   Performance Summary

This study evaluated the performance of prediction models by measuring the prediction accuracy and the processing time spent on training a prediction model. Even though the prediction accuracy is more important than processing time as it directly affects the profits of investment, the processing time has an important influence on whether the system can make adjustments in time to enhance prediction accuracy.

Moreover, to investigate the impact of these five normalization methods, an additional experimental process without data normalization is necessary for comparing and analysis.

5 Experimental Result and Discussion

All the experiments were conducted on a computer with Intel Core i7-3770 CPU, 8 GB RAM and Windows 7 64-bit Enterprise Version. The implementation platform of SVM prediction was carried out via WEKA version 3.7.13, a development environment for knowledge analysis [22], by extending the LibSVM version 1.0.8, which was originally designed by Chang and Lin [23].

In this study, various prediction models were constructed based on training data sets that preprocessed with (or without) different normalization methods. For one data subset with (or without) a normalization procedure, 27 feature subsets that were generated in the feature selection phase and 110 combinations of SVM parameter setting yielded a total of 2970 prediction models for training and testing.

Moreover, it should be noted that, the number of decimal places of normalized data is six, and two for the original data.

5.1 Prediction Accuracy

In terms of the prediction accuracy, Table 2 compares the optimal SVM prediction models which were constructed by different normalization methods respectively. Here, the optimal SVM model for a normalization method is the model which has the best prediction accuracy among the 2970 models. Meanwhile, the bold font and grey background color highlight the highest accuracy in a row obtained for a specified data subset. It can be seen from the results that, it's possible for any types of data to construct an optimal SVM model with the highest prediction accuracy, no matter the data set is normalized or not. For instance, the Decimal Scaling normalization achieved the highest accuracy with 65.38 % in data subset 6, however, it produced the accuracy 55.72 %, which was lower than the highest accuracy 60.70 % for excluding normalization methods in data subset 1. An analysis of the information displayed in Table 2,

Table 2. The prediction accuracy (%) of the optimal model for different normalization methods

| Dataset | Without normalization | Normalization method | | | | | DIFF.[a] |
		Min-Max1	Min-Max2	Z-Score	Decimal Scaling	MMAD	
1	**60.70**	56.22	56.72	59.20	55.72	57.71	-1.5
2	60.29	56.37	57.35	**60.78**	57.35	59.31	0.49
3	60.30	57.79	60.30	**61.31**	57.79	59.80	1.01
4	59.22	**60.68**	**60.68**	58.25	58.25	58.74	1.46
5	60.40	59.41	59.90	61.39	60.40	**62.87**	2.47
6	63.46	63.46	63.46	63.46	**65.38**	64.42	1.92
7	**64.18**	**64.18**	**64.18**	63.18	63.18	63.68	0
8	66.67	65.69	66.18	65.69	65.20	**67.65**	0.98
9	66.17	66.17	66.17	66.17	64.18	**67.66**	1.49

[a]DIFF. represents the difference between the maximum accuracy of "Normalization method" and the accuracy of "Without normalization"

illustrates that the differences between the maximum accuracy of normalization methods and the accuracy of excluding normalization methods is varying from −1.5 % to 2.47 %. It means that these five normalization methods had the possibility to increase the prediction accuracy by at most 2.47 %, which is a very small improvement. Compared to Min-Max1 normalization, Min-Max2 normalization has a slightly better prediction on accuracy.

5.2 Processing Time

The effect on the averaged training time with respect to normalization methods and the number of selected features is presented in Table 3. The effect is measured as a percentage change by comparing the averaged training time spent on the original datasets. The bold font and grey background color highlight the negative value, which means that the corresponding normalization method can speed up the training time of a prediction model. According to Table 3, there is an obvious negative correlation between the number of selected features and the accelerating effect of normalization methods. The more features are adopted to train a prediction model, the smaller the accelerating effect are made by the normalization methods. This acceleration would gradually disappear until the number of selected feature is increased to eight. If the number of selected features is greater than eight, all five types of normalized data will definitely slow down the process of training prediction models, and require more computation time than the original data.

Comparing the accelerating effect of data normalization methods, the obtained results show that, the shortest processing time is when the Decimal Scaling normalization data set is trained, and followed by Min-Max1, Min-Max2, Z-Score and MMAD normalization data sets.

Table 3. The effect of averaged training time with respect to the normalization methods and the number of selected features

Number of selected features	Min-Max1	Min-Max2	Z-Score	Decimal Scaling	MMAD
1	−88.65 %	−81.00 %	−68.39 %	−93.61 %	−58.18 %
2	−84.97 %	−72.29 %	−29.12 %	−91.99 %	−9.75 %
3	−76.05 %	−51.41 %	−1.74 %	−86.29 %	−5.87 %
4	−59.02 %	−17.71 %	−3.89 %	−74.81 %	−4.46 %
5	−29.59 %	−7.25 %	0.47 %	−56.15 %	−4.35 %
6	−10.32 %	−1.65 %	−1.60 %	−31.29 %	−3.15 %
7	−1.90 %	−0.49 %	−0.52 %	−18.40 %	1.17 %
8	3.59 %	5.06 %	8.39 %	−6.16 %	9.17 %
9	8.41 %	12.66 %	14.97 %	10.05 %	17.06 %
10	14.90 %	17.97 %	14.73 %	18.78 %	16.28 %
11	15.96 %	18.44 %	15.74 %	21.61 %	21.69 %
12	31.19 %	33.48 %	38.44 %	40.89 %	44.60 %

(*Continued*)

Table 3. (*Continued*)

Number of selected features	Min-Max1	Min-Max2	Z-Score	Decimal Scaling	MMAD
13	41.94 %	42.24 %	49.75 %	50.67 %	55.62 %
14	52.43 %	54.85 %	62.08 %	62.17 %	62.59 %
15	55.79 %	62.63 %	64.42 %	60.07 %	65.35 %
16	67.95 %	71.61 %	76.57 %	70.21 %	76.81 %
17	82.37 %	83.98 %	83.41 %	81.34 %	87.34 %
18	80.97 %	81.97 %	80.99 %	82.58 %	86.70 %
19	84.00 %	90.07 %	81.88 %	87.95 %	92.07 %
20	88.47 %	91.22 %	87.54 %	89.76 %	94.02 %
21	87.11 %	87.40 %	85.85 %	86.83 %	89.35 %
22	83.86 %	85.10 %	85.42 %	89.31 %	91.23 %
23	79.54 %	80.44 %	85.74 %	86.90 %	88.46 %
24	82.42 %	86.37 %	93.01 %	90.22 %	95.74 %
25	82.47 %	86.48 %	87.59 %	87.69 %	93.92 %
26	78.16 %	82.17 %	90.50 %	84.07 %	92.17 %
27	81.77 %	83.03 %	91.37 %	85.12 %	97.96 %

6 Conclusion

This study is to investigate the impact of various data normalization methods on using SVM and technical indicators to predict stock market index. Five methods of data normalization were discussed; Min-Max1, Min-Max2, Z-Score, Decimal Scaling, and Median and Median Absolute Deviation normalization.

According to the experimental results, it indicates that data normalization has the possibility to construct an SVM prediction model that can achieve the highest prediction accuracy. However, compared to the optimal models without normalization, this improvement of accuracy is very small (2.4 % at most). For improving the prediction accuracy, not one of these five normalization methods is superior over all others in the long term. In terms of the training time, the obtained findings show that, there is an obvious negative correlation between the number of selected features and the accelerating effect of data normalization. And for reducing the processing time of training, the best method that the SVM model prefer is Decimal Scaling normalization, and next preference is Min-Max1 normalization.

All in all, this experimental work can be concluded that, the prediction system based on SVM and technical indicators, should carefully choose an appropriate data normalization method so as to avoid its negative influence on prediction accuracy and the processing time on training.

Acknowledgement. The authors of this paper would like to thank Research and Development Administrative Office of the University of Macau, for the funding support of this project which is called "Building Sustainable Knowledge Networks through Online Communities" with the project code MYRG2015-00024-FST.

Appendix A. The Descriptions and Definition of Input Features

1. OSCP_SMA: the difference between 5-day and 10-day simple moving average

$$OSCP\_SMA_{n,m}(t) = SMA_n(C_t) - SMA_m(C_t) \tag{1}$$

Where n and m are the length of time period, $SMA_n(C_t) = \left(\sum_{i=1}^{n} C_{t-i+1}\right)/n$, C_t is the closing price at time t.

2. OSCP_WMA: the difference between 5-day and 10-day weighted moving average

$$OSCP\_WMA_{n,m}(t) = WMA_n(C_t) - WMA_m(C_t) \tag{2}$$

Where $WMA_n(C_t) = \left(\sum_{i=1}^{n}((n-i+1)\cdot C_{t-i+1})\right)/\sum_{i=1}^{n} i$.

3. OSCP_EMA: the difference between 5-day and 10-day exponential moving average

$$OSCP\_EMA_{n,m}(t) = EMA_n(C_t) - EMA_m(C_t) \tag{3}$$

Where $EMA_n(C_t) = (1-\alpha)\cdot EMA_n(C_{t-1}) + \alpha \cdot C_t$, and $\alpha = 2/(n+1)$.

4. BIAS_5: 5-day Bias Ratio. It measures the divergence of current price from the moving averages over an observation period.

$$BIAS_n(t) = 100 \times \frac{C_t - SMA_n(C_t)}{SMA_n(C_t)} \tag{4}$$

5. CCI_20: 20-day Commodity Channel Index. It measures the variation of a security's price from its statistical mean.

$$CCI_n(t) = \frac{TYP_t - SMA_n(TYP_t)}{0.015 \times MAD_n(TYP_t)} \tag{5}$$

Where, $MAD_n(TYP_t) = (\sum_{i=1}^{n}|TYP_{t-i+1} - SMA_n(TYP_t)|)/n$, $TYP_t = (H_t + L_t + C_t)/3$, H_t is the high price at time t and L_t is the low price at time t.

6. PDI_14: 14-day Positive Directional Index. It summarizes upward trend movement.

$$PDI_n(t) = 100 \times \frac{MMA_n(PDM_t)}{MMA_n(TR_t)} \tag{6}$$

Where

$$PDM_t = \begin{cases} H_t - H_{t-1}, \ \text{if } H_t - H_{t-1} > 0 \ \text{and } H_t - H_{t-1} > L_{t-1} - L_t \\ 0, \text{Otherwise} \end{cases}$$

$$TR_t = \max(|H_t - L_t|, |H_t - C_{t-1}|, |L_t - C_{t-1}|)$$

$$MMA_n(TR_t) = \frac{(n-1) \cdot MMA_n(TR_{t-1}) + TR_t}{n}$$

7. NDI_14: 14-day Negative Directional Index. It summarizes downward trend movement.

$$NDI_n(t) = 100 \times \frac{MMA_n(NDM_t)}{MMA_n(TR_t)} \tag{7}$$

Where

$$NDM_t = \begin{cases} L_{t-1} - L_t, \ \text{if } L_{t-1} - L_t > 0 \ \text{and } H_t - H_{t-1} < L_{t-1} - L_t \\ 0, \text{Otherwise} \end{cases}$$

8. DMI_14: 14-day Directional Moving Index. It measures the difference between positive and negative directional index.

$$DMI_n(t) = PDI_n(t) - NDI_n(t) \tag{8}$$

9. slow%K_3: 3-day Stochastic Slow Index K. It's a simple moving average of 10-day Stochastic Index K, which showed the relationship of the differences of today's closing price with the period lowest price and the trading range

$$slow\%K_n(t) = SMA_n(\%K_m(t)) \tag{9}$$

Where $\%K_m(t) = 100 \times (C_t - LV_m)/(HV_m - LV_m)$, LV_m is the lowest low price (in previous m days period), HV_m is the highest high price.

10. %D_3: 3-day Stochastic Index D. It's a simple moving average of Stochastic Slow Index K.

$$\%D_n(t) = SMA_n(slow\%K_n(t)) \tag{10}$$

11. Stochastic_DIFF: the difference between Stochastic Slow Index K and Stochastic Index D.

$$Stochastic\_DIFF_n(t) = slow\%K_n(t) - \%D_n(t) \tag{11}$$

12. PSY_10: 10-day Psychological Line.

$$PSY_n(t) = 100 \times \left(\frac{UD_n}{n}\right) \tag{12}$$

Where UD_n is the number of closing price upward days during previous n days.

13. MTM_10: 10-day Momentum. It measures the pace at which a trend is accelerating or decelerating.

$$MTM_n(t) = C_t - C_{t-n+1} \tag{13}$$

14. WMTM_9: 9-day simple moving average of MTM_10.

$$WMTM_m(t) = SMA_m(MTM_n(t)) \tag{14}$$

15. MTM_DIFF: the difference between MTM_10 and WMTM_9.

$$WMTM_{n,m}(t) = MTM_n(t) - WMTM_m(t) \tag{15}$$

16. ROC_12: 12-day Price Rate of Change. It measures the percent change of the current price and the price a specified period ago.

$$ROC_n(t) = 100 \times (C_t - C_{t-n+1})/C_{t-n+1} \tag{16}$$

17. RSI_14: 14-day Relative Strength Index. It reflects the price strength by comparing upward and downward close-to-close movements over a predetermined period.

$$RSI_n(t) = 100 - \frac{100}{1 + RS_n(t)} \tag{17}$$

Where

$$RS_n(t) = \frac{SMA_n(Close\_Up_t)}{SMA_n(Close\_Down_t)}$$

$$Close\_Up_t = \begin{cases} C_t - C_{t-1}, & if\ C_t - C_{t-1} > 0 \\ 0, & Otherwise \end{cases}$$

$$Close\_Down_t = \begin{cases} C_{t-1} - C_t, & if\ C_{t-1} - C_t > 0 \\ 0, & Otherwise \end{cases}$$

18. W%R_14: 14-day William's Oscillator Percent R. It helps highlight the overbought/oversold levels by measuring the latest closing price in relation to the highest price and lowest price of the price range over an observation period

$$W\%R_n(t) = -100 \times \frac{HV_n - C_t}{HV_n - LV_n} \tag{18}$$

19. CMF_21: 21-day Chaikin Money Flow. It compares the summation of volume to the closing price and the daily peaks and falls, so that determine whether money is flowing in or out during a definite period.

$$CMF_n(t) = 100 \times \frac{\sum_{i=1}^{n} MFV_{t-i+1}}{\sum_{i=1}^{n} V_{t-i+1}} \tag{19}$$

Where V_t is the trading volume at time t, and $MFV_t = V_t \times (2 \times C_t - L_t - H_t)/(H_t - L_t)$.

20. MFI_14: 14-day Money Flow Index. It measures the strength of money flowing in and out of a security.

$$MFI_n(t) = 100 - \frac{100}{1 + MR_t} \tag{20}$$

Where

$$MR_n = \frac{\sum_{i=1}^{n} +MF_{t-i+1}}{\sum_{i=1}^{n} -MF_{t-i+1}}$$

$$+MF_t = \begin{cases} TPY_t \times V_t, & \text{if } TYP_t > TYP_{t-1} \\ 0, & \text{Otherwise} \end{cases}$$

$$-MF_t = \begin{cases} TPY_t \times V_t, & \text{if } TYP_t < TYP_{t-1} \\ 0, & \text{Otherwise} \end{cases}$$

21. RVI_14: 14-day Relative Volatility Index. It was used to indicate the direction of volatility, and it's similar to the RSI, except that it measures the 10-day standard deviation (STD) of prices changes over a period rather than the absolute price changes.

$$RVI_{n,m}(t) = 100 \times \frac{SMA_m(u_n(t))}{SMA_m(u_n(t)) + SMA_m(d_n(t))} \tag{21}$$

Where

$$u_n(t) = \begin{cases} STD_n(C_t), & \text{if } C_t > C_{t-1} \\ 0, \text{Otherwise} \end{cases}, \quad d_n(t) = \begin{cases} STD_n(C_t), & \text{if } C_t < C_{t-1} \\ 0, \text{Otherwise} \end{cases}$$

$$STD_n(C_t) = \sqrt{\frac{\sum_{i=1}^{n}(C_{t-i+1} - SMA_n(C_t))^2}{n}}$$

22. AROON_25: 25-day Aroon Oscillator. It can help determine whether a security is trending or not and how strong the trend is.

$$Aroon_n(t) = 100 \times (HD_n(t) - LD_n(t))/n \tag{22}$$

Where $HD_n(t)$ is the number of days since the highest high price during the n observation days, and $LD_n(t)$ is the number of days since the lowest low price during the n observation days.

23. EMV: Ease of Movement. It demonstrates how much volume is required to move prices.

$$EMV_t = 1000000 \times (H_t - L_t)\frac{(H_t - H_{t-1}) + (L_t - L_{t-1})}{2 \times V_t} \tag{23}$$

24. WEMV_14: 14-day simple moving average of Ease of Movement.

$$WEMV_n(t) = SMA_n(EMV_t), \tag{24}$$

25. MACD: Moving Average Convergence and Divergence. It's represented as the difference between 12-day and 26-day exponential moving average of closing price.

$$MACD_{n1,n2}(t) = EMA_{n1}(C_t) - EMA_{n2}(C_t) \tag{25}$$

26. MACDSIG: 9-day simple moving average of MACD

$$MACDSIG_{n3}(t) = EMA_{n3}(MACD_{n1,n2}(t)) \tag{26}$$

27. MACD_DIFF: the difference between MACD and MACD_SIG

$$MACD\_DIFF_t = MACD_{n1,n2}(t) - MACDSIG_{n3}(t) \tag{27}$$

References

1. Jacinta, C.: Financial Times Guide to Technical Analysis: How to Trade like a Professional. Pearson, UK (2012)
2. Atsalakis, G.S., Valavanis, K.P.: Surveying stock market forecasting techniques–Part II: soft computing methods. Expert Syst. Appl. **36**(3), 5932–5941 (2009)
3. Chen, W.-H., Shih, J.-Y., Wu, S.: Comparison of support-vector machines and back propagation neural networks in forecasting the six major Asian stock markets. Int. J. Electron. Finan. **1**(1), 49–67 (2006)
4. Kara, Y., Boyacioglu, M.A., Baykan, Ö.K.: Predicting direction of stock price index movement using artificial neural networks and support vector machines: the sample of the Istanbul Stock Exchange. Expert Syst. Appl. **38**(5), 5311–5319 (2011)
5. Vladimir, V.N., Vapnik, V.: The nature of statistical learning theory (1995)
6. Kotsiantis, S.B., Kanellopoulos, D., Pintelas, P.E.: Data preprocessing for supervised leaning. Int. J. Comput. Sci. **1**(2), 111–117 (2006)
7. Malkiel, B.G., Fama, E.F.: Efficient capital markets: a review of theory and empirical work. J. Finan. **25**(2), 383–417 (1970)
8. Malkiel, B.G.: A random walk down Wall Street: the time-tested strategy for successful investing. WW Norton & Company, New York (2007)
9. Kim, K.-J.: Financial time series forecasting using support vector machines. Neurocomputing **55**(1), 307–319 (2003)
10. Kumar, M., Thenmozhi, M.: Forecasting stock index movement: a comparison of support vector machines and random forest. In: Indian Institute of Capital Markets 9th Capital Markets Conference Paper (2006)
11. Huang, W., Nakamori, Y., Wang, S.-Y.: Forecasting stock market movement direction with support vector machine. Comput. Oper. Res. **32**(10), 2513–2522 (2005)
12. Al Shalabi, L., Shaaban, Z.: Normalization as a preprocessing engine for data mining and the approach of preference matrix. In: 2006 International Conference on Dependability of Computer Systems, IEEE (2006)
13. Mustaffa, Z., Yusof, Y.: A comparison of normalization techniques in predicting dengue outbreak. In: International Conference on Business and Economics Research, vol. 1, IACSIT Press (2011)
14. Jayalakshmi, T., Santhakumaran, A.: Statistical normalization and back propagation for classification. Int. J. Comput. Theor. Eng. **3**(1), 89 (2011)
15. Nayak, S.C., Misra, B.B., Behera, H.S.: Impact of data normalization on stock index forecasting. Int. J. Comp. Inf. Syst. Ind. Manag. Appl. **6**, 357–369 (2014)
16. Singh, BK., Verma, K., Thoke, A.S.: Investigations on impact of feature normalization techniques on classifier's performance in breast tumor classification. Int. J. Comput. Appl. **116**(19) (2015)
17. Han, J., Pei, J., Kamber, M.: Data Mining: Concepts And Techniques. Elsevier, Waltham (2011)
18. Achelis, S.B.: Technical Analysis from A to Z. McGraw Hill, New York (2001)
19. Guyon, I., Elisseeff, A.: An introduction to variable and feature selection. J. Mach. Learn. Res. **3**, 1157–1182 (2003)
20. Furey, T.S., et al.: Support vector machine classification and validation of cancer tissue samples using microarray expression data. Bioinformatics **16**(10), 906–914 (2000)

21. Hsu, C-W., Chang, C-C., Lin, C-J.: A practical guide to support vector classification, pp. 1–16 (2003)
22. Hall, M., Frank, E., Holmes, G., Pfahringer, B., Reutemann, P., Witten, I.H.: The WEKA data mining software: an update. ACM SIGKDD Explor. Newsl. **11**(1), 10–18 (2009)
23. Chang, C-C., Lin C.J.: LIBSVM. a library for support vector machines (2012). http://www.csie.ntu.edu.tw/cjlin/libsvm

Interactive Big Data Visualization Model Based on Hot Issues (Online News Articles)

Wael M.S. Yafooz[1,2(✉)], Siti Z.Z. Abidin[1,2], Nasiroh Omar[2], and Shadi Hilles[3]

[1] Advanced Analytics Engineering Center (AAEC), Shah Alam, Malaysia
waelmohamed@hotmail.com
[2] Faculty of Computer and Mathematical Sciences, Universiti Teknologi MARA,
40450 Shah Alam, Selangor, Malaysia
{zaleha,nasiroh}@tmsk.uitm.edu.my
[3] Faculty of Computer and Information Technology, Al Madinah International University,
Shah Alam, Malaysia
shadihilless@gmail.com

Abstract. Big data is a popular term used to describe a massive volume of data, which is a key component of the current information age. Such data is complex and difficult to understand, and therefore, may be not useful for users in that state. News extraction, aggregation, clustering, news topic detection and tracking, and *social* network analysis are some of the several attempts that have been made to manage the massive data in social media. Current visualization tools are difficult to adapt to the constant growth of big data, specifically in online news articles. Therefore, this paper proposes Interactive Big Data Visualization Model Based on Hot Issues (IBDVM). IBDVM can be used to visualize hot issues in daily news articles. It is based on textual data clusters in textual databases that improve the performance, accuracy, and quality of big data visualization. This model is useful for online news reader, news agencies, editors, and researchers who involve in textual documents domains.

Keywords: Big data · Visual analytics · Interactive visualization · Clustering · Information extraction

1 Introduction

The present immense use of computers in daily life has led to the creation of big data. Big data refers to a collection of vast data stored together, and this term is used to describe the exponential growth of structured and unstructured data. Big data has three characteristics: volume, velocity, and variety [1, 2]. Volume refers to the data size, velocity represents the data speed, and variety denotes the data type (e.g., text, video, or music). It is a dataset that cannot be analyzed by relational database tools, statistical analysis tools, and visualization aids that have become popular in the past 20 years, during which the digitized sensor data have begun to develop rapidly. As such, computer scientists, economists, political scientists, bio-informaticists, and other scholars are finding efficient ways to access massive quantities of information and represent these data in meaningful ways.

© Springer Nature Singapore Pte Ltd. 2016
M.W. Berry et al. (Eds.): SCDS 2016, CCIS 652, pp. 89–99, 2016.
DOI: 10.1007/978-981-10-2777-2_8

Handling big data involves four main issues: storage, management, processing, and visualization. For storage, big data requires a huge storage capacity. Big data management, which encompasses data organization, is the most important among these issues. In general, a relational database is typically used for structured and small data. Big data is difficult to process using traditional database techniques because the process of loading the data into a traditional relational database is too time-consuming and costly, indicating the need for additional big data analytic techniques. These techniques are known as the processes of collecting, organizing, and analyzing the data to discover useful information. Such techniques aid managers and decision makers in making appropriate business decisions that can improve operational efficiencies and reduce costs and risks.

Analyzed big data can be useful for many sectors, such as advertising and marketing agencies, logistics, hospitals, web-based businesses, consumer product companies, and government sectors. The most common tools and applications for big data analytics are predictive analytics, data mining, text mining, statistical analysis, and data optimization. During processing, retrieval of big data requires high-performance computing, which is time-consuming, suggesting the need for distributed and parallel computing. Good visualization helps organizations to perform analyses and make decisions more efficiently. However, visualization issues arise when big data is depicted in a graphical representation for analysis.

Many businesses are facing difficulties in extracting and visualizing useful information from big data to make decisions. That is, the challenge does not concern the technology itself, rather the skills that are needed to present big data in the form of useful information to users. The objectives of businesses are to increase efficiencies and to optimize operations. The two main challenges in big data visualization are as follows:

- Human perception: When the number of data entities is very large, users will experience difficulty in extracting meaningful information.
- Screen space limitation: When large data are displayed on the screen in terms of data items, the screen space will be insufficient.

Current visualization tools are difficult to adapt to the growth of big data, specifically in online news articles. Therefore, Interactive Big Data Visualization Model Based on Hot Issues (IBDVM) is proposed in this paper. IBDVM is used to visualize hot issues in daily news articles. The user can collect hot issues in online news on a daily, weekly, or monthly basis, or for a specific period. In this way, the user can obtain useful information from the news articles. The proposed model works in multi-layer visualization, thus giving the user general and detailed views. By contrast, available tools for graphical analysis become problematic when dealing with extremely large amounts of information or various categories of information because of screen space limitation. IBDVM is based on our proposed TVSM model, which uses relational databases that contain the most relevant textual documents and online news agencies.

The rest of this paper is organized as follows: Sect. 2 presents overview on related work, while the system architecture of the proposed model and expected results are described in Sect. 3. Section 4 provides the conclusion.

2   Related Work

Big data often has unstructured formats which must be processed in several steps to be useful for users. These steps include data accessing, preprocessing, data modeling, and data visualization as shown in Fig. 1. These steps can be performed in one or more data analytic tools and visualization. Many data preprocessing, analytics, and visualization tools are available. Therefore, these steps can be performed automatically and manually depending on the approaches and tools employed such as *Datawrapper, Tableau, Timeline, JS D3.js, FusionCharts, Chart.js Python, Scala and R*. However, many challenges arise in big data processing, such as redundancy reduction and data compression, data life cycle management, analytical mechanism, expendability, and scalability. Furthermore, processing and visualizing big data are major challenges that are more difficult than storage or management.

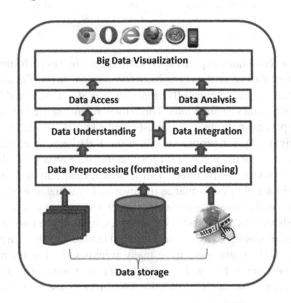

Fig. 1.   General view for big data visualization.

Data visualization (visual analytics) is increasingly becoming an important component of big data analytics. Visual analytics enables organizations to extract raw data and present it in a meaningful way that generates the most value. The visualization is not only static graphic but also interactive which based on selecting, linking, filtering and remapping [3]. Therefore, many researchers have proposed several techniques for visualizing big data, such as pixel-oriented visualization techniques [4], spatial displacement techniques, topological distortion [5], big data reduction [6], parallel coordinates [7] and visual query [8, 9].

In addition, there are several problems for big data visualization such as visual noise, information loss, large image perception and high performance requirements [10, 11].

Many commercial products are available for data analytics and visualization, as summarized in Table 1.

Table 1. Summary of tools for big data visualization

Tools/criteria	User intervention	Type	Dashboard support	Modeling and analytics	Multiple-layer	Open source	Research work
imMens	Yes	Web-Based	Yes	N/A	No	No	[8]
CartoDB	Yes	Web-Based	Yes	N/A	No	No	[12]
Processing.js	Yes	Library	No	N/A	No	Yes	[13]
Weave	Yes	Web-Based	Yes	N/A	No	No	[14]
DataWatch	Yes		Yes	N/A	No	No	[15]
ParaView	Yes	Web-Based	Yes	N/A	No	Yes	[16]
SAS	Yes	Desktop	Yes	Yes	No	No	[17]
Tableau		Desktop	Yes	Yes	No	No	[18]
Datawrapper	Yes	Web-Based	Yes	N/A	No	Yes	[19]
FusionCharts	Yes	Library	No	N/A	No	Yes	[20]
IBDVM	No	Desktop	Yes	Yes	Yes	No	Proposed

However, these techniques require expert knowledge, such as performing processing steps, and they show the output data (graphical form) as one layer. These tools focus on numerical or graphical data, whereas other tools, such as SQL Oracle and IBM, are more concerned on textual data visualizations. These tools require skilled users, prerequisite format, and long processing time. Users must also be knowledgeable enough to shape the data in the right way for proper visualization. Relational database management system (RDBMS) is a robust mechanism for manipulating structured data; therefore, using RDBMS to convert unstructured to structured textual data and visualize it will save time and facilitate high-performance retrieval and visualization of textual data.

There are many tools focus on online news articles visualizations as summarized in Table 2. Such tools used to provide users a good picture about the online news articles. These articles can be considered as the most referred textual data that are produced from daily information [21]. These tools have many weaknesses. The examples include; limited number of dataset, not show the hot issues of online news, required some format as input to application, loss much information and the news are not clustered based on relevant information and hot topics.

Table 2. Online news articles visualization

Tools	Input format	Cluster data	Drill-down	Size	Research work
NewsViews	Web	No	No	Limited	[22]
Contextifier	Web	No	No	Limited	[23]
Newsmap	Web	No	No	Limited	[24]
eagereyes	Web	No	No	Limited	[25]
Newser	Web	No	No	Limited	[26]
Spectra	Web	No	No	Limited	[27]
IBDVM	Web+desktop	Yes	Yes	Unlimited	Proposed

3   System Architecture

This section describes the system architecture of IBDVM, as shown in Fig. 2. IBDVM consists of two phases: input phase and processing interface phase.

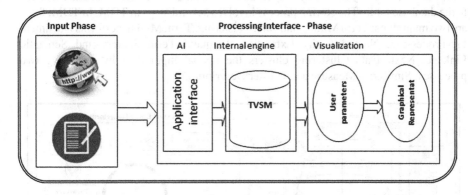

Fig. 2.   General view of BDVM.

3.1   Input Phase

In the input phase, the input data are inserted into the processing interface through the application interface. These data are in textual form. The input tier is the first layer of the conceptual design of IBDVM. It includes the input of textual files that must be stored in the database table. This tier is known as the external layer of the relational database engine. These files usually consist of textual data, which are unstructured documents from different sources, such as news articles, personal data, and files from discussion forums. Dealing with this type of data during the storing process in relational databases involves several steps, which are discussed in the processing interface phase.

3.2   Processing Interface Phase

In second phase is processing interface that consists of three stages: application interface, internal engine and visualization.

1. Application Interface

The processing interface tier, which is known as the internal layer of IBDVM, is the core process of the proposed model. The processing interface tier works as the inter-mediate layer between the input (external) and processing interface.

This application interface stage, which is an external layer of the relational database engine process, handles the interaction between the user and database engine. In this stage, users enter any format of textual data to relational database attributes.

2. Internal Engine

The internal engine is the textual virtual schema model (TVSM) [28, 29]. TVSM performs process of converting textual file from unstructured to structured form by extracting important information. Then, it automatically executes data clustering techniques. TVSM can be used in any relational database such as Oracle, SQL Server and Sybase. TVSM consists of two main components which are Data Acquisition (DA) as the first phase and its second phase is Textual Data Management (TDM) as shown in Fig. 3. In DA, ordinary users enter any format of textual data (structured or unstructured) to the database table. The storing process involves several steps in TDM. In TDM, there are two main steps: Term Mining and Data Clustering. Term Mining performs Document Pre-Processing, Named Entity Extraction, Frequent Term Search, and Semantic Linkage. Next, Data Clustering clusters the textual input data by executing two processes: Similarity Measure and Cluster Description.

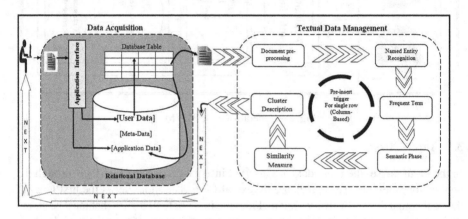

Fig. 3. TVSM system architecture [28, 29].

Document Pre-Processing is a process of cleaning and rearranging the textual data. In document cleaning, Filtration, Stop Words Removal, Stemming and Document Representation are carried out. Filtration removes any format or noise from a textual document such as HTML tags or XML. Stop Words Removal removes the list of stop-words such as 'a', 'an', 'the' according to the standard list [30]. Stemming converts words to source by using the porter algorithm [31]. Document Representation uses the Vector Space Model, which is used for representing a document in the form of words, and its frequencies with added columns for Named Entity and Frequent Term. The output of this process can be used in the Named Entity Extraction process.

The Named Entity Extraction is a process to extract the named entity such as person, organization and place, based on NER-Stanford [32]. The extracted named entity is stored along with its frequencies. The Frequent Term Search process mines and selects the frequent terms from textual data according to its frequencies known as minimum support. All Selected Frequent Terms are used in the next stage to get the semantic.

The Semantic Phase deals with the synonym for selected frequent terms from WordNet database [33] that enables TVSM to discover the hidden semantic relation between textual data (documents). Thus, the synonym of Selected Frequent Terms and Named Entity are used as input for Matching Process to find an existing textual data cluster. If the cluster is found, it is considered as the selected cluster for that textual document. Otherwise, a new textual data cluster is created. The matching process is based on the Similarity Measure, which will be discussed in the next section. The last process is adding the selected frequent terms and named entities to the cluster description. Cluster description is the collection of selected frequent terms and named entities that describe the textual data cluster. It is used in the process of similarity measure for incoming text data.

The results of the experiment show that the performance of TVSM outperforms K-mean, BKM, UPGMA, Oracle text and FICH clustering algorithms. Since TVSM works as an incremental clustering process, there is no need to re-cluster textual data every time. On the other hand, the goodness (quality of clusters) achieves a higher score. Specifically, when a dataset contains news articles, they often hold named entities.

3. Visualization

Visualization, which is also known as visual analytics, is a professional way of delivering information to users to support them with useful information. Such information can help them to make sound decisions based on big data. Visual analytics enables organizations to extract raw data and present it in a meaningful way that generates the most value.

The contribution of the proposed model is to show the hot issues of online news in two steps, namely, user parameters and graphical representations. The user parameters include publication period of news articles, field of the news, area of the news, and news feeds (RSS). Meanwhile, in graphical representation, the figure shows the hot issues in economics, sports, politics, and other fields of news. When a user clicks of one of these topics, the figure shows the details. The graphical representation is based on the visualization process.

The proposed model can significantly improve the problem of noises visualization, loss of information, large image perception and high performance requirements, due to the model developed based on TVSM that achieves a high textual data clustering in relational databases.

The visualization process is based on the textual data clusters executed using TVSM. Visualization has two components: Named Entity Extraction and Frequent Term Search. Named Entity Extraction is the extraction of significant words, such as persons, organizations and places. The extracted word, that is, a named entity, is stored in a database table along with its number of occurrences in a given document. Then, Frequent Term Search is performed to mine and select the frequent terms from textual data based on the chosen number of minimal frequencies, known as minimum support. These two components have already been created through TVSM. The visualization is based on the frequency of both components. Therefore, online news articles can be visualized as in Fig. 4.

Figure 4 shows the dataset of online new articles that contain 744 articles divided into six categories, which have already been clustered using TVSM. A graph for analysis is difficult to create when dealing with extremely large amount of information or various categories of information, such as when a dataset contains million of articles. The best way to resolve this problem is to give the user a way of clicking on the parts of categories and show news articles in sub-categories.

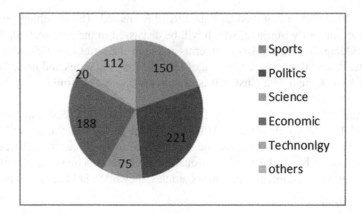

Fig. 4. Categories of online news articles.

For example: If the user clicks on sports, then the detailed view will appear as in Fig. 5.

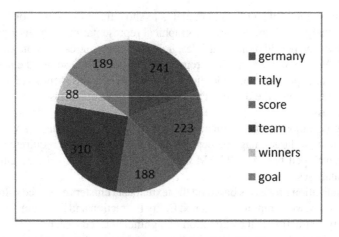

Fig. 5. Detailed view of sport category

Figure 5 shows the frequency of words in the datasets, meaning that these terms are the most frequent words that represent textual document, which in this case are online news articles. By selecting any section of the graphic, the source of these online news articles will appear as in Fig. 6.

Figure 6 shows the source of the news articles. When a user clicks on the online news agencies, all news articles will be listed by title. Visualization can be performed on the input data from a user, such as categories, period of online news published, or hot issues. In this way, the user saves time and obtains accurate high-quality news articles in interactive and excellent graphical representation. All visualization processes are based on data clusters that have already been created using TVSM.

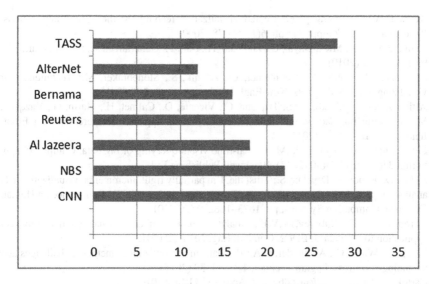

Fig. 6. Source of Sport online news articles.

4 Conclusion

In this paper, we presented the processes of analyzing big data and obtaining useful information. The most important tools and scripting languages used for big data visualization are already presented; more focus should be given to the visualization of big data. The most common stage for textual data is the relational database. Therefore, the system architecture of the IBDVM retrieves data from the relational database and delivers the data for users. IBDVM produces accurate and good-quality graphical representations of visualization. This visualization can show online news articles in a general view (main category) and its sub-categories. Therefore, the performance of visualizations is very fast because the textual data (news articles) are already clustered in relational databases using TVSM. By contrast, traditional approaches and tools encounter many issues during data accessing, which require long processing time and skilled users who are familiar with the tools or scripting languages.

Acknowledgment. The authors would like to thank Universiti Teknologi MARA and Ministry of Education, Malaysia (600-RMI/FRGS 5/3(161/2013)) for the financial support.

References

1. Bizer, C., Boncz, P., Brodie, M.L., Erling, O.: The meaningful use of big data: four perspectives–four challenges. ACM SIGMOD Rec. **40**(4), 56–60 (2012)
2. Sagiroglu, S., Sinanc, D.: Big data: a review. In: 2013 International Conference on Collaboration Technologies and Systems (CTS), pp. 42–47. IEEE (2013)
3. Simon, P.: The Visual Organization: Data Visualization, Big Data, and the Quest for Better Decisions. John Wiley & Sons, Hoboken (2014)

4. Keim, D.A.: Designing pixel-oriented visualization techniques: theory and applications. IEEE Trans. Vis. Comput. Graph. **6**(1), 59–78 (2000)
5. Keim, D.A., Hao, M.C., Dayal, U., Janetzko, H., Bak, P.: Generalized scatter plots. Inf. Vis. **9**(4), 301–311 (2010)
6. DeBrabant, J., Battle, L., Cetintemel, U., Zdonik, S., Stonebraker, M.: Techniques for visualizing massive data sets. New Engl. Database Summit **2**(9) (2013)
7. Artigues, A., Cucchietti, F.M., Tripiana, C., Vicente, D., Calmet, H., Marín, G., Vazquez, M.: Scientific big data visualization: a coupled tools approach. Supercomputing Front. Innovations **1**(3), 4–18 (2015)
8. Liu, Z., Jiang, B., Heer, J.: imMens: real-time visual querying of big data. Comput. Graph. Forum **32**(3–4), 421–430 (2013). Blackwell Publishing Ltd.
9. Fisher, D., Popov, I., Drucker, S.: Trust me, I'm partially right: incremental visualization lets analysts explore large datasets faster. In: Proceedings of the SIGCHI Conference on Human Factors in Computing Systems, pp. 1673–1682. ACM (2012)
10. Gorodov, E.Y.E., Gubarev, V.V.E.: Analytical review of data visualization methods in application to big data. J. Electr. Comput. Eng. **2013**, 22 (2013)
11. Wang, L., Wang, G., Alexander, C.A.: Big data and visualization: methods, challenges and technology progress. Digital Technol. **1**(1), 33–38 (2015)
12. CartoDB (2016). https://cartodb.com/. Accessed 11 July 2016
13. Agrawal, R., Kadadi, A., Dai, X., Andres, F.: Challenges and opportunities with big data visualization. In: Proceedings of the 7th International Conference on Management of Computational and Collective intElligence in Digital EcoSystems, pp. 169–173. ACM (2015)
14. Weave, Institute for Visualization and Perception Research (IVPR) of the University of Massachusetts (2016). https://www.oicweave.org/. Accessed 11 July 2016
15. Gartner, Datawatch, Gartner says Self-Service Data Prep the next Big Disruption in Business Intelligence (2016). http://www.datawatch.com/. Accessed 11 July 2016
16. Alya system, large scale computational mechanics (2016). http://www.bsc.es/es/computer-applications/alya-system. Accessed 11 July 2016
17. Stephanie Robertson, Tracking down answers to your questions about data scientists. http://www.sas.com/en_us/insights/articles/analytics/Tracking-down-answers-to-your-questions-about-data-scientists.html. Accessed 11 July 2016
18. ROSS PEREZ, Next Generation Cloud BI: Tableau Server hosted on Amazon EC2. http://www.tableau.com/learn/whitepapers/next-generation-cloud-bi-tableau-server-hosted-amazon-ec2. Accessed 23 July 2016
19. Aisch, G.: Doing the line chart right. http://academy.datawrapper.de/doing-the-line-chart-right-66523.html. Accessed 23 July 2016
20. Sanket Nadhani, fusioncharts. http://www.fusioncharts.com/company/Not-Just-Another-Pie-In-The-Sky.pdf. Accessed 23 July 2016
21. Yafooz, W.M., Abidin, S.Z., Omar, N.: Challenges and issues on online news management. In: 2011 IEEE International Conference on Control System, Computing and Engineering (ICCSCE), pp. 482–487. IEEE (2011)
22. Gao, T., Hullman, J.R., Adar, E., Hecht, B., Diakopoulos, N.: NewsViews: an automated pipeline for creating custom geovisualizations for news. In: Proceedings of the 32nd Annual ACM Conference on Human Factors in Computing Systems, pp. 3005–3014. ACM (2014)
23. Hullman, J., Diakopoulos, N., Adar, E.: Contextifier: automatic generation of annotated stock visualizations. In: Proceedings of the SIGCHI Conference on Human Factors in Computing Systems, pp. 2707–2716. ACM (2013)
24. Ong, Thian-Huat, Chen, Hsinchun, Sung, Wai-ki, Zhu, Bin: Newsmap: a knowledge map for online news. Decis. Support Syst. **39**(4), 583–597 (2005)

25. Robert Kosara, Eagereyes (2016). https://eagereyes.org/. Accessed 23 July 2016
26. Michael Wolff, Newser (2016). http://www.newser.com/. Accessed 10 July 2016
27. NBC News, Spectra (2016). http://spectramsnbc.com/. Accessed 10 July 2016
28. Yafooz, W.M., Abidin, S.Z., Omar, N., Halim, R.A.: Dynamic semantic textual document clustering using frequent terms and named entity. In: 2013 IEEE 3rd International Conference on System Engineering and Technology (ICSET), pp. 336–340. IEEE (2013)
29. Yafooz, W.M., Abidin, S.Z., Omar, N., Halim, R.A.: Model for automatic textual data clustering in relational databases schema. In: Herawan, T., et al. (eds.) Proceedings of the First International Conference on Advanced Data and Information Engineering (DaEng-2013), Part I. LNEE, vol. 285, pp. 31–40. Springer, Singapore (2014)
30. Wilbur, W.J., Sirotkin, K.: The automatic identification of stop words. J. Inf. Sci. **18**(1), 45–55 (1992)
31. Porter, M.F.: An algorithm for suffix stripping. Program **14**(3), 130–137 (1980)
32. Finkel, J.R., Manning, C.D.: Joint parsing and named entity recognition. In: Proceedings of Human Language Technologies: The 2009 Annual Conference of the North American Chapter of the Association for Computational Linguistics, pp. 326–334. Association for Computational Linguistics (2009)
33. Miller, G.A.: WordNet: a lexical database for English. Commun. ACM **38**(11), 39–41 (1995)

Metabolites Selection and Classification of Metabolomics Data on Alzheimer's Disease Using Random Forest

Mohammad Nasir Abdullah[1]([📧]), Bee Wah Yap[2], Yuslina Zakaria[3],
and Abu Bakar Abdul Majeed[3,4]

[1] Department of Statistics, Faculty of Computer and Mathematical Sciences,
Universiti Teknologi MARA, Perak Branch, 35400 Tapah, Perak, Malaysia
nasir916@perak.uitm.edu.my
[2] Advanced Analytic Engineering Centre, Faculty of Computer
and Mathematical Sciences, Universiti Teknologi MARA,
40000 Shah Alam, Selangor, Malaysia
beewah@tmsk.uitm.edu.my
[3] Faculty of Pharmacy, Universiti Teknologi MARA,
42300 Bandar Puncak Alam, Selangor, Malaysia
yuslina@puncakalam.uitm.edu.my,
abubakar@salam.uitm.edu.my
[4] Brain Degeneration and Therapeutics Group, Pharmaceutical and Life Science
Community of Research, Universiti Teknologi MARA,
40450 Shah Alam, Selangor, Malaysia

Abstract. Alzheimer's disease (AD) is neurodegenerative disorder character-ized by the gradual memory loss, impairment of cognitive functions and pro-gressive disability. It is known from previous studies that symptoms of AD are due to synaptic dysfunction and neuronal death in the area of the brain, which performs memory consolidation. Thus, the investigation of deviations in various cellular metabolite linkages is crucial to advance our understanding of early disease mechanism and to identify novel therapeutic targets. This study aims to identify small sets of metabolites that could be potential biomarkers of AD. Liquid chromatography/mass spectrometry-quadrupole time of flight (LC/MS-QTOF)-based metabolomics data were used to determine potential biomarkers. The metabolic profiling detected a total of 100 metabolites for 55 AD patients and 55 healthy control. Random forest (RF), a supervised classification algo-rithm was used to select the important features that might elucidate biomarkers of AD. Mean decrease accuracy of .05 or higher indicates important variables. Out of 100 metabolites, 10 were significantly modified, namely N-(2-hydroxyethyl) icosanamide which had the highest Gini index followed by X11-12-dihyroxy (arachidic) acid, N-(2-hydroxyethyl) palmitamide, phytosph-ingosine, dihydrosphingosine, deschlorobenzoyl indomenthacin, XZN-2-hydroxyethyl (icos) 11-enamide, X1-hexadecanoyl (sn) glycerol, trypthophan and dihydroceramide C2.

Keywords: Alzheimer's disease · Biomarkers · Metabolite · Random forest

© Springer Nature Singapore Pte Ltd. 2016
M.W. Berry et al. (Eds.): SCDS 2016, CCIS 652, pp. 100–112, 2016.
DOI: 10.1007/978-981-10-2777-2_9

1 Introduction

Alzheimer's disease (AD) is a neurodegenerative disorder characterized by the gradual progression of memory loss, impairment of cognitive functions and progressive disability [1–3]. The end result of the disease is dementing disorder which mostly happens in the elderly people [4–6]. AD incurs tremendous social and economic costs not only to the patient but also to caregivers and family members. Genes associated with the development of AD might have an effect on chemicals known as neurotransmitters, which allow message to be communicated between nerve cells in the brains [7]. Previous studies have reported that the symptoms of AD are synaptic dysfunction and neuronal death in the area of the brain which performs memory consolidations [8]. Genetic factors are also involved in the pathogenesis of AD [9, 10]. The disease could also result from a combination of environmental and lifestyle factors as the life expectancy of the population expands [11]. Several studies have shown that the genetic factor for AD is $\varepsilon 4$ allele of the gene encoding apolipoprotein E (APOE) [4, 12, 13].

The prevalence of AD is 27 million person worldwide [5], and it is predicted to affect more than 80 million people by the year 2050 [5, 9]. In the USA, the number of disease is expected to increase to 13 million while 4 million in the European Union (EU) by the year 2050 [14]. In Asia, approximately 5.9 % of all individuals in China aged 65 years and above have developed AD [14]. In Malaysia, it has been estimated that there are currently about 60 thousand people having the disease and most of them have not been diagnosed [15].

Hitherto, there is no treatment that can effectively reverse AD [7, 16]. An accurate early diagnosis of AD is still difficult because early symptoms of the disease are shared with a variety of disorders, which reflect common neuropathological features [9].

Lately, the field of metabolomics has become a prevailing profiling method to determine novel biomarkers and to recognize the state of the disease especially AD [17]. In addition, metabolomics can provide unbiased identification of small molecules systematically. By definition, metabolomics basically cover a wide-range of small molecules such as glucose, cholesterol, adenosine triphosphate (ATP), biogenic amine neurotransmitters, signalling molecules among several classes of compounds within a biological system [18]. Metabolomics or metabonomics participate in metabolic reaction required for growth, maintenance and normal function of an organism [11, 18]. It can also display changes at higher hierarchical levels such as proteome, transcriptomics and genomics [19].

The aim of this study was to identify small set of metabolites that could be used as biomarkers for AD using the random forest supervised learning method. This paper is organized as follows. Section 2 reviews and presents some related studies on identifying potential biomarkers for AD. The methodology is explained in Sect. 3 while the results and discussions are given in Sect. 4. Some discussion of results and the conclusion are given in Sect. 5.

2 Literature Review

Liang et al. [17] obtained representative fingerprints of low molecular weight metabolites from serum samples to distinguish between AD and control group. They found 9 potential biomarker candidates where 6 of them were up-regulated (sphinganine-1-phosphate, 7-ketocholesterol, 3-methoxytyrosine, deoxyribose-5-phosphate, P-phenyllactic acid, LysoPC(15:0)) and the remaining 3 (phenylalanine, omithine, L-glutamic acid) were down-regulated. Meanwhile, [20] used ultra-performance liquid chromatography/electrospray ionization quadrupole-time-of-flight-high-definition mass spectrometry (UPLC-Q/TOF-MS/MS-based) metabolomics approach to identify AD and the protective effect of Kai-Xin-San (KXS). Their study found 48 potential biomarkers and the first 10 potential metabolites biomarkers were Lysine, Allantoic acid, 5'-deoxyadenosine, uridine (down-regulated), N6-acetyl-L-lysine, glycerpphosphocholine, IDP, leucine (down-regulated), formylanthranilate (down-regulated) and adenine (down-regulated) Another study, which employed an autopsy-confirmed AD patients found that NE, methionine and alpha-tocopherol have negative correlations with tangle formation or amyloid deposition, while 5-hyroxytrytophan and tyramine have positive correlation with amyloid deposition [21]. Recently, a study which employed salivary specimens on 22 AD, 25 MCI and 35 normal persons found that TyrAsn-ser was upregulated in AD [22]. On other view, a study which also employed salivary techniques on AD was found seven potential metabolic biomarkers that up-regulated which are cytidine, sphinganine-1-phosphate, phenyllactic acid, pyroglutamic acid, L-glutamic acid, omithine, L-tryptophan and hypoxanthine [23]. In addition, [23] also found that three metabolic biomarkers that down regulated which are 3-dehydrocamitine, inosine and hypoxanthine.

3 Methodology

3.1 Dataset

The sample consisted of 55 AD patients and another 55 healthy controls (HC). These subjects were recruited from the Memory and Geriatric clinic of the Universiti Malaya Medical Centre (UMMC). The inclusion criteria for AD patients were age more than 65 years old, fulfill the criteria for probable AD based on the Revise National Institute of Neurological and Communication disorders – Alzheimer Disease and Related Disorders Association criteria, and diagnosis made by neurologist or geriatrician. The inclusion criteria also include mini mental state examination (MMSE) score of less than or equal to 26 and the most important part is informed consent provided by the subject and caregiver. While the exclusion criteria were age less than 65 years old, patients functionally independent as measured by Katz Basic Activities of Daily Living (ADL) and Lawton Instrumental ADL (BADL) scales, the Mini Mental State Examination (MMSE) score of more than 27 and patients were able to communicate and give informed consent.

Metabolic profiling involves the multi-component analysis of biological fluids, tissue and cell extracts using NMR spectroscopy and/or mass spectrometry [24]. The data were collected based on blood sample of the subjects. Then the blood sample was prepared for extraction using liquid chromatography mass spectrometry quadrupole-time-of-flight (LC-MS-QTOF) and the metabolic profiling was obtained. The dataset contains only 100 metabolites selected based on prior analysis using t-test with correction of p-value less than 0.05 [25] which controls for False Discovery Rate. The 100 metabolites are listed in Table 1.

3.2 Data Pre-processing and Preliminary Analysis

Data were analyzed using R software, an open source statistical software [26]. All the observations were checked and cleaned. Data exploration was carried out using descriptive statistics such as minimum and maximum value, mean, range, standard deviation and box plots to observe if there were outliers and to view the distribution of the data.

The basic characteristics of the patients such as gender, ethnicity, smoking history and alcohol history were also recorded. The chi-squared test was used to test the association between AD and demographic variables.

3.3 Random Forest

In this paper, we choose Random Forest (RF) as a classifier to selects potential biomarkers for AD after done thorough review on several classifiers performance and found that RF is one of the top classifier that can classify biomarkers from high dimensional data.

RF is a classification and regression technique which involves constructing a multitude of decision trees in the training phase. To be more specific, it is an ensemble method of trees developed from a training data set and validated internally to achieve an accurate prediction of the target variable from the predictors for the purpose of future observations [27]. RF will create multiple classification and regression trees (CART) based on the bootstrap sample from the original training data. It also randomly search the features to determine the splitting point [28] for growing a tree. In addition, the RF does not over fit as the number of trees increases but it will produce a restrictive value on the generalization error [29].

In this paper, we choose RF as a classifier to select potential biomarkers for AD after done thorough review on several classifiers performance and found that RF is one of the top classifier that can classify biomarkers from high dimensional data.

The procedure for growing a single tree was based on the CART methodology where the tree started at single node (N) in the training set (D). Next, if all observations in D are in the same class, then node N will become a leaf. If they are in different classes, then the feature selection method using Gini index was applied (at this stage it is called as splitting point). These steps were repeated for other features. The stopping criterion for splitting point is when all observations in D belong to the same class.

In CART, the measure used for feature selection (splitting point) method used was Gini index, where $Gini(D) = 1 - \sum_{i=1}^{m} p_i^2$. The p_i is the probability of a feature in D belong to class $(C_i \, where \, i = 0, 1)$ and it was estimated using $\frac{|C_i D|}{|D|}$. The Gini index

Table 1. List of metabolomics elements ($p = 100$)

2 - thiopheneacrylic acid	ritodrine glucuronide	24-nor-5beta-cholane-3alpha,7alpha,12alpha,22,23-pentol	4,7,10,13,16-docosapentaenoic acid	cyclopenta[c]pyrrole, benzenesulfonamide deriv.
9R-(2-cyclopentenyl)-1-nonanol	11,12-dihydroxy arachidic acid	4-methyl-undecanoic acid	deschlorobenzoyl Indomethacin	cis-11-hexadecenal
dihydroceramide C2	3-keto palmitic acid	rimiterol	methyl 4-[2-(2-formyl-vinyl)-3-hydroxy-5-oxo-cyclopentyl]-butanoate	N-oxide gliclazide
(Z)-N-(2-hydroxyethyl)icos-11-enamide	perindoprilat lactam A	GPCho(18:3(9Z,12Z,15Z)/0:0) (1-nonadecanoyl-2-tricosanoyl-sn-glycero-3-phosphoethanolamine)	11-methyl-octadecanoic acid	b-d-glucopyranosiduronic acid
13,14-dihydro-19(R)-hydroxyPGE1	3-hydroxypromazine sulfoxide	5-s-cysteinyldopamine	nalorphine	2-ethylacrylylcarnitine
cholest-5-ene	cetylpyridinium	1-(9Z,12Z-octadecadienoyl)-rac-glycerol	droperidol	cinnamaldehyde
lauric acid	9,12,14-octadecatrienoic acid	2-mercapto-octadecanoic acid	20alpha-dihydroprogesterone glucuronide	furafylline
2-hydroxy-nonadecanoic acid	9,12-octadecadienal	norcodeine	24-nor-5beta-cholane-3alpha,7alpha,12alpha,23-tetrol	1-indanone
indoleacrylic acid	5,8-dihydroxy-3,4-dihydrocarbostyril	7E,9Z-dodecadien-1-ol	3-nitrotyrosine	octadecanamide
GPCho(O-15:0/2:0[U]) - (1-hexadecanoyl-2-(9Z-octadecenoyl)-3-(7Z,10Z,13Z,16Z,19Z-docosapentaenoyl)-sn-glycerol)	embelin	1-hexadecanoyl-sn-glycerol	methacholine	B-ureidoisobutyric acid
cis-5-tetradecenoylcarnitine	5a-tetrahydrocorticosterone	1-acetylindole	inosine	3-hydroxy-n-glycyl-2,6-xylidine (3-hydroxyglycinexylidide)
tryptophan	N-palmitoylsphingosine	L-glutamic acid dibutyl ester	N-arachidonoyl d-serine	L-phenylalanine n-butyl ester
6-[3]-ladderane-1-hexanol	N-gamma-acetyl-n-2-formyl-5-methoxykynurenamine	3,11-dihydroxy myristoic acid	GPCho(O-18:2(9Z,12Z)/2:0[U]) - (1-(9Z-hexadecenoyl)-2-(9Z,12Z,15Z-octadecatrienoyl)-3-eicosanoyl-sn-glycerol)	GPEtn(10:0/11:0)[U] - (1-a-linolenoyl-2-myristoleoyl-sn-glycero-3-phosphoethanolamine)
dihydrosphingosine	3E,5E-tridecadienoic acid	2H-indol-2-one, 4-[2-(dipropylamino)ethyl]-1,3-dihydro-7-hydroxy-glucuronide	1-octadecanamine	neuraminic acid
10-hydroxy-2E,8E-decadiene-4,6-diynoic acid	3-hydroxylidocaine	GPCho(15:0/0:0) - (1-arachidonoyl-2-myristoleoyl-sn-glycero-3-phosphocholine)	paraxanthine	warfarin
9-hexadecen-1-ol	phytosphingosine	corticosterone	guanosine	proline
N-(2-hydroxyethyl)icosanamide	2-hydroxy-3-(4-methoxyethylphenoxy)-propanoic acid	loganin	2-oxo-docosanoic acid	2-methylbutyrylglycine
procaterol	3-hydroxy-eicosanoic acid	2E,5Z,8Z,11Z,14Z-eicosapentaenoic acid	salsolidine	methyl 6-hydroperoxy-4,8,11,14,17-eicosapentaenoate
1-cyclohexene-1-acrylic acid, 2,6,6-trimethyl-3-oxo-	N-(2-hydroxyethyl)palmitamide	17-phenyl-trinor-PGF2alpha isopropyl ester	TG(17:2(9Z,12Z)/20:2(11Z,14Z)/22:6(4Z,7Z,10Z,13Z,16Z,19Z))[iso6]	SM(d18:1/0:0) - ((2S,3S,4R)-2-amino-3,4-dihydroxyoctadecyl dihydrogen phosphate)
o-hexanoyl-r-carnitine	15S-hpEDE	allocortol	cyclopenta[c]pyrrole, benzoic acid deriv	lipoamide

considers binary split for each feature to determine the best split on N. The decision of splitting point was based on the lowest Gini impurity value computed within m features. Therefore, in order to obtain the value of Gini impurity, the weighted sum of the impurity of each resulting partition was calculated using $Gini_A(D) = \sum_{j=1}^{2} \frac{|D_j|}{|D|} \times Gini(D)$. Then the Gini impurity was defined as $\Delta Gini(A) = Gini(D) - Gini_A(D)$.

The number of input features to be selected for each splitting is denoted as *mtry* where it is the square root of number of variables in D [30].

The RF algorithm consists the element of bagging (bootstrap aggregation where the sample of features is taken with replacement) and randomization of CART [31]. It produces numbers of trees based on randomized CART from D and sub-samples of size for *mtry* based on bootstrapping from D. In addition, [29] (the founder of RF) have proved that the RF does not over-fit when the number of trees increase. However, if we were energetically grow as much as many trees, it would increase computational cost. [32] reported that as the number of trees increased it did not constantly show that the mean performance is significantly better than fewer trees. In this study, we performed a simulation study to determine the ideal number of trees to grow for the dataset.

The variable importance (VI) measures an association between a feature and the target (AD or HC). To estimate the VI for features j, the out of bag sample (which the sample that not included in the bootstrap sample to construct tree) were pass down the tree and the accuracy of the performance was measured [28]. The VI for features j can be defined as $VI(X_j) = \frac{1}{M}\sum_t \left(err\widetilde{OOB}_t^j - errOOB_t \right)$, where M is the number of trees in RF and t is the sum of all trees [33].

The analysis of RF were done using R package which was described in [34] focusing on the randomForest procedure. The procedure of executing RF is shown in Fig. 1.

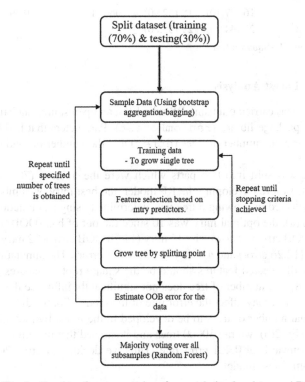

Fig. 1. Random forest procedure for metabolomics data on AD.

4 Results

The preliminary analyses using Chi-Square test of association shows no significant association between AD and HC with gender, ethnicity and alcohol history. However, there are significant associations of AD with smoking history. The Chi-Square test of association results are presented in Table 2. These results could not be generalized due to the small sample size.

Table 2. Demographic profile of subjects and Chi-square test of association results

Characteristics	HC (n (%))	AD (n (%))	Chi-Square test (df)	p-value[a]
Gender				
Male	28 (56.00)	22 (44.00)	1.320 (1)	0.251
Female	27 (45.00)	33 (55.00)		
Ethnicity				
Chinese	32 (52.46)	29 (47.54)	0.3312 (1)	0.565
Non-Chinese	23 (46.94)	26 (53.06)		
Smoking History				
Yes	14 (63.60)	8 (36.40)	3.989 (1)	0.046
No	25 (39.10)	39 (60.90)		
Alcohol History				
Yes	16 (47.10)	18 (52.90)	0.066 (1)	0.797
No	23 (44.20)	29 (55.80)		

[a]Pearson Chi-Squared test.

4.1 Random Forest Analysis

The RF analysis was carried out using RStudio [26], an open source statistical software. We used R RF package library (randomForest). Parameters that had been used in RF algorithm were the number of trees (ntree) and the number of randomly selected features (mtry).

The dataset was split into two parts which were the training (70 %) and testing (30 %) samples. Before developing the RF model, the best mtry or number of random features to be selected in growing a tree was determined using a simulation procedure. Based on Fig. 2(a), the optimal mtry was 10 since the out of bag (OOB) error was the lowest (11.74 OOB error), followed by 15 mtry (11.78 OOB error), 22 mtry (11.92 OOB error), 33 mtry (12.26 error) and 49 mtry (12.59 OOB error). The simulation confirmed that the number of selected features should be the square root of features [34, 35].

Based on [28], the number of tree does not significantly influence the classification outcome. However, it may affect the error rate of the model. So, we did a simulation of searching optimal number of trees to be developed in the forest by looking at the error rate. Based on Fig. 2(b), we ran 50000 trees and it showed that the error rate stabilized at 18000 trees onwards at 0.221 error rate. Thus we decided to grow 20000 trees to build the random forest model.

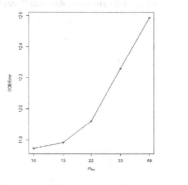

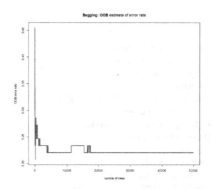

(a) Simulation on optimization of

mtry

(b) Bagging: OOB estimate of error

rate

Fig. 2. Simulation to find optimal parameters in RF.

Next we determined the important biomarkers in metabolomics dataset for AD. In RF, variable importance (VI) is based on two measures: mean decrease in accuracy and mean decrease in Gini. Based on [36], the mean decrease accuracy is the result of the permutation of the average over all trees and it is used to measure the importance of the variables in RF. Figure 3 shows the 30 metabolites from a total of 100 metabolites. RF results show that N-(2-hydroxyethyl) icosanamide is the most important biomarker for AD since the mean decrease in accuracy was 0.0161. This means removing N-(2-hydroxyethyl) icosanamide would result in increasing the misclassification by 0.0161. The next two important biomarkers are 11, 12–dihydroxy arachidic acid and N–(2-hydroxyethyl) palmitamide with mean decrease accuracy of 0.0100 and 0.0092 respectively. Furthermore, seven biomarkers have more than 0.005 and less than 0.008 mean decrease accuracy.

The performance of the RF model is summarized in Table 3, where measures of diagnostic accuracy for training set and test set were reported. For the training set, the sensitivity of the model was acceptable (72.50 %) with specificity of 83.78 %. In addition, the area under the curve was 79.39 % with an estimated OOB error rate of 7.80 %. Next, for the test set, the sensitivity was reported to be 80.0 %, the sensitivity was 72.22 %, the area under the curve was 81.48 % with an estimated OOB error rate of 2.00 %.

Table 3. Summary of RF model performance on metabolomics data for AD.

Measure	Percentage (training set)	Percentage (test set)
Sensitivity	72.50 %	80.0 %
Specificity	83.78 %	72.22 %
Area under curve	79.39 %	81.48 %
OOB Error rate	7.80 %	2.00 %

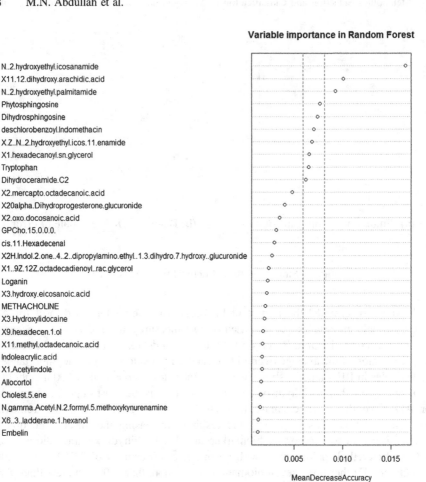

Fig. 3. Variable importance in RF based on decrease in accuracy.

5 Discussion and Conclusion

This study found that N-(2-hydroxyethyl) icosanamide could be an important bio-marker since the metabolite was involved in the regulation of the inflammatory immune response and acted as a protector from neuronal death [37]. In addition, the mean intensity (MI) of N-(2-hydroxyethyl) icosanamide in AD (mean(sd): 12.97(6.57)) was higher compared to HC (mean(sd): 5.20 (7.50)). Next, the 11, 12–dihydroxy arachidic acid and N–(2-hydroxyethyl) palmitamide appeared to be subsequent biomarkers for AD. To date, there has been no publication reported on the relation between 11, 12–dihydroxy arachidic acid and AD. Similarly, the N–(2-hydroxyethyl) palmitamide is not known to be associated with AD but it has been reported that the compound is related to anti-inflammatory agent [38]. In addition, by looking at the MI value of these two biomarkers, we found that for 11, 12–dihydroxy arachidic acid the MI was higher (mean(sd): 4.22 (4.67)) in AD compared to HC (mean(sd): 1.54 (2.86)). Similarly,

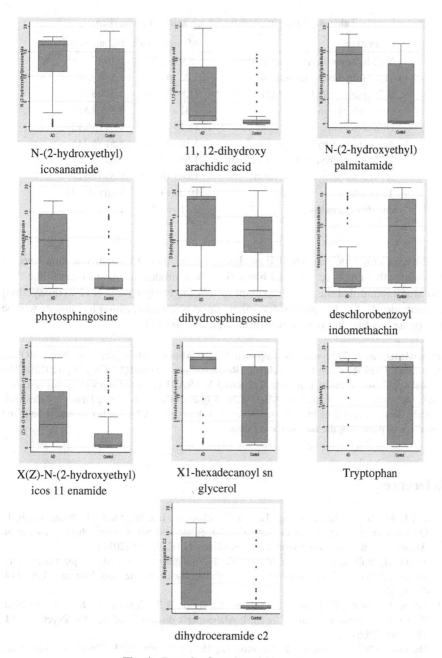

Fig. 4. Box plot for selected biomarkers.

N-(2-hydroxyethyl) palmitamide MI for AD (mean(sd): 11.80 (6.00)) was also higher compared to HC (mean(sd): 4.52 (6.05)). The MI of the selected biomarkers is summarized in Table 4. Boxplots in Fig. 4 show that the MI is higher in AD compared to HC for all biomarkers except deschlorobenzoyl indomenthacin.

Table 4. Summary statistic for ten selected biomarkers

Biomarker	AD mean (sd)	HC mean (sd)
N-(2-hydroxyethyl) icosanamide	12.97 (6.57)	5.20 (7.50)
11,12-dihydroxy arachidic acid	4.22 (4.67)	1.54 (2.86)
N 2 hydroxyethyl palmitamide	11.80 (6.00)	4.52 (6.05)
Phytosphingosine	8.31 (6.45)	2.89 (4.88)
Dihydrosphingosine	14.15 (7.36)	11.15 (6.18)
deschlorobenzoyl indomenthacin	3.10 (4.87)	8.51 (6.11)
XZN-2-hydroxyethyl (icos) 11-enamide	4.66 (4.13)	2.03 (3.30)
X1-hexadecanoyl (sn) glycerol	13.85 (6.64)	8.35 (7.12)
Trypthophan	19.89 (3.74)	13.65 (9.23)
dihydroceramide c2	7.11 (6.37)	1.76 (3.83)

We can conclude that the RF analysis can be used to identify a small set of metabolites that could be potential biomarkers in neurodegenerative diseases like AD. This study found that the most important metabolite compound is N-(2-hydroxyethyl) icosanamide. As for the future work, other supervised classification techniques will be investigated to find the most powerful biomarkers for AD.

Acknowledgement. This work is supported by the Ministry of Higher Education (MOHE) under Long Term Research Grant Scheme (Reference no: 600-RMI/LRGS 5/3 [3/2012]). The authors would like to thank Universiti Teknologi MARA (UiTM) for supporting this research. Lastly, we would like to thank Che Adlia Enche Ady at Collaborative Drug Discovery Research (CDDR) Group, Faculty of Pharmacy, Universiti Teknologi MARA for her technical help and generosity in providing the metabolomics data.

References

1. Lv, J., Ma, S., Zhang, X., Zheng, L., Ma, Y., Zhao, X., Lai, W., Shen, H., Wang, Q., Ji, J.: Quantitative proteomics reveals that PEA15 regulates astroglial Abeta phagocytosis in an Alzheimer's disease mouse model. J. Proteomics **110C**, 45–58 (2014)
2. Motta, M., Imbesi, R., Di Rosa, M., Stivala, F., Malaguarnera, L.: Altered plasma cytokine levels in Alzheimer's disease: correlation with the disease progression. Immunol. Lett. **114**, 46–51 (2007)
3. Sillen, A., Forsell, C., Lilius, L., Axelman, K., Bjork, B.F., Onkamo, P., Kere, J., Winblad, B., Graff, C.: Genome scan on Swedish Alzheimer's disease families. Mol. Psychiatry **11**, 182–186 (2006)
4. Darawi, M.N., Ai-Vyrn, C., Ramasamy, K., Hua, P.P., Pin, T.M., Kamaruzzaman, S.B., Majeed, A.B.: Allele-specific polymerase chain reaction for the detection of Alzheimer's disease-related single nucleotide polymorphisms. BMC Med. Genet. **14**, 27 (2013)

5. Doecke, J.D., Laws, S.M., Faux, N.G., Wilson, W., Burnham, S.C., Lam, C.P., Mondal, A., Bedo, J., Bush, A.I., Brown, B., De Ruyck, K., Ellis, K.A., Fowler, C., Gupta, V.B., Head, R., Macaulay, S.L., Pertile, K., Rowe, C.C., Rembach, A., Rodrigues, M., Rumble, R., Szoeke, C., Taddei, K., Taddei, T., Trounson, B., Ames, D., Masters, C.L., Martins, R.N., Alzheimer's Disease Neuroimaging, I., Australian Imaging, B., Lifestyle Research, G.: Blood-based protein biomarkers for diagnosis of Alzheimer disease. Arch. Neurol. **69**, 1318–1325 (2012)
6. Silva, A.R., Grinberg, L.T., Farfel, J.M., Diniz, B.S., Lima, L.A., Silva, P.J., Ferretti, R.E., Rocha, R.M., Filho, W.J., Carraro, D.M., Brentani, H.: Transcriptional alterations related to neuropathology and clinical manifestation of Alzheimer's disease. PLoS ONE **7**, e48751 (2012)
7. Cayton, H., Graham, N., Warner, J.: Alzheimer's and Other Dementias. Class Publishing, London (2008)
8. Zetterberg, H., Mattsson, N., Shaw, L.M., Blennow, K.: Biochemical markers in Alzheimer's disease clinical trials. Biomark. Med. **4**, 91–98 (2010)
9. Humpel, C.: Identifying and validating biomarkers for Alzheimer's disease. Trends Biotechnol. **29**, 26–32 (2011)
10. Sala, G., Galimberti, G., Canevari, C., Raggi, M.E., Isella, V., Facheris, M., Appollonio, I., Ferrarese, C.: Peripheral cytokine release in Alzheimer patients: correlation with disease severity. Neurobiol. Aging **24**, 909–914 (2003)
11. Ibanez, C., Simo, C., Barupal, D.K., Fiehn, O., Kivipelto, M., Cedazo-Minguez, A., Cifuentes, A.: A new metabolomic workflow for early detection of Alzheimer's disease. J. Chromatogr. A **1302**, 65–71 (2013)
12. Mayeux, R., Saunders, A.M., Shea, S., Mirra, S., Evans, D., Roses, A.D., Hyman, B.T., Crain, B., Tang, M.X., Phelps, C.H.: Utility of the apolipoprotein E genotype in the diagnosis of Alzheimer's disease. Alzheimer's Disease Centers Consortium on Apolipoprotein E and Alzheimer's Disease. N. Engl. J. Med. **338**, 506–511 (1998)
13. Whiley, L., Sen, A., Heaton, J., Proitsi, P., Garcia-Gomez, D., Leung, R., Smith, N., Thambisetty, M., Kloszewska, I., Mecocci, P., Soininen, H., Tsolaki, M., Vellas, B., Lovestone, S., Legido-Quigley, C., AddNeuroMed, C.: Evidence of altered phosphatidyl-choline metabolism in Alzheimer's disease. Neurobiol. Aging **35**, 271–278 (2014)
14. Wang, D.C., Sun, C.H., Liu, L.Y., Sun, X.H., Jin, X.W., Song, W.L., Liu, X.Q., Wan, X.L.: Serum fatty acid profiles using GC-MS and multivariate statistical analysis: potential biomarkers of Alzheimer's disease. Neurobiol. Aging **33**, 1057–1066 (2012)
15. Alzheimer's Disease Foundation: About Alzheimer's (2014)
16. Ibáñez, C., Simó, C., Cifuentes, A.: Metabolomics in Alzheimer's disease research. Electrophoresis. n/a–n/a (2013)
17. Liang, Q., Liu, H., Zhang, T., Jiang, Y., Xing, H., Zhang, A.: Discovery of serum metabolites for diagnosis of progression of mild cognitive impairment to Alzheimer's disease using an optimized metabolomics method. RSC Adv. **6**, 3586–3591 (2016)
18. Quinones, M.P., Kaddurah-Daouk, R.: Metabolomics tools for identifying biomarkers for neuropsychiatric diseases. Neurobiol. Dis. **35**, 165–176 (2009)
19. Armitage, E.G., Kotze, H.L., Williams, K.J.: Correlation-Based Network Analysis of Cancer Metabolism: A New Systems Biology Approach in Metabolomics. Springer, New York (2014)
20. Chu, H., Zhang, A., Han, Y., Lu, S., Kong, L., Han, J., Liu, Z., Sun, H., Wang, X.: Metabolomics approach to explore the effects of Kai-Xin-San on Alzheimer's disease using UPLC/ESI-Q-TOF mass spectrometry. J. Chromatogr. B **1015–1016**, 50–61 (2016)

21. Kaddurah-Daouk, R., Rozen, S., Matson, W., Han, X., Hulette, C.M., Burke, J.R., Doraiswamy, P.M., Welsh-Bohmer, K.A.: Metabolomic changes in autopsy-confirmed Alzheimer's disease. Alzheimer's Dement. **7**, 309–317 (2011)
22. Sapkota, S., Tran, T., Huan, T., Lechelt, K., Macdonald, S., Camicioli, R., Li, L., Dixon, R. A.: Metabolomics analyses of Salivary sample discriminate normal aging, mild cognitive impairment and Alzheimer's disease groups and produce biomarkers predictive of neurocognitive performance. JALZ **11**, P654 (2016)
23. Liang, Q., Liu, H., Li, X., Zhang, A.-H.: High-throughput metabolomics analysis discovers salivary biomarkers for predicting mild cognitive impairment and Alzheimer's disease. RSC Adv. **6**, 75499–75504 (2016)
24. Beckonert, O., Keun, H.C., Ebbels, T.M.D., Bundy, J., Holmes, E., Lindon, J.C., Nicholson, J.K.: Metabolic profiling, metabolomic and metabonomic procedures for NMR spectroscopy of urine, plasma, serum and tissue extracts. Nat. Protoc. **2**, 2692–2703 (2007)
25. Benjamini, Y., Hochberg, Y.: Controlling the false discovery rate: a practical and powerful approach to multiple testing (1995)
26. RStudio Team: RStudio: Integrated Development for R (2015). http://www.rstudio.com/
27. Boulesteix, A.L., Janitza, S., Kruppa, J., König, I.R.: Overview of random forest methodology and practical guidance with emphasis on computational biology and bioinformatics. Wiley Interdiscip. Rev. Data Min. Knowl. Discov. **2**, 493–507 (2012)
28. Khalilia, M., Chakraborty, S., Popescu, M.: Predicting disease risks from highly imbalanced data using random forest. BMC Med. Inform. Decis. Mak. **11**, 51 (2011)
29. Breiman, L.: Random forests. Mach. Learn. **45**, 5–32 (2001)
30. Díaz-Uriarte, R., Alvarez de Andrés, S.: Gene selection and classification of microarray data using random forest. BMC Bioinform. **7**, 3 (2006)
31. Pudlo, P., Marin, J.-M., Robert, C.P., Cornuet, J.-M., Estoup, A.: ABC model choice via random forests. Mol. Biol. Evol. **32**, 28 (2015)
32. Oshiro, T.M., Perez, P.S., Baranauskas, J.A.: How many trees in a random forest? In: Perner, P. (ed.) MLDM 2012. LNCS, vol. 7376, pp. 154–168. Springer, Heidelberg (2012)
33. Genuer, R., Poggi, J., Tuleau-malot, C.: Variable selection using random forests. Pattern Recogn. Lett. **31**, 2225–2236 (2010)
34. Liaw, A., Wiener, M.: Classification and regression by randomForest. R News **2**, 18–22 (2002)
35. Breiman, L., Cutler, A.: setting up, using, and understanding random forests V4. 0. University of California, Department of Statistics (2003)
36. Hastie, T., Tibshirani, R., Friedman, J.: The Elements of Statistical Learning: Data Mining, Inference and Prediction. Springer, New York (2009)
37. Brand, B., Hadlich, F., Brandt, B., Schauer, N., Graunke, K.L., Langbein, J., Repsilber, D., Ponsuksili, S., Schwerin, M.: Temperament type specific metabolite profiles of the prefrontal cortex and serum in cattle. PLoS ONE **10**, e0125044 (2015)
38. Kuehl, F.A., Jacob, T.A., Galey, O.H., Ormond, R.E., Meisinger, M.A.P.: The identification of N-(2-hydroxyethyl)-palmitamide as a naturally occurring anti-inflammatory agent. J. Am. Oil Chem. Soc. **79**, 5577–78 (1957)

A Multi-objectives Genetic Algorithm Clustering Ensembles Based Approach to Summarize Relational Data

Rayner Alfred[1(✉)], Gabriel Jong Chiye[1], Yuto Lim[2],
Chin Kim On[1], and Joe Henry Obit[1]

[1] Faculty of Computing and Informatics, Universiti Malaysia Sabah, Kota Kinabalu, Malaysia
{ralfred,kimonchin,joehenry}@ums.edu.my,
gabrieljongums@gmail.com
[2] School of Information Science, Japan Advanced Institute of Science and Technology,
Nomi, Japan
ylim@jaist.ac.jp

Abstract. In learning relational data, the Dynamic Aggregation of Relational Attributes algorithm is capable to transform a multi-relational database into a vector space representation, in which a traditional clustering algorithm can then be applied directly to summarize relational data. However, the performance of the algorithm is highly dependent on the quality of clusters produced. A small change in the initialization of the clustering algorithm parameters may cause adverse effects to the clusters quality produced. In optimizing the quality of clusters, a Genetic Algorithm is used to find the best combination of initializations in order to produce the optimal clusters. The proposed method involves the task of finding the best initialization with respect to the number of clusters, proximity distance measurements, fitness functions, and classifiers used for the evaluation. Based on the results obtained, clustering coupled with Euclidean distance is found to perform better in the classification stage compared to using clustering coupled with Cosine similarity. Based on the findings, the cluster entropy is the best fitness function, followed by multi-objectives fitness function used in the genetic algorithm. This is most probably because of the involvement of external measurement that takes the class label into consideration in optimizing the structure of the cluster results. In short, this paper shows the influence of varying the initialization values on the predictive performance.

Keywords: Relational data mining · k-means · Clustering · Ensembles · Genetic algorithm · Multi-objectives

1 Introduction

Data mining is the process of discovering interesting patterns and knowledge from large amounts of data [1, 2]. It involves several stages, first is the data preprocessing, which relevant data is selected and retrieved from the databases, cleaning to remove noise and handle missing data, integration to combine data from multiple sources. Second, data transformation, where features selection and transformation is applied, to produce an appropriate representation which represent the databases. Third, data mining process in

© Springer Nature Singapore Pte Ltd. 2016
M.W. Berry et al. (Eds.): SCDS 2016, CCIS 652, pp. 113–122, 2016.
DOI: 10.1007/978-981-10-2777-2_10

which intelligent methods are used to extract patterns, such as rules, clusters and etc. Finally, pattern evaluation and presentation, where interesting patterns extracted are presented in knowledge which is easily human understandable.

From database point of view, a typical relational database consists of multiple relations known as tables [3, 4]. These relations are connected via semantic links such as entity relationship links. Many useful traditional data mining algorithms work on single table form only and cannot be applied directly in learning multi-relational databases [5]. Flattening the multi-relational databases leads will lead to inaccurate clustering and results in wrong business decision.

Thus, several approaches have been proposed to learn multi-relational databases. Inductive Logic Programming [6], Propositionalization [7] and Dynamic Aggregation for Relational Attributes [8] are popular approaches in learning relational datasets. Dynamic Aggregation of Relational Attributes is designed based on a clustering technique [9, 10] in which it is capable of transforming multi-relational databases to vector space representation, where traditional clustering algorithms can be applied directly to learn the multi-relational data.

Traditional k-means clustering algorithm is a hill climbing algorithm which is sensitive to its initialization [9, 11, 12]. The predefined number of clusters, k which is usually assumed to be known priori and the initialization of initial centroids of cluster, will heavily affect the quality of final clusters. The quality of clustering and the time complexity in terms of the number of iterations required to converge also depends on the initial selection of centroids. As different initialization in k-means clustering algorithm will produce different results, which affect the performance of Dynamic Aggregations of Relational Attributes, it is essential to combine and generalize these results produced by different initialization to form one single consensus result of higher stability using clustering ensembles.

Thus, the objectives of this paper are, to design and propose the complete framework of the ensemble approach to learning multi-relational data; to implement a Genetic Algorithm based clustering algorithm in order to find the best initialization for each k-means clustering applied; and to assess the performance of the clustering ensembles by using a C4.5 classifier and Naïve Bayes classifier based on different distance metrics and different fitness function (i.e., single and multi-objectives). The rest of this paper is organized as followed. Section 2 explains some related works. Section 3 discusses the Multi-Objectives Genetic Algorithm Hybridization used in this paper. Section 4 discusses the experimental setup and Sect. 5 discusses the experimental results obtained. Section 6 concludes this paper.

2   Related Works

2.1   Multi-relational Data Mining

A typical multi-relational database consists of multiple relations as tables and these tables are associated to each other through primary and foreign keys. In a one-to-many relationship set of tables, a record stored in the target table corresponding to multiple records stored in the non-target table. One of the famous approaches in learning multi-relational data is known as inductive logic programming [13]. The term for inductive logic programming was first introduced by Muggleton [14]. It uses logic for representation. Based on the background knowledge provided, it constructs predicates for logic representations and make hypotheses base on the syntactic predicates [6]. However, most of the inductive logic programming based approaches are usually inefficient for databases with complex schemas, not appropriate for continuous values and not capable of dealing with missing values [15].

Another common approach is propositionalization [7] that captures and stores relational representation in propositional form. These propositional forms are known as new features and are usually stored as attributes in a vector form. Although propositionalization approaches have some inherent limitations, such as learning recurrent definitions, it has advantages over traditional inductive logic programming based approaches by allowed the use of propositional algorithms including probabilistic algorithms while learning relational representations [7].

Alfred recently introduced Dynamic Aggregation for Relational Attributes approach to summarize the entire contents of non-target tables before the target table can be processed for knowledge discovery [8]. It contains three stages i.e. data preprocessing, data transformation and data summarization. Data are first preprocessed by discretizing continuous values into categorical values. Then, a theory from information retrieval is borrowed, to transform the multi-relational databases. Records from non-target table are transformed into "bag of words", so that frequency-inverse document frequency matrix can be produced.

2.2 Clustering

After where data is transformed from multi-relational data representation to vector space representation, a traditional clustering algorithm can then be applied. Data mining can be generalized to supervised learning and unsupervised learning. Clustering is an unsupervised learning algorithm where class labels are not available during training phase [16]. Clustering can be categorized to two main categories: hierarchy and partition [10, 11, 17, 18]. Hierarchy clustering will produce nested clusters tree. In contrast with hierarchical clustering, a record can be grouped from one cluster to another cluster in different iterative of partition clustering. k-means clustering is a partition clustering algorithm [19]. The number of clusters, k is predefined. For each N record instances, we will calculate the vector distance between the records to each of the centroid of clusters using Euclidean distance, $dist$, as follows,

$$dist = \sqrt{\sum_{i=1}^{d} \left(x_i - y_i\right)^2} \tag{1}$$

where d is the maximum feature dimension, x is the record instance and y is the centroid of interest. The record instance will be assigned to the nearest centroid, which is the centroid with lowest Euclidean distance.

Alternatively, the distance between the records to each of the centroid of clusters can also be calculated using the cosine similarity, as follows,

$$\cos(x, y) = \frac{x \cdot y}{\|x\| \|y\|} \tag{2}$$

where $\cdot$ indicates the vector dot product, $x \cdot y = \sum_{k=1}^{n} x_k y_k$, and $\|x\|$ is the length of vector x, $\|x\| = \sqrt{\sum_{k=1}^{n} x_k^2} = \sqrt{x \cdot x}$. Through finite number of iterations, k-means algorithm is guaranteed to converge to a local optimum. k-means algorithm is sensitive to the centroids initialization. It is common to run k-means clustering with different number of clusters and different centroids initialization. After clustering algorithm is applied, clustering validity can be used to evaluate the quality of the final clustering result. The clustering validity index has been defined as combination of compactness and separation [20]. Compactness is the measurement of distance between records of the same cluster. Lower distance between records of the same cluster should have better validity. Separation is the measurement of distance between records of different cluster. Records of different cluster should separate as far as possible with each other. The Sum of Squared Error (SSE) is the summation of squared distance for each record in the data set with other records [21]. It is defined as,

$$SSE(X, Y) = |X - Y| \tag{3}$$

where X and Y are vector of two records in the vector space. Davies-Bouldin Index (DBI) is an internal validity index [20]. It is defined as,

$$DBI = \frac{1}{c} \sum_{i=1}^{c} \max_{i \neq j} \left\{ \frac{d(X_i) + d(X_j)}{d(c_i, c_j)} \right\} \tag{4}$$

where c is the total number of clusters, i and j are the cluster labels in comparison, $d(X_i)$ and $d(X_j)$ are records in cluster i and j to their respective centroids, $d(c_i, c_j)$ is distance between centroid i and j. Lower value of DBI indicates better clustering result. Entropy (Ent) is an external validity index [20]. It is the measurement of purity of clusters' class labels. Lower value of Entropy indicates lower impurities within the same cluster. First, we calculate the class distribution of objects in each cluster as follows,

$$E_j = \sum_i p_{ij} \log(p_{ij}) \tag{5}$$

where j is the cluster, i is the class. p_{ij} is the probability of occurrences of class i in cluster j. Total Entropy is calculated as follows,

$$E = \sum_{j=1}^{m} \frac{n_j}{n} E_j \tag{6}$$

where n_j is the size of cluster j, m is the number of clusters, and n is the total number of records in the data set.

2.3 Multi-objectives Optimization

Real life optimizations are mostly multi-objectives and these objectives are mostly contradictory. The goal of multi-objectives optimizations is to search for solutions that optimize a number of functions to satisfy more than one constraint (Wahid, Gao, & Peter, 2014). Traditional genetic algorithms [22–24] can be accommodated with specialized multi-objectives function in order to provider higher solutions diversity. There are generally two approaches to multi-objectives optimization. The first general approach is either combines individual functions into one composite function or moves all objectives to one constraint set. Combining individual functions is also known as weighted sum approaches (Wen, Li, Gao, Wan, & Wang, 2013), where a weight w_i is assigned for each normalized objective function, $f_i(x)$ so that the multi-objectives problem is converted to a single objective problem:

$$f(x) = w_1 f_1(x) + w_2 2(x) + \ldots + w_i f_i(x) \qquad (7)$$

In clustering for instance, there are many functions that can be used to validate the cluster quality such as SSE, DBI and entropy as described in Sect. 2.2 above. These individual cluster quality validity functions can be normalized and then combined using equal weighted sum to convert it to a single minimization function.

3 Multi-objectives Genetic Algorithm Hybridization

Since k-means algorithm is sensitive to initialization and would fall into a local optima, rather than running it for multiple times using random initialization, it has been hybridized with a genetic algorithm [22–24] in order to find the global optima. In this paper, a binary encoding is used with chromosome length = N where N is the total number of records. Each gene within the chromosome represents an instance of records of the data set. The value of each gene represents whether that particular instance of records is an initial centroid of clusters. For each number of clusters, k the population size of 100 is used and is let to evolve for 100 generations. Each chromosome represents a possible solution, which is a set of initial centroid of clusters. After k-means algorithm run is completed for each chromosome in the first generation, the chromosome quality will be evaluated using the individual fitness function previously and multi-objectives function as described below.

Individual fitness functions described in Sect. 2.2, can also be normalized, combined according to weighted sum approach into one optimization objective. The normalization of each of the single objective function is done based on per generation basis in the genetic algorithm,

$$f_{normalized}(x) = \frac{f_i(x) - f_{min}(x)}{f_{max}(x) - f_{min}(x)} \tag{8}$$

where $f_i(x)$ is the fitness value of the individual, $f_{max}(x)$ and $f_{min}(x)$ is the maximum and minimum fitness value in the generation respectively. The multi-objectives fitness function can then be defined as,

$$f^*(x) = w_1 f_1(x) + w_2 f_2(x) + w_3 f_3(x) \tag{9}$$

where $f^*(x)$ is the computed weighted sum multi-objectives fitness value of the individual, $f_1(x)$, $f_2(x)$ and $f_3(x)$ are normalized value for sum of squared error, Davies-Bouldin index and cluster entropy respectively, and w_1, w_2 and w_3 are the weight applied to respective normalized value, where $w_1 = w_2 = w_3 = 1$.

4   Experimental Setup

The main objective of the experiment that is conducted involves the task of finding the best initialization with respect to the number of clusters, proximity distance measurements, fitness functions, and classifiers used for the evaluation. There are two main parts in this experiment. In the first part, the data will be clustered based on given number of clusters, k. The main task here is to determine the best number of clusters in order to summarize the transformed data. Let the number of rows in a data set be n then the ranges of k will be from 2 to $2\sqrt{n}$, with increment of 1. For example, there are 189 rows of data within Mutagenesis dataset, 2 times of square root of 189 is 27. We further extended 27 to 30 to allow better view of the output results in term of accuracy. The clustering results are then appended into the target table as new additional feature. Then the table is fed into the classifiers (e.g., C4.5 and Naïve Bayes classifiers with 10-fold CV) in order to evaluate the predictive performances.

In the second part of the experiment, the results from the first part of the experiment with number of clusters, k from 2 to maximum test range is collected. The cluster ID which is determined from the first part of this experiment will become the features of the second part of the experiment. The second part of the experiment consists purely of categorical data as cluster ID is categorical value. Same range of k is used. After clustering, the results are fed into WEKA using the same classifier. Results of the second part are then compared with the first part of the experiment.

For each number of clusters, k, a population size of 100 is used and there is 100 generations. Each chromosome represents a set of initial centroid of clusters. During reproduction, the chromosomes will undergo uniform crossover and mutation. For each gene of the chromosome, they will be tested with a crossover rate whether to exchange the information. After the offspring is produced, each of the genes of the offspring chromosome will be tested with mutation rate whether to flip the bit. To maintain the number of true bits which represent the number of clusters, when a bit is flipped, another bit of opposite value is also randomly chosen and flipped.

The experiment is first executed with genetic algorithm using sum of squared error (SSE) as measurement of fitness function which will be used for selection.

The experiment is then repeated using Davis-Bouldin Index (DBI) and Cluster Entropy (Ent) as fitness function. After that, the experiment is repeated again using a normalized multi-objectives (MO) fitness function, which is the combination of all previous three fitness function earlier.

Three Mutagenesis and Three Hepatitis datasets were used, namely B1, B2, B3 and H1, H2, H3 respectively. The number of clusters, k used in Mutagensis datasets starts from $k = 2$ to $k = 30$. The number of clusters, k used in Hepatitis datasets starts from $k = 2$ to $k = 50$. Each of the dataset was executed in the experiment using 3 different single fitness function and 1 combined and normalized multi-objectives fitness function.

5   Results and Discussion

Six different datasets i.e. B1, B2 and B3 from mutagenesis and H1, H2 and H3 from hepatitis have been used. Each dataset is then summarized using different combination of distance measurement (e.g., Euclidean distance and Cosine) and genetic algorithm fitness function (e.g., SSE, DBI, Ent, MO) to produce sets of 48 results. Each result set are then fed into two different classifier, C4.5 and Weka Naïve-Bayes classifiers, which consequently produce the 96 average predictive accuracy increment tabulated in Table 1.

Table 1. Increment of average predictive accuracy in percentage generated by different from classifiers using summarized datasets with clustering ensembles of different combination of experimental settings.

Weka classifier	Dataset \ Fitness Function, Distance measurement	Sum of squared error		Davies-Bouldin index		Cluster entropy		Multi-objectives	
		Euclidean	Cosine	Euclidean	Cosine	Euclidean	Cosine	Euclidean	Cosine
C4.5	B1	1.85	1.62	−0.46	0.97	3.59	0.28	2.09	2.60
	B2	0.83	−0.07	−0.02	1.78	2.44	2.84	0.00	1.41
	B3	−0.38	−0.44	4.51	0.30	3.93	3.19	4.12	0.33
	H1	2.15	2.76	−9.4	6.43	2.44	2.03	2.20	2.97
	H2	−0.02	−0.01	−0.09	0.00	0.41	1.25	0.00	0.66
	H3	−0.08	0.00	−0.09	0.00	1.46	0.07	0.07	0.15
Naïve Bayes	B1	0.11	10.11	1.08	−0.49	−0.49	0.97	0.11	1.24
	B2	0.38	−0.97	0.49	−0.81	2.37	1.35	1.73	−0.38
	B3	0.32	0.27	0.97	0.16	3.18	2.21	0.91	0.60
	H1	3.59	1.96	2.45	4.33	2.35	3.59	2.75	5.77
	H2	3.62	1.10	3.59	0.71	8.31	3.76	6.91	3.54
	H3	6.18	−0.30	5.13	0.39	9.10	−1.00	9.55	0.70

In term of classifier used to generate predictive accuracy, Weka C4.5 classifier is able to achieve increment in average predictive accuracy as high as 6.43 % with 32 results set of positive increment and 16 results set of no improvement or decreased average predictive accuracy, which is able to conclude that 66.66 % of the summarized data can perform better on a Weka C4.5 classifier. On the other hand, Weka Naïve Bayes classifier is able to achieve increment in average predictive accuracy as high as 10.11 %

with 41 results set of positive increment and 7 results set of no improvement or decreased average predictive accuracy, which brings to a conclusion that 85.41 % of the summarized data can perform better on a Weka Naïve Bayes classifier. From this point of view, Weka Naïve Bayes classifier will benefits from a summarized data more than a Weka C4.5 classifier does (Table 2).

Table 2. Average predictive accuracy increment in percentage based on different classifier used and fitness function used in the genetic algorithm.

Fitness function \ Classifier	Weka C4.5	Weka Naïve Bayes	Overall
Sum of squared error	0.67	2.19	1.43
Davies-Bouldin index	0.32	1.5	0.91
Cluster entropy	1.99	2.97	2.48
Multi-objectives	1.38	2.78	2.08

The table above concluded the results from point of view of performance of different fitness function used in the genetic algorithm to optimize the initialization of initial centroids of the k-means algorithm. From the results, it can be concluded that cluster entropy has the highest performance in optimizing the initial centroids of the k-means clustering algorithm, with overall average increment of predictive accuracy of 2.48 %, followed by multi-objectives with 2.08 %. This is most probable caused by that class label has been used as part of evaluation in the cluster entropy fitness function, which favors both classifiers.

The table above concluded the results using another perspective, to compare which summarized datasets have better performance in term of average predictive accuracy increment according to different classifier used. It can be concluded that summarized mutagenesis datasets have better performance for Weka C4.5 classifier but summarized hepatitis datasets have better performance for Weka Naïve Bayes classifier (Table 3).

Table 3. Average predictive accuracy increment in percentage based on different classifier used and different summarized dataset used.

Dataset \ Classifier	Weka C4.5	Weka Naïve Bayes
Mutagenesis (B1, B2, B3)	1.55	1.05
Hepatitis (H1, H2, H3)	0.63	3.67

6   Conclusion

In this paper, a framework of ensemble approach to learning relational data is proposed and designed. A genetic algorithm based k-means clustering algorithm has been implemented in clustering relational data to find the optimal set of solutions for each number of clusters, k. It can be concluded that the combination of Euclidean distance has the better performance over cosine similarity for mutagenesis datasets and Weka C4.5 classifier, but cosine similarity has better performance over Euclidean distance for hepatitis datasets and Weka C4.5 classifier. On the other hand, it was found that for Weka

Naïve Bayes classifier, where cosine similarity has better performance for mutagenesis dataset and Euclidean distance has better performance for hepatitis dataset. The weight for each individual objective function in the multi-objectives optimization process in this work is not properly tuned as it is not an easy task even for expert with domain knowledge. To address this problem in the future, the weights of each individual function can also be optimized using evolutionary algorithms. The purpose of doing so is to find the optimum combination of weights for each individual function along the evolutionary algorithms of multi-objectives optimization in order to find the best set of initial centroids. Pareto approaches can also be implemented to compare the performance of cluster ensemble with a weighted sum approach.

In future works, the processes in the transformation stage may be integrated with the summarization stage, such as feature aggregation or feature selection using an evolutionary approach, together with the optimization of initial centroids of k-means clustering; so that the best set of features contributing to the highest information gain can be selected, thus indirectly ignoring the insignificant tuples in the non-target table.

References

1. Cattral, R., Oppacher, F., Graham, K.J.L.: Techniques for evolutionary rule discovery in data mining. In: Conference on Evolutionary Computation, pp. 1737–1744 (2009)
2. Xu, L., Jiang, C., Wang, J., Yuan, J., Ren, Y.: Information security in big data: privacy and data mining. In: IEEE 2014, pp. 1149–1176 (2014)
3. Dzeroski, S.: Relational data mining. In: Maimon, O., Rokach, L. (eds.) Data Mining and Knowledge Discovery Handbook, pp. 887–911. Springer US (2010)
4. Ling, P., Rong, X.: Double-Phase Locality Sensitive Hashing of neighborhood development for multi-relational data. In: 13th UK Workshop on Computational Intelligence (UKCI), pp. 206–213 (2013)
5. Mistry, U., Thakkar, A.R.: Link-based classification for Multi-Relational database. In: Recent Advances and Innovations in Engineering (ICRAIE), pp. 1–6 (2014)
6. Zhang, W.: Multi-relational data mining based on higher-order inductive logic. In: WRI Global Congress in Intelligent Systems, Xiamen, pp. 453–458 (2009)
7. Roth, D., Yih, W.-T.: Propositionalization of relational learning: an information extraction case study. In: 17th International Joint Conference on Artificial Intelligence, Seattle (2001)
8. Nguyen, T.-S., Duong, T.-A., Kheau, C.S., Alfred, R., Keng, L.H.: Dimensionality reduction in data summarization approach to learning relational data. In: Selamat, A., Nguyen, N.T., Haron, H. (eds.) ACIIDS 2013, Part I. LNCS, vol. 7802, pp. 166–175. Springer, Heidelberg (2013)
9. Lu, B., Ju, F.: An optimized genetic K-means clustering algorithm. In: International Conference on Computer Science and Information Processing, pp. 1296–1299 (2012)
10. Li, T., Chen, Y.: A weight entropy k-means algorithm for clustering dataset with mixed numeric and categorical data. In: Fifth International Conference on Fuzzy Systems and Knowledge Discovery, Shandong, pp. 36–41(2008)
11. Cui, X., Potok, T.E., Palathingal, P.: Document clustering using particle swarm optimization. In: IEEE Swarm Intelligence Symposium 2005, pp. 185–191(2005)
12. Abdel-Kader, R.F.: Genetically improved PSO algorithm for efficient data clustering. In: 2nd International Conference on Machine Learning and Computing (ICMLC), pp. 71–75 (2010)

13. Bharwad, N.D., Goswami, M.M.: Proposed efficient approach for classification for multi-relational data mining using Bayesian Belief Network. In: 2014 International Conference on Green Computing Communication and Electrical Engineering, pp. 1–4 (2004)
14. Muggleton, S.: Inductive Logic Programming. New Gener. Comput. **8**(4), 295–318 (1991)
15. Guo, J., Zheng, L., Li, T.: An efficient graph-based multi-relational data mining algorithm. In: International Conference on Computational Intelligence and Security, pp. 176–180 (2007)
16. Dutta, D., Dutta, P., Sil, J.: Data clustering with mixed features by multi objective generic algorithm. In: 12th International Conference on Hybrid Intelligent Systems, Pune, pp. 336–341 (2012)
17. Shah, N., Mahajan, S.: Document clustering: a detailed review. Int. J. Appl. Inf. Syst. **4**, 30–38 (2012)
18. Chen, C.-L., Tseng, F.S.C., Liang, T.: An integration of WordNet and fuzzy association rule mining for multi-label document clustering. Data Knowl. Eng. **69**(11), 1208–1226 (2010)
19. Pettinger, D., Di Fatta, G.: Space partitioning for scalable k-means. In: 9th International Conference in Machine Learning and Apps (ICMLA), pp. 319–324 (2010)
20. Rendon, E., Abundez, A.A.I., Quiroz, E.M.: Internal versus External cluster validation indexes. Int. J. Comput. and Commun. **5**(1), 27–32 (2011)
21. Bilal, M., Masud, S., Athar, S.: FPGA design for statistics-inspired approximate sum-of-squared-error computation in multimedia applications. IEEE Trans. Circ. Syst. II: Express Briefs **59**(8), 506–510 (2012)
22. Mitchell, M.: An Introduction to Genetic Algorithms. MIT Press, London (1999)
23. Razali, N.M., Geraghty, J.: Genetic algorithm performance with different selection strategies in solving TSP. In: Proceedings of the World Congress on Engineering 2011, London, vol. II (2011)
24. Wahid, A., Gao, X., Peter, A.: Multi-view clustering of web documents using multi-objective genetic algorithm. In: 2014 IEEE Congress Evolutionary Computation (CEC), pp. 2625–2632 (2014)
25. Wen, X., Li, X., Gao, L., Wan, L., Wang, W.: Multi-objective genetic algorithm for integrated process planning and scheduling with fuzzy processing time. In: 2013 Sixth International Conference on Advanced Computational Intelligence (ICACI), pp. 293–298 (2013)
26. Konak, A., Coit, D.W., Smith, A.E.: Multi-objective optimization using genetic algorithms: a tutorial. Reliab. Eng. Syst. Saf. **9**(9), 992–1007 (2006)
27. Ismail, F.S., Yusof, R., Waqiyuddin, S.M.M.: Multi-objective optimization problems: method and application. In: 2011 4th International Conference on Modeling, Simulation and Applied Optimization (ICMSAO), pp. 1–6 (2011)
28. Zeghichi, N., Assas, M., Mouss, L.H.: Genetic algorithm with pareto fronts for multi-criteria optimization case study milling parameters optimization. In: 2011 5th International Conference on Software, Knowledge Information, Industrial Management and Applications (SKIMA), Benevento, pp. 1–5 (2011)
29. Atashkari, K., NarimanZadeh, N., Ghavimi, A.R., Mahmoodabadi, M.J., Aghaienezhad, F.: Multi-objective optimization of power and heating system based on artificial bee colony. In: International Symposium on Innovations in Intelligent Systems and Applications (INISTA), Istanbul, pp. 64–68 (2011)

Automated Generating Thai Stupa Image Descriptions with Grid Pattern and Decision Tree

Sathit Prasomphan[✉], Panuwut nomrubporn, and Pirat Pathanarat

Department of Computer and Information Science, Faculty of Applied Science, King Mongkut's University of Technology North Bangkok, 1518 Pracharat 1 Road, Wongsawang, Bangsue, Bangkok, 10800, Thailand
ssp.kmutnb@gmail.com, panuwut@gmail.com, pirat@gmail.com

Abstract. This research presents a novel algorithm for generating descriptions of stupa image such as stupa's era, stupa's architecture and other description by using information inside image which divided into grid and learning stupa description from the generated information with decision tree. In this paper, we get information inside image by divided image into several grid patterns, for example 10 × 10 and use data inside that image to submit to the decision tree model. The proposed algorithm aims to generate the descriptions in each stupa image. Decision tree was used for being the classifier for generating the description. We have presented a new approach to feature extraction based on analysis of information in image by using the grid information. The algorithms were tested with stupa image dataset in Phra Nakhon Si Ayutta province, Sukhothai province and Bangkok. The experimental results show that the proposed framework can efficiently give the correct descriptions to the stupa image compared to using the traditional method.

Keywords: Automated generating image descriptions · Decision tree · Grid pattern · Feature extraction

1 Introduction

The long history of Siam or Thailand which has been developed for becoming a civilization country since Sukhothai era, Ayutthaya era and Rattanakosin era shows the civilization of that era through the archaeological site for example, temple, palace etc. From the past through present which has a long decade ago, there are too many things effect to the archaeological site. Accordingly, there remains some part of that place. From that reason, if we can show the descriptions of archaeological site especially the component of stupa in each era from the image content, it will show the understanding of that place to each other in the present.

Several researchers have developed an image content retrieval algorithms for generating descriptions from that image [4–6]. Andrej Karpathy and Li-Fei-Fei [1] proposed the research named "Deep Visual-Semantic Alignments for Generating Image Descriptions" which showed the model for generating language for image and put the generated description inside that image. They proposed the rule for generating relationship

© Springer Nature Singapore Pte Ltd. 2016
M.W. Berry et al. (Eds.): SCDS 2016, CCIS 652, pp. 123–135, 2016.
DOI: 10.1007/978-981-10-2777-2_11

between language and image by using recurrent neural network. In their research, the sentence was generated from image by using image from several databases such as Flickr8K, Flickr30K and MSCOCO. Richard Socher et al. [2] proposed the research with title "Grounded Compositional Semantics for Finding and Describing Images with Sentences". This research, introduced a relationship between image and sentence by using recursive neural network [3] for showing image descriptions by using convolution neural network. Ali Farhadi et al. [8] introduced "Every Picture Tells a Story: Generating Sentences from Images". They showed model for generating sentence from image by using the scoring method for calculating the relationship between sentence and image. They suggested the relationship between image space, meaning space, and sentence space. The problem occurs during used the above algorithms. There are some confusion to create the relationship between the image space, meaning space, and sentence space. From those problems, in this research we interested in developing algorithms for generating description of archaeological site image. We interested in content retrieval in stupa image such as the era, architecture and the importance feature. These descriptions are generated by using information inside the grid pattern. After that, the generated information is used in the learning process for generating description by using decision tree model. We have presented a new approach to feature extraction based on analysis grid pattern of an image.

The remaining of this paper is organized as follows. At first, we show the characteristics of stupa. Next, the process for generating image description based on grid pattern and decision tree is discussed in Sect. 2. Results and discussion are discussed in Sect. 3. Finally, the conclusions of the research are presented in Sect. 4.

2   Proposed Algorithms

The following section, we first briefly review the fundamental of characteristic of Thailand's stupa architecture. Next, the fundamental of grid pattern is proposed. After that, image content description generation is introduced based on recognizing information inside each grid. Finally, decision tree is shown for learning the image description and generating that description to the stupa image.

2.1   Characteristics of Stupa

The Chedi or Stupa is a reliquary tower, sometimes referred to in the west as a Pagoda. Inside it are enshrined relics, sometimes of the Buddha (bones, teeth etc.) sometimes of kings or other important people, sometimes with images of the Buddha. The shape of the Chedi probably originated in the shape of an ancient Indian burial mound (stupa) and is multi-symbolic: of the Buddha himself, seated cross-legged on a pedestal or lotus throne, of the different levels of the Buddhist Cosmology from under the earth through to the heavens, and of Mount Meru itself, the mountain at the center of the Hindu and Buddhist cosmos [7]. In Thailand, there are too many style of stupa or chedi. In general, three styles of stupa in Thailand which is divided by the era of that architecture are presented. For example, Sukhothai era, Ayutthaya era and Rattanakosin era. Some architecture in each era

Fig. 1. Example of stupa in Sukhothai era. (a) the lotus blossom style (b) the bell-shaped style (c) the Prang style (d) the Chomhar style [7].

Fig. 2. Example of stupa in Ayutthaya era. [7].

Fig. 3. Example of stupa in Rattanakosin era. [7]

will be overlaps with other era. The stupa's architecture divided by these eras can be described with the following details. The stupa in Sukhothai era can be classified into these categories: the lotus blossom style, the bell-shaped style, the Prang style, etc. The example of stupa in Sukhothai era can be shown in Fig. 1. The stupa in Ayutthaya era can be classified into these categories: the bell-shaped style, the Prang style, etc. The example of stupa in Ayutthaya era can be shown in Fig. 2. The stupa in Rattanakosin era can be classified into these categories: the square wooden pagoda style, the Prang style, etc. The example of stupa in Rattanakosin era can be shown in Fig. 3.

Fig. 4. Example of stupa in (a) the Prasat style (b) the bell-shaped style (c) the Prang style (d) the lotus blossom style (e) Yohmum style [7].

Several researches classified the stupa architecture into 5 styles which are: Prasat style, Prang style, the bell-shaped style, the lotus blossom style, the Yohmum style. Examples of these styles are shown in Fig. 4.

2.2 Grid Pattern

Grid pattern is a model for applying a zoning inside an image. This model was done by dividing image into grid patterns for considering ratio in each image. We use information in each grid for creating classification model. The grid pattern has 2 style which is parallel or linear grid pattern and cross hatch grid pattern as shown in Fig. 5.

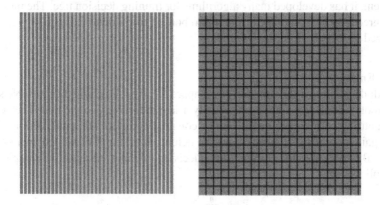

Fig. 5. Grid in linear grid pattern and cross hatch grid pattern

2.3 Decision Tree for Data Classification

Decision tree learning is a model for discrete-value function by using tree model which composed of set of rule with if-then command for easily to human to understanding the decision tree.

In machine learning, decision tree is a mathematic model for forecasting type of object by considering attribute of object. In inner node of tree will display variable, the branch of tree will display a possible value of variable, inside leaf node will display a type of object. A decision tree usually use in risk management. It is a part of decision theory and a basic theory for data mining.

2.3.1 Characteristics of Decision Tree

The benefit of decision tree is to group or to classify the imported datasets for each case or instance. Each node of the decision tree is a variable or attribute of a data set. For example, to decide whether to play sport or not, its variable to consider is the scenery, wind, humidity, temperature, etc., and an expected variable of the tree is to decide whether to play or not. Each attribute will have its own value which is a set of attribute-value pair such as scenery variable it may be raining or sunny, the decision whether to play variable is yes or no. The decision in decision tree starts from the root node by testing the test parameters inside each node. Then continue to test in the branches of a tree to go to the next node. This test is done until it finds the leaf node which shows the predicted results. The example of using decision trees to predict whether to play a sport or not, by considering three variables, namely weather, humidity and winds. So if the day is displayed as a feature vector as [weather = sunny, humidity = high] to predict whether or not to play a sport, it starts from the root node by testing a variable "weather", which is equal to the "sunny" then to test the parameters "humidity" which is equal to "high" then the conclusion to play sport is "No".

2.3.2 The Algorithm to Create a Decision Tree

At present, it has developed many algorithms for training decision tree. The most basic way is greedy search algorithm from top to bottom (top-down) called ID3, which was developed in 1986 by John Ross Quinlan.

2.3.3 Entropy

ID3 built decision tree from the top down approach, by asking whether variable should be the roots of this decision and repeatedly asked recursively to the tree. They choose the best attribute for becoming the root by considering information gain. Prior to information gain knowledge it must known the definition of purity of tainted data by using entropy. By definition, the entropy of the decision tree in a subset of the sample S is E (S) as follows.

$$E(S) - \sum_{j=1}^{n} ps(j) \log_2 ps(j)$$

where,

- S is a sample,
- $ps(j)$ is a ratio in case of attribute has a j value.

Actually, decision tree with a boolean attribute will have entropy as follows.

$$E(S) = -p_{yes} log_2(p_{yes}) - p_{no} log_2(p_{no})$$

Entropy has value between 0–1 in which value will be 0 if the result has only one output for example, yes or no. It will increase the entropy value if it has different values. In other words, entropy will have increasing value if dataset has impurity.

2.3.4 Information Gain

From above definition of entropy, we can define the nature of the good variables to be selected to be root node as follows, A will be a good variable to be selected if only if this variable is well divided the sample data to a possible data only one. Moreover, the average entropy of the partition has a lowest value. This value called expected value of the reduction of entropy after the data is divided by A. So the gain knowledge of A defines by,

$$Gain(S, A) = E(S) - \sum_{v=value(A)} \frac{|S_v|}{S} E(S_v)$$

where,

- S is sample which composed of many attribute,
- E is entropy of data sample,
- A is attribute,
- value (A) is set of possible value of A,
- S_v is all of sample which A attribute has v value.

It is shown that, if more gain value of A, the split sample S with A attributes, its entropy will nearly zero which making decisions to close more. Gain knowledge is a great value to tell the good of variables to be considered.

2.4   Image Content Generator System

In this section, we describe the process of image content generator system. The following steps will perform for retrieving the description of interested image:

Image input: The training stupa image is stored in database. The testing stupa images which want to know there's descriptions will be taken from camera and will be compared with the training stupa image dataset.

Pre-processing: After the stupa image was taken, the preprocessing process was performed. First, transforms the RGB image to gray scale image. Next, use the image enhancement algorithms to improve the quality of image.

Edge detection: This process is used for checking the edge that pass through or near to the interested point by measure the different of intensity of nearest points or finding the line surrounding the object inside image. If the edge was known, the image description was generated. The process for finding the accurate edge is not easy especially to finding the edge in the low quality image which has the low difference between foreground and background or the brightness is not cover to all of image. The edge is formed by the difference of intensity from one point to other points. If it has high different we can see the accurate edge, if not, the edge will blurred. The method to finding edge has so many algorithms, for example, Sobel, Roberts, Prewitt, and Canny. In our research we used Laplacian for becoming the main algorithms to detect edge of image (Fig. 6).

Feature extraction: Feature extraction is the process for retrieving the identity of each image to be vector for using in the training and testing processes. A grid pattern algorithm is used for being the process to get information inside image.

Image segmentation: Image segmentation is the process for dividing image into several sections.

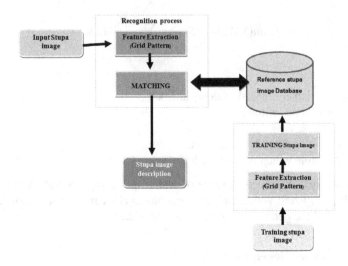

Fig. 6. The process of image content generator system.

Matching: This is the process for matching the most similar image between input image and the reference image. It is the process which was used after localization process. If the wrong used of localization process, the learning process for matching image to get description will not accurate. The process for generating description after matching process, the algorithms will use information inside grid of that image inside the database to show and set description to the input image. Accordingly, inside the image database, instead of contains only the reference image, it will contain the descriptions of the reference image too. Information inside database is stupa's architecture, stupa's era and other importance details of stupa.

In this paper, decision tree have been created in order to distinguish for information in each grid of an image. The formulation of decision tree was shown as the previous section.

Use input attribute from the grid which is the summation of the 1 value in each grid generated from grid pattern algorithms for becoming an input to decision tree. If the grid is divided with 10*10, we use 100 attributes to be input attribute.

Train on only the training set by setting the stopping criteria and the parameters. The difference between our proposed algorithms and traditional method is using the semi-supervised technique. The process in each step is to find the information gain and entropy in each node and select the best node to be the root node as shown in the previous section. The algorithms will adjust a tree until it give the best model for predicting class. This will be done from the direction of the upper layer towards the input layer. This algorithm will adjust weight from initial weight until it gives the satisfied mean square error. After the matching process by using decision tree, the system will generate the image description which taken from database to show in the target image. Example of the decision tree model is shown in Fig. 7 which generated by calculating entropy and gain in each attribute by using information inside grid.

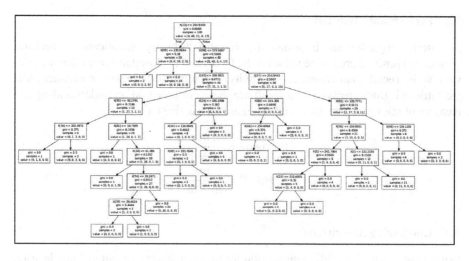

Fig. 7. Decision tree architecture.

3   Experimental Setup

3.1   Experimental Setup

In this research, datasets were collected from the archaeological site at Phra Nakhon Si Ayutta province, Sukhothai province and Bangkok province which have the following constraints.

1. An image which is used in the experiments is the stupa image in Ayutthaya era, Sukhothai era and Rattanakosin era.
2. Each image must have the stupa in landscape direction which coverage at least 80 percent.
3. Image must contain full of stupa and used only .jpg or .png file.

The number of image using in the research was shown in Table 1.

Table 1. Number of samples in the Experiments

Architecture	Number of image
Prasat style	100
Bell-shaped style	100
Prang style	100
Lotus Blossom style	100
Yohmum style	100
All	**500**

3.2 Performance Indexed

In pattern recognition and information retrieval with binary classification, **precision** (also called positive predictive value) is the fraction of retrieved instances that are relevant, while **recall** (also known as sensitivity) is the fraction of relevant instances that are retrieved. Both precision and recall are therefore based on an understanding and measure of relevance. Precision and recall are then defined as:

$$\text{Precision} = \frac{tp}{tp + fp}$$

$$\text{Recall} = \frac{tp}{tp + fn}$$

3.3 Comparing Algorithms

In this paper we used the following algorithms for comparing: grid pattern with decision tree, grid pattern with regression, grid pattern with Euclidean distance, and line contour technique. We compared each of this method with our proposed algorithms which used to predict the architecture for example: Prasat style, bell-shaped style, Prang style, lotus blossom style, Yohmum style. We used the combination of feature extraction technique with grid pattern and decision tree algorithms.

4 Experimental Results and Discussions

4.1 Experimental Results

The experimental results for stupa image content generator will be shown by using the classification results by using the proposed algorithms to classify architecture of that stupa. Accordingly, the accuracy of our image description generator depends on the classification accuracy. In order to confirm the performance of that accuracy we used precision and recall to measure. The results of classification were shown in Table 2. The experimental results shown that by using the proposed method they give these results: Table 2 shows the performance index of the proposed algorithms which is the combination between grid pattern with size of 10 × 10 and decision tree, the performance indexed of the algorithms by using grid pattern with regression, the performance indexed of the algorithms by using grid pattern with Euclidean distance. The experimental results confirms that using information inside grid of image which generated from the grid pattern algorithms and using the decision tree for training to get an architecture of image can be successfully used for generating the description of image as shown that in table. The proposed algorithms can be used to classify the era and architecture of the stupa image which can provide the accuracy about 70–80 % in average.

Table 2. Performance indexed of the proposed algorithms by using grid pattern with different technique.

Stupa's architecture	Performance indexed of the proposed algorithms					
	Grid pattern with decision tree		Grid pattern with regression		Grid pattern with Euclidean distance	
	Precision	Recall	Precision	Recall	Precision	Recall
Prasat style	0.1666	0.2000	0.0000	0.0000	0.000	0.000
bell-shaped style	0.6460	0.5640	0.5000	0.5100	0.5830	0.5490
Prang style	0.6660	0.7770	0.6190	0.8125	0.6660	0.8240
lotus blossom style	0.1250	0.1250	0.2500	0.2000	0.2500	0.3330
Yohmum style	0.2353	0.2500	0.2350	0.1818	0.2350	0.1818

Example of using SIFT algorithms for generate information for matching the architecture of stupa between testing image and image in database with different comparing algorithms for example, Euclidean distance, neural network, and KNN can be shown in Table 3. The result of using line contour algorithms for generating information for matching the architecture of stupa between testing image and image in database can be shown in Table 4. As we can see from these algorithms, the similarity of tested image and image in database has a little similarity. Accordingly our proposed algorithms can give the higher result for classify the architecture. So, it can be used to generate the description from that architecture of an image.

Table 3. Example of using SIFT algorithms for generate key point for matching the architecture of stupa.

Algorithms	Example
SIFT with Euclidean distance	
SIFT algorithms with nearest neighbors	
SIFT algorithms with neural network	

Table 4. Example of using line contour algorithms for matching the architecture of stupa.

Matching Shape	similarity distance
	0.7882
	0.5548

5   Conclusion

This research presents a novel algorithm for generating descriptions of stupa image such as stupa's era, stupa's architecture and other description by using information inside image which divided into grid and learning stupa description from the generated information with decision tree. We get information inside image by divided image into several grid patterns, for example 10×10 and use data inside that image to submit to the decision tree model. The proposed algorithm aims to generate the descriptions in each stupa image. Decision tree was used for being the classifier for generating the description. The experimental results show that the proposed framework can efficiently give the correct descriptions to the stupa image compared to using the traditional method.

Acknowledgment. This research was funded by King Mongkut's University of Technology North Bangkok. Contract no. KMUTNB-59-GEN-048.

References

1. Karpathy, A., Fei-Fei, L.: Deep visual-semantic alignments for generating image descriptions. In: The IEEE Conference on Computer Vision and Pattern Recognition (CVPR), pp. 3128–3137 (2015)
2. Socher, R., Karpathy, A., Le, Q.V., Manning, C.D., Ng, A.Y.: Grounded compositional semantics for finding and describing images with sentences. TACL **2**, 207–218 (2014)
3. Zaremba, W., Sutskever, I., Vinyals, O.: Recurrent neural network regularization. arXiv preprint arXiv:1409.2329 (2014)

4. Young, P., Lai, A., Hodosh, M., Hockenmaier, J.: From image descriptions to visual denotations: new similarity metrics for semantic inference over event descriptions. TACL **2**, 67–78 (2014)
5. Hodosh, M., Young, P., Hockenmaier, J.: Framing image description as a ranking task: data, models and evaluation metrics. J. Artif. Intell. Res. **47**, 853–899 (2013)
6. Su, H., Wang, F., Yi, L., Guibas, L.J.: 3D-assisted image feature synthesis for novel views of an object. In: International Conference on Computer Vision (ICCV), Santiago (2015)
7. Temple Architecture: Available online at http://www.thailandbytrain.com/TempleGuide.html
8. Farhadi, A., Hejrati, M., Sadeghi, M.A., Young, P., Rashtchian, C., Hockenmaier, J., Forsyth, D.: Every picture tells a story: generating sentences from images. In: Daniilidis, K., Maragos, P., Paragios, N. (eds.) ECCV 2010, Part IV. LNCS, vol. 6314, pp. 15–29. Springer, Heidelberg (2010)
9. Lowe, D.G.: Object recognition from local scale-invariant features. In: Proceedings of the 7th International Conference on Computer Vision (ICCV 1999), Corfu, Greece (1999)
10. Lowe, D.G.: Distinctive image features from scale-invariant key points. Int. J. Comput. Vis. **60**(2), 91–110 (2004)

Imbalance Effects on Classification Using Binary Logistic Regression

Hezlin Aryani Abd Rahman[1]([✉]) and Bee Wah Yap[2]

[1] Faculty of Computer and Mathematical Sciences, Centre of Statistical
and Decision Science Studies, Universiti Teknologi MARA,
40450 Shah Alam, Selangor, Malaysia
hezlin@tmsk.uitm.edu.my
[2] Faculty of Computer and Mathematical Sciences, Advanced Analytics
Engineering Centre, Universiti Teknologi MARA,
40450 Shah Alam, Selangor, Malaysia
beewah@tmsk.uitm.edu.my

Abstract. Classification problems involving imbalance data will affect the performance of classifiers. In predictive analytics, logistic regression is a statistical technique which is often used as a benchmark when other classifiers, such as Naïve Bayes, decision tree, artificial neural network and support vector machine, are applied to a classification problem. This study investigates the effect of imbalanced ratio in the response variable on the parameter estimate of the binary logistic regression via a simulation study. Datasets were simulated with controlled different percentages of imbalance ratio (IR), from 1 % to 50 %, and for various sample sizes. The simulated datasets were then modeled using binary logistic regression. The bias in the estimates was measured using MSE (Mean Square Error). The simulation results provided evidence that imbalance ratio affects the parameter estimates where severe imbalance (IR = 1 %, 2 %, 5 %) has higher MSE. Additionally, the effects of high imbalance (IR $\leq$ 5 %) will be more severe when sample size is small (n = 100 & n = 500). Further investigation using real dataset from the UCI repository (Bupa Liver (n = 345) and Diabetes Messidor, n = 1151)) confirmed the imbalanced ratio effect on the parameter estimates and the odds ratio, and thus will lead to misleading results.

Keywords: Imbalance data · Parameter estimates · Logistic regression · Simulation, predictive analytics

1 Introduction

Classification problems with class imbalance, whereby one class has more observations than the other, emerge in many data mining applications, ranging from medical diagnostics [1–5], finance [6–8], marketing [9], manufacturing [10] and geology [11]. Due to their practical importance, the class imbalance problem have been widely studied by many researchers [12–21].

Logistic regression (LR) is a conventional statistical method and often used in predictive analytics as a benchmark when other classifiers are used. However, the imbalance situation creates a challenge for LR, whereby the focus of the many classifiers

© Springer Nature Singapore Pte Ltd. 2016
M.W. Berry et al. (Eds.): SCDS 2016, CCIS 652, pp. 136–147, 2016.
DOI: 10.1007/978-981-10-2777-2_12

is normally on the overall optimization without taking into account the relative distribution between the classes [22]. Hence, the classification results obtained from the classifiers tends to be biased towards the majority class. Unfortunately, this imbalance problem is prominent as majority of the real dataset, regardless of field, suffers from some imbalance in nature [23]. This leads to the fact that the class with fewer observations is often misclassified into the majority classes [12, 14, 24].

Past studies either reported the effect of imbalance dataset on classifiers such as decision tree (DT), Support Vector Machine (SVM), and artificial neural network (ANN) [25–27] or uses LR on actual datasets to reveal the impact of IR in terms of predictions [2, 7, 19, 28]. However, no simulation study has been performed to determine the impact of imbalanced ratio on the estimate of the LR parameter (β) and thus the odds ratio (e^{β}) of the LR model.

The aim of this study is to investigate the effects of different IR on the parameter estimate of the binary LR model via a simulation study. This paper is organized as follows: Sect. 2 covers some previous studies on imbalanced data and applications of LR. The simulation methodology is explained in Sect. 3 and the results are presented in Sect. 4. Some discussions and the conclusion are given in Sect. 5.

2   Literature Review

Learning in the presence of imbalance data offers a great challenge to data scientists around the world. Techniques such as DT, ANN and SVM, performed well for balanced data but will not classify well when applied to imbalanced datasets [29]. [30] illustrated how overlapping within the dataset in the presence of imbalance data made it difficult for any classifier to predict observations correctly. A study by [31] demonstrated using actual datasets that the performance of a classifier (C4.5) is affected by the percentage of imbalance ratio.

Most studies reported the effect of IR towards the performance of standard classifiers using actual datasets with different ratio of imbalances [2, 17, 19, 25, 31]. [2] introduced REMED (Rule Extraction Medical Diagnosis), a 3-step algorithm using simple LR to select attributes for the model and adjust the percentage of the partition to improve accuracy of the model and evaluated it using real datasets. Although REMED is a competitive algorithm that can improve accuracy, it is restricted only to its domain, which is medical diagnostics. [17] reported that IR, dataset size and complexity, all contribute to the predictive performance of a classifier. They categorized the IR into three categories (balance, small, large), dataset size into four categories (very small, small, medium, and large) and complexity of the dataset into four categories (small, medium, large and very large) and experimented on different classifier (k-nearest neighbor (KNN), C4.5, SVM, multi-layered perceptron (MLP), naïve bayes (NB), and Adaboost (AB)). They concluded that higher IR has more influence on the performance of all the classifiers. [19] performed an extensive experiment on 35 dataset with different ratio of imbalance (1.33 %−34.90 %) using different sampling strategies (random oversampling (ROS), random undersampling (RUS), one-sided selection (OSS), cluster-based oversampling (CBOS), Wilson's editing (WE), SMOTE (SM), and borderline-SMOTE (BSM) on different classifiers (NB, DT C4.5, LR, random forest

(RF), and SVM). Their study significantly shows that sampling strategy improves the performance of the chosen classifiers and different classifiers works best with different sampling strategy. [25] observed that actual datasets do not provide the best distribution for any classifiers and sampling strategy is needed to improve the predictive performance of classifiers. Their experiment involves five actual dataset with C4.5 as the classifier. The study shows that SMOTE improve the performance of the overall classifier better than other sampling strategies and RUS works better than ROS with replication. [31] also experimented on 22 real dataset with different IR on different classifiers (C4.5, C4.5Rules, CN2 and RIPPER, Back-propagation Neural Network, NB and SVM). Using different sampling strategies (ROS, SMOTE, borderline-SMOTE, AdaSyn, and MetaCost), they concluded that the most affected in terms of accuracy (AUC) is the rule-based algorithm (C4.5Rule, RIPPER) and the least affected is the statistical algorithm (SVM).

From the studies mentioned above, we found that the performance of different standard classifiers such as NB, DT, ANN, and SVM were compared and each study concluded differently as to which classifier performed better [2, 17, 28, 32].

In predictive analytics, LR is an important classifier as it provides important information about the effect of an independent variable (IV) on the dependent variable (DV) through the odds ratio, which is the exponential value of the regression parameter [33]. LR is a statistical model and it is often used as a benchmark when other classifiers are used. However, with the presence of imbalance, the predictive models seemingly underestimate the class probabilities for minority class, despite evidently good overall calibration [34]. Significant odds ratio indicates a significant relationship between the DV and IVs.

The simulation study by [35] found that when sample size is large which is at least 500, the parameter estimates accuracy for LR improves. Their simulation study shows that the estimation of LR parameters is severely affected by types of covariates (continuous, categorical, and count data) and sample size. Meanwhile, [28] did a simulation study to evaluate the performance of six types of classifiers (ANN, Linear Discriminant Analysis (LDA), RF, SVM and penalized logistic regression (PLR)) on highly imbalanced data. However, their results show that the PLR with ROS method, failed to remove the biasness towards the majority class.

Most studies investigated the issue of IR by applying various classifiers to several datasets and thus there is no conclusive evidence as to which classifier is the best as it depends on the sample size, severity of imbalanced and types of data. Furthermore, no simulation study has been performed to determine the impact of IR on performance of classifiers. Simulation studies can provide empirical evidence on the impact of IR on the β-value and the odds ratio of the LR model.

3 Methods

There are two unknown parameters, β_0 and β_1 for a simple binary LR model. The parameters are estimated using the maximum likelihood method. The likelihood function by assuming observations to be independent is given by the following equation [33]:

$$L(\beta_0, \beta_1) = \prod_{i=1}^{n} \pi(x_i)^{y_i} [1 - \pi(x_i)]^{1-y_i} \tag{1}$$

The maximization of the likelihood function is required in order to estimate β_0 and β_1. Equivalently, the maximization of the natural logarithm of the likelihood function can be denoted by the following:

$$\log[L(\beta_0, \beta_1)] = \sum_{i=1}^{n} \{y_i \log[\pi(x_i)] + (1 - y_i) \log[1 - \pi(x_i)]\} \tag{2}$$

Referring to simple LR equation model, the equation in (2) can be expressed as follows [33]:

$$\log[L(\beta_0, \beta_1)] = \sum_{i=1}^{n} y_i(\beta_0 + \beta_1 x_i) - \sum_{i=1}^{n} \log[1 + \exp(\beta_0 + \beta_1 x_i)] \tag{3}$$

To maximize (3), one of the approaches is by differentiating $\log[L(\beta_0, \beta_1)]$ with respect to β_0 and β_1, and set the result of these two equations equal to zero as such:

$$\sum_{i=1}^{n} [y_i - \pi(x_i)] = 0 \quad \text{and} \quad \sum_{i=1}^{n} x_i[y_i - \pi(x_i)] = 0 \tag{4}$$

One of the methods that can be used to solve Eq. (4) is using iterative computing methods. The maximum likelihood estimates of β_0 and β_1, are denoted by $\hat{\beta}_0$ and $\hat{\beta}_1$. The maximum likelihood estimate of probability that event occurs, $\pi(x_i)$ is for case i denoted by:

$$\hat{\pi}(x_i) = \frac{e^{\hat{\beta}_0 + \hat{\beta}_1 x_i}}{1 + e^{\hat{\beta}_0 + \hat{\beta}_1 x_i}} \tag{5}$$

$\hat{\pi}(x_i)$ is also known as fitted or predicted value and the sum of $\hat{\pi}(x_i)$ is equal to the sum of the observed values:

$$\sum_{i=1}^{n} y_i = \sum_{i=1}^{n} \hat{\pi}(x_i) \tag{6}$$

The estimated logit function is written as follows:

$$\hat{g}(x_i) = \log\left[\frac{\hat{\pi}(x_i)}{1 - \hat{\pi}(x_i)}\right] = \hat{\beta}_0 + \hat{\beta}_1 x_i \tag{7}$$

This study assesses the effect of different percentages of IR on estimation of parameter coefficients for binary LR model. The fitted model was compared with the true logistic model in order to assess the effect on the parameter estimation. The simulations were performed using R, an open source programming software.

We fit a binary LR model to the simulated data and obtain the estimated coefficients, $\hat{\beta}$. The study focuses on a model with a single continuous covariate. For a single covariate, the value of the regression coefficient (β_1) for the logistic model was set at 2.08 which gives a significant odds ratio (OR) of 8.004 for X ($OR = e^{2.08} = 8.004$).

We considered eight ratios: 1 %, 2 %, 5 %, 10 %, 20 %, 30 %, 40 %, and 50 % where IR 5 % or less represents severe imbalance in the response variable. However, the complexity of generating the simulated dataset specifically for fixed percentages of IR requires β_0 values to vary for different IR percentages. The full LR model is thus set as the following:

$$g(x) = \beta_{0k} + 2.08x_k \tag{8}$$

where β_{0k} is determined by the IR and $k = 1, 2, 3, \ldots 8$.

The distribution of the covariate (X) considered in this study is the standard normal distribution, $N(0,1)$. The sample sizes generated ranges from 100, 500, 1000, 1500, 2000, 2500, 3000, 3500, 4000, 4500, and 5000. The simulation involves 10,000 replications. The R code developed for this simulation is made available at https://github.com/hezlin/simulationx1.git. It is also provided in the Appendix.

4 Simulation Results

Table 1 summarizes the results of parameter estimates for different sample sizes and IR. The results show that the estimates of β_0 and β_1 are far from the true parameter values for smaller sample size (n = 100, 500) especially with high IR value (IR = 1 %, 2 %, 5 %).

However, the estimates of the LR coefficient improve when the sample size increases to 1000 or more irrespective of IR. Figure 1 illustrates effect of IR and sample size. The red line is the set parameter value which is $\beta_1 = 2.08$.

Referring to Fig. 1, the effect on parameter estimate is most severe for small sample (n = 100). For sample size n = 500, at IR = 20 %, the value of the estimate is closer to the actual value. For sample size n = 1000 and n = 1500, at IR = 5 % the estimated β_1 is close enough to 2.08. Results show that for larger sample size (n = 1000 and above), the estimations are biased if the IR is less than 10 %.

The clustered boxplot in Fig. 2 shows clearly the effect of IR for various sample sizes. The estimates get closer to the true value when the sample size increases. The dispersion (standard deviation) of all $\hat{\beta}_1$ decreases as sample size and IR increases.

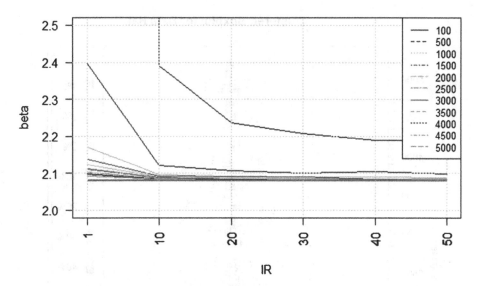

Fig. 1. Parameter estimates, $\hat{\beta}_1$, for different IR and sample size. This illustrates effect of different percentages of imbalance (IR = 1 %, 2 %, 5 %, 10 %, 20 %, 30 %, 40 %, 50 %) and sample size (n = 100, 500, 1000, 1500, 2000, 2500, 3000, 3500, 4000, 4500, 5000). The red line is the set parameter value, $\beta_1 = 2.08$. (Color figure online)

5   Application to Real Dataset

In this section, we illustrate the IR effect using two real datasets. The Bupa Liver and the Diabetes Messidor datasets were selected from the UCI repository. From the original dataset, we used stratified sampling to obtain the IR accordingly. The parameter estimates obtained are averaged over 1000 replications. Table 2 summarizes the results for this experiment.

The Bupa Liver Disorder dataset [36], consists of 345 observations, a response variable, *selector* (1 = *Yes* (58 %) and 0 = *No* (42 %)) and 6 covariates. We selected *alkaphos* (alkaline phosphate) as the independent variable as it is a continuous covariate. Results in Table 2 show that the value of $\hat{\beta}_0$ is far from the actual value as imbalance increases (that is when the percentage of IR gets smaller).

Although, the value of $\hat{\beta}_1$ and *OR* seems to be close enough to the true value from the original dataset, the p-value increases as IR increase (i.e. when imbalance decreases). Additionally, the confidence interval (CI) for $\hat{\beta}_1$ becomes wider for severe imbalance. For IR = 2 % and 1 %, the true value (-0.042) is no longer within the CI and thus the covariate will be conclude as not significant. These results show that IR affects the p-values for both estimates of $\hat{\beta}_0$ and $\hat{\beta}_1$.

The Diabetes Messidor dataset [37], consist of 1151 observations. The response variable selected is *DR status* (1 = with DR (53 %) and 0 = without DR (47 %)) and

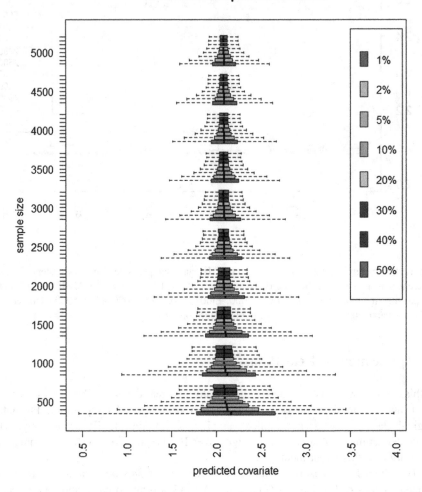

Fig. 2. Horizontal clustered boxplots for $\hat{\beta}_1$.

the dataset has 16 covariates. The covariate selected for this study is the *dmtroptdisc* (diameter of the optical disc). Results in Table 2 show that the $\hat{\beta}_0$ and $\hat{\beta}_1$ values becomes larger as the IR increases. The p-values for $\hat{\beta}_1$ increases as imbalance become more severe. The most obvious biased can be seen in the values of the odds-ratios which increase tremendously there is severe imbalance (at 5 % or less). These two applications confirm the results of the simulation study that imbalance data will lead to wrong conclusion on the effect of the independent variable on the response variable.

Table 1. Parameter Estimates for Different IR and Sample Size

Size	IR	$\hat{\beta}_1$	CI (lower)	CI (upper)	β_0	$\hat{\beta}_0$
100	1	82.173	94.183	70.163	-6.627	-193.893
	2	47.084	63.954	30.213	-5.764	-96.990
	5	4.822	6.061	3.582	-4.672	-9.117
	10	2.390	2.500	2.281	-3.566	-3.881
	20	2.236	2.247	2.224	-2.304	-2.403
	30	2.208	2.219	2.198	-1.427	-1.478
	40	2.190	2.199	2.180	-0.718	-0.720
	50	2.191	2.200	2.181	-0.274	-0.118
500	1	2.398	2.518	2.278	-6.649	-7.259
	2	2.189	2.199	2.178	-5.760	-5.920
	5	2.146	2.153	2.139	-4.548	-4.621
	10	2.121	2.127	2.115	-3.505	-3.543
	20	2.106	2.111	2.102	-2.284	-2.303
	30	2.101	2.105	2.097	-1.420	-1.427
	40	2.104	2.108	2.100	-0.688	-0.691
	50	2.098	2.102	2.095	0.001	0.000
1000	1	2.170	2.179	2.160	-6.594	-6.729
	2	2.133	2.140	2.127	-5.814	-5.738
	5	2.110	2.115	2.105	-4.567	-4.529
	10	2.096	2.100	2.092	-3.501	-3.514
	20	2.092	2.096	2.089	-2.285	-2.292
	30	2.091	2.094	2.088	-1.418	-1.423
	40	2.091	2.093	2.088	-0.685	-0.687
	50	2.089	2.091	2.086	-0.002	-0.001
1500	1	2.138	2.145	2.131	-6.569	-6.662
	2	2.116	2.122	2.111	-5.731	-5.781
	5	2.101	2.105	2.097	-4.528	-4.554
	10	2.092	2.095	2.089	-3.499	-3.509
	20	2.090	2.092	2.087	-2.283	-2.288
	30	2.089	2.092	2.087	-1.417	-1.422
	40	2.085	2.087	2.083	-0.684	-0.685
	50	2.087	2.089	2.084	-0.001	0.000
2000	1	2.123	2.129	2.117	-6.565	-6.629
	2	2.100	2.105	2.095	-5.726	-5.752
	5	2.093	2.096	2.090	-4.525	-4.541
	10	2.090	2.092	2.087	-3.498	-3.507
	20	2.084	2.087	2.082	-2.281	-2.284
	30	2.087	2.089	2.085	-1.417	-1.420
	40	2.085	2.087	2.083	-0.684	-0.685
	50	2.087	2.089	2.085	-0.002	0.000
2500	1	2.112	2.117	2.107	-6.558	-6.605
	2	2.100	2.104	2.095	-5.726	-5.751
	5	2.089	2.091	2.086	-4.524	-4.535
	10	2.088	2.090	2.086	-3.498	-3.505
	20	2.085	2.087	2.083	-2.282	-2.285
	30	2.083	2.085	2.082	-1.418	-1.419
	40	2.084	2.085	2.082	-0.683	-0.684
	50	2.083	2.085	2.082	0.000	0.000
3000	1	2.111	2.116	2.106	-6.552	-6.601
	2	2.096	2.099	2.092	-5.720	-5.743
	5	2.088	2.091	2.086	-4.524	-4.535
	10	2.088	2.090	2.086	-3.498	-3.505
	20	2.085	2.087	2.083	-2.281	-2.284
	30	2.084	2.086	2.083	-1.419	-1.420
	40	2.084	2.085	2.082	-0.683	-0.684
	50	2.083	2.084	2.081	-0.048	-0.020
3500	1	2.103	2.107	2.098	-6.551	-6.586
	2	2.094	2.098	2.091	-5.712	-5.738
	5	2.088	2.091	2.086	-4.525	-4.534
	10	2.085	2.087	2.083	-3.497	-3.502
	20	2.083	2.084	2.081	-2.281	-2.283
	30	2.083	2.085	2.082	-1.417	-1.419
	40	2.082	2.084	2.081	-0.685	-0.685
	50	2.083	2.084	2.081	0.000	0.000
4000	1	2.098	2.102	2.094	-6.547	-6.575
	2	2.090	2.094	2.087	-5.710	-5.731
	5	2.088	2.091	2.086	-4.524	-4.534
	10	2.085	2.086	2.083	-3.496	-3.501
	20	2.084	2.084	2.082	-2.282	-2.285
	30	2.082	2.084	2.081	-1.417	-1.418
	40	2.082	2.084	2.081	-0.683	-0.684
	50	2.083	2.085	2.082	0.000	0.000
4500	1	2.097	2.101	2.093	-6.547	-6.573
	2	2.093	2.096	2.090	-5.718	-5.736
	5	2.086	2.088	2.084	-4.526	-4.531
	10	2.083	2.085	2.081	-3.496	-3.499
	20	2.082	2.084	2.082	-2.281	-2.283
	30	2.083	2.085	2.081	-1.418	-1.419
	40	2.082	2.084	2.080	-0.685	-0.685
	50	2.083	2.085	2.081	0.000	0.001
5000	1	2.092	2.096	2.089	-6.541	-6.564
	2	2.088	2.091	2.085	-5.713	-5.727
	5	2.088	2.091	2.085	-4.524	-4.533
	10	2.083	2.085	2.081	-3.497	-3.500
	20	2.081	2.082	2.080	-2.282	-2.282
	30	2.083	2.084	2.082	-1.418	-1.419
	40	2.082	2.084	2.080	-0.685	-0.685
	50	2.083	2.084	2.084	0.000	0.000

Table 2. Real dataset applications results

Dataset	Data/IR	$\hat{\beta}_0$, [p-value] C.I (lower, upper)	$\hat{\beta}_1$, p-value C.I (lower, upper)	*Odds-Ratio (OR)* C.I (lower, upper)
Bupa Liver (select status)	Original (145:200)	4.143, [0.069] (4.1433,4.1435)	−0.042, [0.093] (−0.04244, −0.042236)	0.959 (0.9584, 0.9585)
	40 % (133:200)	4.192, [0.077] (4.162, 4.221)	−0.042, [0.110] (−0.04226, −0.04161)	0.959 (0.9586, 0.9593)
	30 % (85:200)	4.530, [0.119] (4.454, 4.606)	−0.041, [0.205] (−0.042, −0.040)	0.960 (0.959, 0.961)
	20 % (50:200)	4.931, [0.179] (4.814, 5.048)	−0.039, [0.326] (−0.041, −0.038)	0.962 (0.960, 0.963)
	10 % (22:200)	5.773, [0.277] (5.573, 5.973)	−0.039, [0.447] (−0.042, −0.037)	0.962 (0.960, 0.964)
	5 % (10:200)	6.549, [0.361] (6.253, 6.845)	−0.039, [0.500] (−0.042, −0.036)	0.963 (0.960, 0.966)
	2 % (4:200)	7.436, [0.469] (6.987, 7.886)	−0.038, [0.554] (−0.043, −0.033)	0.965 (0.961, 0.970)
	1 % (2:200)	8.692, [0.503] (8.059, 9.325)	−0.044, [0.550] (−0.051, −0.037)	0.963 (0.957, 0.970)
Diabetes Messidor (DR status)	Original (540:611)	0.498, [0.170] (0.4975, 0.4977)	−3.449, [0.295] (−3.44852, −3.44854)	0.032 (0.031, 0.032)
	40 % (407:611)	0.771, [0.063] (0.766, 0.776)	−3.365, [0.372] (−3.411, −3.319)	0.079 (0.074, 0.085)
	30 % (261:611)	1.218, [0.019] (1.209, 1.227)	−3.381, [0.44] (−3.466, −3.297)	0.69 (0.419, 0.961)
	20 % (152:611)	1.752, [0.01] (1.738, 1.765)	−3.307, [0.482] (−3.431, −3.183)	104.45 (−52.948, 261.849)
	10 % (68:611)	2.514, [0.014] (2.491, 2.536)	−2.881, [0.511] (−3.089, −2.674)	116816.15 (218.528, 233413.773)
	5 % (35:611)	3.177, [0.025] (3.145, 3.209)	−2.821, [0.516] (−3.117, −2.524)	1.528E + 11 (−1.45E + 11, 4.51E + 11)
	2 % (13:611)	4.146, [0.069] (4.094, 4.199)	2.443, [0.517] (−2.929, −1.958)	1.535E + 20 (−1.47E + 20, 4.54E + 20)
	1 % (7:611)	4.664, [0.126] (4.593, 4.735)	−1.269, [0.513] (−1.926, −0.612)	8.043E + 26 (−6.63E + 26, 2.27E + 27)

6 Conclusion

This simulation study shows that the imbalance ratio in the response variable will affect the parameter estimates of the binary LR model. Imbalance ratio will lead to imprecise estimates and the bias is more severe if sample size is small. The imbalance ratio also affects the p-value of the parameter estimates and a covariate may be inaccurately reported as not significant. There are approaches suggested for handling imbalance data such as ROS, RUS, SMOTE, and clustering. Future work will include simulation study involving categorical covariates and the classification performance of other classifiers such as SVM, DT, ANN and NB.

Acknowledgements. Our gratitude goes to the Research Management Institute (RMI) Universiti Teknologi MARA and the Ministry of Higher Education (MOHE) Malaysia for the funding of this research under the Malaysian Fundamental Research Grant, 600- RMI/FRGS 5/3 (16/2012). We also thank Prof. Dr. Haibo He (Rhodes Island University), Prof. Dr. Ronaldo Prati (Universidade Federal do ABC), Dr. Pam Davey and Dr. Carolle Birrell (University of Wollongong) for sharing their knowledge and providing valuable comments for this study.

Appendix

```
#fitting the model
set.seed(54321)
while(n<=nrep)
{ #set bnot value
  for(i in seq(start,end,0.0001))
  { x1 <- rnorm(ndata,0,1)        # some continuous variables
    z <- (i + 2.08*x1)   # linear combination with a bias
    pr1 <- 1/(1+exp(-z))
# pass through an inv-logit function
    ry<-runif(ndata,0,1) #generate value of y
    u <-as.matrix(ry) #generate value of y
    y <- ifelse((u<=pr1),1,0)
    y<-cbind(y)
    m_y <- (mean(y)*100)
    if(m_y == perc && n <=nrep)
    {dt<-data.frame(y=y,x=x1)
      mod<-glm(y~x1,data=dt,family="binomial")
#fit binary logistic regression
      beta0Hat[n]<-as.numeric(mod$coef[1])
      beta1Hat[n]<-as.numeric(mod$coef[2]) #store coefficient value
      resulty[n]<-as.numeric(sum(y))
      betanot[n]<-as.numeric(i)
      n <- n + 1 }
  }
}
mean(beta1Hat)
ci.b1 <- CI(beta1Hat,ci=0.95)
MSEbeta1Hat <- round(sum((beta1Hat-2.08)^2/nrep),3)
```

References

1. Datir, A.A., Wadhe, A.P.: Review on need of data mining techniques for biomedical field. Int. J. Comput. Inf. Technol. Bioinforma. **2**, 1–5 (2014)
2. Mena, L., Gonzalez, J.A.: Machine learning for imbalanced datasets: application in medical diagnostic. In: Proceedings of the Nineteenth International Florida Artificial Intelligence Research Society Conference (FLAIRS 2006), pp. 574–579. AAAI Press (2006). http://www.informatik.uni-trier.de/~ley/db/conf/flairs/flairs2006.html

3. Oztekin, A., Delen, D., Kong, Z.J.: Predicting the graft survival for heart-lung transplantation patients: an integrated data mining methodology. Int. J. Med. Inform. **78**, e84–e96 (2009)
4. Sathian, B.: Reporting dichotomous data using logistic regression in medical research: the scenario in developing countries. Nepal J. Epidemiol. **1**, 111–113 (2011)
5. Uyar, A., Bener, A., Ciray, H., Bahceci, M.: Handling the imbalance problem of IVF implantation prediction. IAENG Int. J. Comput. Sci. **37** (2010)
6. Akbani, R., Kwek, S.S., Japkowicz, N.: Applying support vector machines to imbalanced datasets. In: Boulicaut, J.-F., Esposito, F., Giannotti, F., Pedreschi, D. (eds.) ECML 2004. LNCS (LNAI), vol. 3201, pp. 39–50. Springer, Heidelberg (2004)
7. Burez, J., Van den Poel, D.: Handling class imbalance in customer churn prediction. Expert Syst. Appl. **36**, 4626–4636 (2009)
8. Ogwueleka, F.: Data mining application in credit card fraud detection system. J. Eng. Sci. Technol. **6**, 311–322 (2011)
9. Nikulin, V., McLachlan, G.J.: Classification of imbalanced marketing data with balanced random sets. In: JLMR: Workshop and Conference Proceedings, vol. 7, pp. 89–100 (2009). http://jmlr.csail.mit.edu/proceedings/papers/v7/nikulin09/nikulin09.pdf
10. Sobran, N., Ahmad, A., Ibrahim, Z.: Classification of Imbalanced Dataset Using Conventional Naïve Bayes Classifier in 35–42 (2013). http://worldconferences.net/proceedings/aics2013/toc/papers_aics2013/A021-NURMAISARAHMOHDSOBRAN-ClassificationofImabalanceddatasetusingconventionalnaivebayesclassifier.pdf
11. Thogmartin, W.E., Knutson, M.G., Sauer, J.R.: Predicting regional abundance of rare grassland birds with a hierarchical spatial count model. Condor **108**, 25–46 (2006)
12. Chawla, N.V., Japkowicz, N., Kotcz, A.: Editorial: special issue on learning from imbalanced data sets. ACM SIGKDD Explor. Newsl. **6**, 1 (2004)
13. Drummond, C., Holte, R.: Severe class imbalance: why better algorithms aren't the answer. In: Gama, J., Camacho, R., Brazdil, P.B., Jorge, A.M., Torgo, L. (eds.) ECML 2005. LNCS (LNAI), vol. 3720, pp. 539–546. Springer, Heidelberg (2005)
14. He, H., Garcia, E.E.A.: Learning from imbalanced data. IEEE Trans. Knowl. Data Eng. **21**, 1263–1284 (2009)
15. Japkowicz, N., Stephen, S.: The class imbalance problem: a systematic study. Intell. Data Anal. **6**, 429–449 (2002)
16. Japkowicz, N.: Learning from imbalanced data sets: a comparison of various strategies. In: AAAI Workshop on Learning from Imbalanced Data Sets 0–5 (2000). doi:10.1007/s13398-014-0173-7.2
17. Lemnaru, C., Potolea, R.: Imbalanced classification problems: systematic study, issues and best practices. In: Zhang, R., Zhang, J., Zhang, Z., Filipe, J., Cordeiro, J. (eds.) ICEIS 2011. LNBIP, vol. 102, pp. 35–50. Springer, Heidelberg (2012)
18. Longadge, R., Dongre, S.S., Malik, L.: Class imbalance problem in data mining review. Int. J. Comput. Sci. Netw. **2**, 83–87 (2013)
19. Van Hulse, J., Khoshgoftaar, T.M., Napolitano, A.: Experimental perspectives on learning from imbalanced data. In: Proceedings of 24th International Conference on Machine Learning, pp. 935–942 (2007). doi:10.1145/1273496.1273614
20. Visa, S., Ralescu, A.: Issues in mining imbalanced data sets - a review paper. In: Proceedings of the Sixteen Midwest Artificial Intelligence and Cognitive Science Conference, MAICS-2005, pp. 67–73 (2005)
21. Weiss, G.M.: Foundations of imbalanced learning. In: He, H., Ma, Y. (eds.) Imbalanced Learning: Foundations, Algorithms, Applications, pp. 13–42. Wiley & IEEE Press (2013). http://storm.cis.fordham.edu/gweiss/papers/foundations-imbalanced-13.pdf

22. Dong, Y., Guo, H., Zhi, W., Fan, M.: Class imbalance oriented logistic regression. In: 2014 International Conference Cyber-Enabled Distributed Computing and Knowledge Discovery, pp. 187–192 (2014). doi:10.1109/CyberC.2014.42
23. Goel, G., Maguire, L., Li, Y., McLoone, S.: Evaluation of sampling methods for learning from imbalanced data. Intell. Comput. Theor. **7995**, 392–401 (2013)
24. Weiss, G.M., Provost, F.: Learning when training data are costly: the effect of class distribution on tree induction. J. Arti. Intell. Res. **19**, 315–354 (2003)
25. Chawla, N.V.: C4. 5 and imbalanced data sets: investigating the effect of sampling method, probabilistic estimate, and decision tree structure. In: Proceedings of the International Conference on Machine Learning, Workshop Learning from Imbalanced Data Set II (2003). https://www3.nd.edu/∼dial/papers/ICML03.pdf
26. Cohen, G., Hilario, M., Sax, H., Hugonnet, S., Geissbuhler, A.: Learning from imbalanced data in surveillance of nosocomial infection. Artif. Intell. Med. **37**, 7–18 (2006)
27. Galar, M., Fernandez, A., Barrenechea, E., Bustince, H., Herrera, F.: A review on ensembles for the class imbalance problem bagging, boosting, and hybrid-based approaches. IEEE Trans. Syst. Man Cybern. Part C Appl. Rev. **99**, 1–22 (2011)
28. Blagus, R., Lusa, L.: Class prediction for high-dimensional class-imbalanced data. BMC Bioinform. **11**, 523 (2010)
29. Anand, A., Pugalenthi, G., Fogel, G.B., Suganthan, P.N.: An approach for classification of highly imbalanced data using weighting and undersampling. Amino Acids **39**, 1385–1391 (2010)
30. Batista, G., Prati, R.C., Monard, M.C.: A study of the behavior of several methods for balancing machine learning training data. ACM SIGKDD Explor. Newsl. **6**, 20 (2004)
31. Prati, R.C., Batista, G.E.A.P.A., Silva, D.F.: Class imbalance revisited: a new experimental setup to assess the performance of treatment methods. Knowl. Inf. Syst. 45, 247–270 (2014)
32. Sarmanova, A., Albayrak, S.: Alleviating class imbalance problem in data mining. In: Signal Processing and Communications Applications Conference, pp. 1–4 (2013)
33. Hosmer, D.W., Lemeshow, S.: Applied Logistic Regression Second Edition. Applied Logistic Regression (2004). doi:10.1002/0471722146
34. Wallace, B.C., Dahabreh, I.J.: Class Probability Estimates are Unreliable for Imbalanced Data (and How to Fix Them). *ICDM* (2012). http://www.cebm.brown.edu/static/papers/wallace-dahabreh-icdm-12-preprint.pdf
35. Hamid, H.A., Yap, B.W., Xie, X.-J., Abd Rahman, H.A.: Assessing the Effects of Different Types of Covariates for Binary Logistic Regression. **425,** 425–430 (2015)
36. Forsyth, R.S.: BUPA Liver Disorders (1990). https://archive.ics.uci.edu/ml/datasets/Liver+Disorders
37. Antal, B., Hajdu, A.: An ensemble-based system for automatic screening of diabetic retinopathy. Knowl. Based Syst. **60**, 20–27 (2014)

Weak Classifiers Performance Measure in Handling Noisy Clinical Trial Data

Ezzatul Akmal Kamaru-Zaman[1,2(✉)], Andrew Brass[2], James Weatherall[3],
and Shuzlina Abdul Rahman[1]

[1] Faculty of Computer and Mathematical Sciences, Universiti Teknologi MARA, Shah Alam,
Selangor, Malaysia
ezza@tmsk.uitm.edu.my
[2] School of Computer Science, University of Manchester, Manchester, UK
[3] Advanced Analytics Centre, Astra Zeneca R&D, Alderley Park, Chesire, UK

Abstract. Most research concluded that machine learning performance is better
when dealing with cleaned dataset compared to dirty dataset. In this paper, we
experimented three weak or base machine learning classifiers: Decision Table,
Naive Bayes and k-Nearest Neighbor to see their performance on real-world,
noisy and messy clinical trial dataset rather than employing beautifully designed
dataset. We involved the clinical trial data scientist in leading us to a better data
analysis exploration and enhancing the performance result evaluation. The classifiers performances were analyzed using Accuracy and Receiver Operating
Characteristic (ROC), supported with sensitivity, specificity and precision values
which resulted to contradiction of conclusion made by previous research. We
employed pre-processing techniques such as interquartile range technique to
remove the outliers and mean imputation to handle missing values and these
techniques resulted to; all three classifiers work better in dirty dataset compared
to imputed and clean dataset by showing highest accuracy and ROC measure.
Decision Table turns out to be the best classifier when dealing with real-world
noisy clinical trial.

Keywords: Clinical trial · Classifier · Decision Table · k-Nearest Neighbor ·
Machine learning · Naïve bayes · Noisy data

1 Introduction

Clinical trial resides in applied data science where it involves biomedical and health
informatics, statistical innovation and scientific computing for analysis and data prediction. Clinical data is a real-world data problem where it can be noisy and messy. Rather
than using conventional prediction, machine learning as a part of analysis using scientific
computing has become another popular method in making prediction of drug release
and survival projection.

Machine learning is defined as study of employing scientific computational techniques to systemize the process of knowledge acquisition by learning and adapting on
patterns of data recorded [1, 2]. The capability of machine learning depends on how it
appropriately handles various types of noisy data.

© Springer Nature Singapore Pte Ltd. 2016
M.W. Berry et al. (Eds.): SCDS 2016, CCIS 652, pp. 148–157, 2016.
DOI: 10.1007/978-981-10-2777-2_13

Ultimately, noisy data becomes a drawback in the ability to draw accurate hypotheses. Thorough procedures such as sorting out detection of noise from hypothesis construction need to be carried out so that noisy data do not affect during the building up of hypothesis process [3]. An indelicate evaluation may be made when dealing with missing data, imbalanced and extreme value data. Several researches have concluded that a cleaner and less noisy dataset may likely to build a better predictor.

Based on research done by [4], there are two major types of noise comprises of attribute noise and class noise. Attribute noise refers to falsify values of one or more attributes. Missing data is categorized as attribute noise where it is defined as unavailable value of variable in a dataset. In term of clinical trial dataset, missing data problem may occur due to discontinuation of treatment of subject. Thus, the evaluation data of the patient may be left unfilled. Missing data value also can arise from lack of systematic clinical trial planning and design. Handling missing data becomes a serious issue to clinical trial as it compromised inferences made [5].

Meanwhile, class noise is the noise occurs when an instance is not labelled appropriately. Outlier originally defined by [6] as data point that lies out of the normal dataset range falls under this class noise type. There is no robust computation technique to determine whether the outlier going to be interpreted as something important or just inaccurate data that will give a drawback to data analysis. This statement also was supported by [7], where in some cases, machine learning treated noisy instances as mislabelled disregard the real meaning of it that probably arises from noisy error in attribute values or erroneous in class labels. That is why assistance by the physician to exploit outlier's meaning in clinical trial dataset is crucial when employing machine learning process.

2   Handling Noisy Data Using Weak Classifiers

Noisy data exist in almost any real-world dataset. Different learners behave differently to the presence of noise. There is also a wide variation in the relative performances of the algorithms [8]. Frequently, noise from different attributes performs differently with the system performance. It is not necessary for a noise handling mechanism to handle every attribute because handling noise on noise-sensitive attributes would be more crucial [9].

Learning process will slow down if too much noisy data with irrelevant information [10]. Hence, noisy data cleaning has to be done. For higher levels of noise, removing noise from attribute information may decrease the predictive accuracy of the resulting classifier if the same attribute noise is present when the classifier is subsequently used, as demonstrated by [11]. However, for class noise, cleaning the training data will produce a classifier with a higher predictive accuracy. Noise can still drastically affect the accuracy even if the use of pruning and learning ensembles partially addresses the problem.

Machine learning technique has been used in many clinical trial analysis and prediction. Naive Bayes and decision tree are compared in research by [12] to see the recurrence of prostate cancer. The research resulted to nomogram performs better than decision tree. The drawback of decision tree is that only certain variable is selected while nomogram goes through all variables.

Research made by Kalapanidas [13] has implemented 11 machine learning techniques on the presence of real pollution real noise dataset. He divided the techniques into two groups, regression type and classification type. From all analysis made, he concluded in his paper that linear regression can handle increment of noise better than other techniques. Meanwhile, decision table has appeared the second best as it is less sensitive even when noise is controlled and increase.

In this paper, we experimented on few weak classifiers. Weak classifiers as defined by Vannuci [14] are single learners that constructed by low complexity algorithm. Recent researches have been made to boost the performance of weak classifiers by combining a group of weak classifiers and created new classifier method known as Ensemble Methods (EM). [15–17]. Nonetheless, EM has been agonized by the real world problem dataset especially when dealing with noisy data. Hence, in this paper, we take an approach to evaluate the weak classifiers performance when handling noisy data disregard to any ensemble techniques. The following subsection explains three types of weak classifier, which are Decision Table, Naïve Bayes and k-Nearest Neighbor (k-NN).

2.1 Decision Table

Decision Table is defined as a class for building and using a simple decision table majority classifier. Kohavi in [18] defined Decision Table consist of two term:

1. Schema consists of optimal feature.
2. Body, are sets of instances with a values in every features and class label.

Decision Table select schema through maximization of cross-validation. In order to predict the label for each unlabeled instances I, computation is denoted as follow:

Let $X =$ the sets of labeled instances that match the given I;
With the condition where only features in schema are matched, leaving out other features.

If $X =$ null, majority class in Decision Table will be returned,
Else majority class in X will be returned.

Wets [19] enhanced Kohavi's experiment by defining Decision Table into actions and conditions. Action is the class label meanwhile conditions is the highly correlated features. In his enhancement of Decision Table, he tabulates every condition to be mapped into the one and only actions. The mapped values can be recursively referred back as lookup table in order to make the prediction [19]. Since table structure is not altered every time new entry is added or removed, cross-validation method is crucial for Decision Table to evaluate the feature hence, leave-out-one cross validation is set in this classifier implementation.

2.2 Naïve Bayes

Naive Bayes is proposed by [20] that implemented naive assumption or precisely defined as 'nonparametric density approximation' statistical methods to alleviate the issue of handling continuous variable when constructing a probability distribution model. The algorithm is defined as follow:

For each y defined as class variable, given an instance to be classified, defined as vector x from 1 to n features: $x = (x_1, \ldots, x_n)$. Bayes theorem denotes the following:

$$P(y|x_1, \ldots, x_n) = \frac{P(y)P(x_1, \ldots, x_n|y)}{P(x_1, \ldots, x_n)}$$

Or in simpler way, it is defined as follow:

$$posterior = \frac{prior \, x \, likehood}{evidence}$$

Naive independence assumption is implemented where:

$$P(x_1|y, x_1, \ldots, x_{i-1}, x_{i+1}, \ldots, x_n) = P(x_1|y)$$

And for all i denoted, we decomposed relationship to:

$$P(y|x_1, \ldots, x_n) = \frac{P(y) \prod_{i-1}^{n} P(x_1|y)}{P(x_1, \ldots, x_n)}$$

Therefore, we created the model of classification rule as follow since vector x is constant:

$$P(y|x_1, \ldots, x_n) \propto P(y) \prod_{i-1}^{n} P(x_1|y)$$

Where we make assumptions based on the distribution of. Estimator classes are used by Naive Bayes to make classification where the numeric values of estimator precision are selected based on training data analysis.

2.3 K-Nearest Neighbour (KNN)

KNN algorithm works in training dataset by searching k-most similar features in k-neighbourhood of search subspaces. KNN algorithm employed Euclidean distance to find k-nearest neighbourhood in real set of data. Any predicted similar instances are summarized and prediction of unseen instances will then be made based on this summary.

In WEKA, classifier of KNN used instance-based learning (IBk) algorithm constructed by Aha [21]. The prediction attribute of the most similar instances is summarized and returned as the prediction for the unseen instance. For real-valued data, the Euclidean distance can be used. Other types of data such as categorical or binary data, Hamming distance can be used. In the case of regression problems, the average of the predicted attribute may be returned. In the case of classification, the most prevalent class may be returned.

3 Experimental Framework

3.1 Clinical Trial Dataset

Prostate cancer clinical trial data acquired from Project Data Sphere was used for our study. The dataset representation and information have been analysed to understand the underlying facts of the real data with the assistance of Astra Zeneca data scientist team. This data was donated by Sponsor Amgen, Inc named as Amgen Protocol Number (Denosumab) 20050103 and it was used to study the effect of Denosumab and Zoledronic Acid (ZOMETA) medicine to treat bone metastases in men with hormone-refractory prostate cancer. It is a rich dataset of 756 patients' records with three years of the study period and analysis on survival of the cancer has been done one this study and we defined it as the class label or what to predict.

3.2 Data Design and Analysis

70 % of the time of experiment focusses on data preprocessing which employed scrutinizing the data design and making data analysis. With the help of the data scientist, we recognized this dataset is made up of 16 continuous features and 17 categorical features including class label which sum up to 33 features altogether. Meanwhile, the class label is made up of 49.86 % of Yes classification of survival and 50.14 % of No classification of survival hence no imbalanced issues are perceived. When employed in WEKA for missing data analysis, about 85 % of missing values reported in two out of 33 features. Interquartile range method was used in this dataset to detect outlier and extreme values. This dataset is considered as balanced dataset based on its balance distribution and data design.

These data; defined as dirty data were then pre-processed to get the imputed and cleaned dataset. Process of imputation was done by identifying the missing values and replacing the values with mean. Heading towards to get a clean dataset, the imputed dataset was employed with interquartile range method to detect data noise such as outliers and extreme values using **Interquartile Range** filter built in WEKA. Once detection has been made, the outliers and extreme values were removed using **RemoveWithValues** filter. The three condition of dataset (dirty, imputed and cleaned) are then fed into machine learning classifiers. In this paper, we are focusing on implementing weak classifiers consist of Decision Table, Naive Bayes and KNN.

3.3 Machine Learning Parameter Setting

In this experiment, Decision Table was customized to use Best-First-Search as searching method to find the highly related attribute combination. In general, Best-First-Search can be explained by using greedy hill climbing method in order to explore the attribute subset spaces. The direction of searching was set to be forward as it started with an empty set and incrementally adds new attributes subsets. In order to terminate the search, searchTermination was set to 5. If there are five non-improving nodes found in sequentially, searching will be stopped. For Naive Bayes, the classifier used standard deviation estimator classifier instead of kernel classifier. Meanwhile, for KNN setting, we chose linear nearest neighbor search algorithm that implements brute force search algorithm and Euclidean Distance as the distance function. Through cross-validation, we had set the value of initial neighborhood to be 1.

3.4 Performance Measure

The performances of these three classifiers were evaluated by making observation on correctly classified instance or also known as accuracy based on survival prediction. Not only that, a confusion matrix that contributes to performance evaluation was also obtained to get the Specificity (True Negative Rate) and Sensitivity (True Positive Rate). Precision is also recorded to evaluate the number of correct labels is in positive class.

Specificity and sensitivity values are crucial in plotting ROC Curve in order to measure the ability of the classifier in making the correct classification. From this, we can analyze the performance of machine learning techniques at which some can work better when dealing with cleaned dataset and some of them perform the same disregard handling noisy or cleaned dataset. In a Receiver Operating Characteristic (ROC) curve, the Sensitivity is plotted in function of the false positive rate (1-Specificity) from different cut points. Each point on the ROC curve will represent a sensitivity or specificity pair respective to a specific decision threshold. Hence, the nearer the ROC curve is to the upper left corner, the higher the whole accuracy of the test [22].

4 Results and Analysis

In this study, we are weighing the accuracy of machine learning methods in predicting survival classification. Table 1 displays the accuracy of three analyzed weak classifier approaches in three different dataset conditions that are dirty dataset, imputed dataset and cleaned dataset. In Dirty dataset, Decision Tables produces the greatest accuracy of 86.77 %. There is a little significant difference of Decision Table compared to KNN and Naive Bayes with accuracies of less than 80 %. The variances are nearly 10 % differences. The worst technique that we can conclude here is Naive Bayes with 73.15 % accuracy.

As explained in Sect. 3.2, dirty dataset is then imputed and cleaned. During imputation, missing values were filled in with mean values of overall instances. Meanwhile, cleaned dataset consists of removal of outliers and extreme values. Accordingly, the performances of those three weak classifiers were evaluated. Similar performance differences was observed in Decision Table as it still outperformed the other two techniques with highest accuracy of 83.33 % in imputed dataset and 84.20 % in cleaned dataset compared to KNN and Naive Bayes as shown in Table 1.

Table 1. Performance Measure of the Classifiers

		Decision table	KNN	Naïve Bayes
Dirty data	Accuracy (%)	**86.77**	79.37	73.15
	ROC	**0.94**	0.79	0.83
	Sensitivity	0.87	0.79	0.73
	Specificity	0.13	0.21	0.27
	Precision	0.87	0.79	0.75
Imputed data	Accuracy (%)	**83.33**	76.32	73.68
	ROC	**0.93**	0.76	0.81
	Specificity	0.83	0.76	0.74
	Sensitivity	0.167	0.24	0.26
	Precision	0.83	0.76	0.75
Cleaned data	Accuracy (%)	**84.20**	76.92	72.59
	ROC	**0.93**	0.76	0.81
	Specificity	0.86	0.77	0.73
	Sensitivity	0.16	0.23	0.27
	Precision	0.84	0.77	0.74

Similarly, when we analyzed ROC as shown in Table 1, Decision Table expressively beats the other two classifiers, followed by Naive Bayes and KNN. We can infer that for all of the classifiers performance is better when dealing with dirty dataset as ROC values are higher compared to ROC values evaluated in cleaned dataset.

Precision of all classifiers are higher in dirty data as compared to cleaned data and can be referred in Table 1 as well. Decision Table's sensitivity and precision outperforms other two classifiers and has lowest specificity value supporting the fact it is best classifier.

Overall, Decision Table performs the best in all state of dataset as a weak classifier. It is quite interesting to see simple technique such as Decision Table exceed the performance of other techniques. Hall [23] has stated that sometimes simple classifier gives better accuracy based on the approach of classification algorithm. However, he also indicated that simple techniques may over fit and surpass other techniques because of data used are not informative. Even though Li [24] in his study indicated that noise is one of the factor for inconsistency of decision table, we have proved that decision table is high tolerate with noise, supported by the result found by Kalapanidas [13].

The additional remarkable outcome of this evaluation is Naive Bayes is the worst techniques for all three conditions of datasets with less than 75 % of accuracy and significantly different from others. This is probably because Naïve Bayes treat the attributes independently while they may be correlated to each other hence, this may affect the result of prediction [25].

Another useful evaluation we can observe from this outcome is most of the techniques yielded improved accuracy when dealing with dirty data compared to cleaned dataset. In overall, the techniques perform approximately more than 1 % better in dirty dataset condition. This inference may relate to the preprocessing process in which we probably have overcleaned the data.

Even though the performance result of the three classifiers is not too significant, we realized that KNN differ significantly compared to Decision Table with accuracy less than 80 %. This may be due to the performance of KNN are based on initialization of values of k or initial value of neighborhood, as explained in parameter setting subsection. Proper setting of k need to be considered when implements KNN in different dataset [26]. Thus, due to complicated setting of k, we can conclude that KNN should be avoided when dealing with some dataset.

After three months of association with Astra Zeneca, we have met the panel of experts in Astra Zeneca to evaluate our findings on the study. Experts have accepted this study although a poorer result of machine learning was produced while treating a cleaner data compared to dirty data. Astra Zeneca expert noticed that there are few reasons why the result of this project contradicted to past works done. Indicating the imputation methods used in data pre-processing procedure is one of the reasons. The usage of this simple method leads to our diverse result findings from others. Overall, the experts agreed that cleaning are not necessarily been made meticulously as it will lead to over cleaning. Simple classifiers lead to better prediction and last but not least, they suggested that different imputation methods work differently depending on the missing-ness nature.

5 Conclusion

In term of analyzing robust machine learning classifiers, we have determined that sometimes simple methods work better. This statement has been established from accuracy analysis of Decision Table as it works the best when handling balance dataset, regardless of how big or small the number of data used for training. Meanwhile, the enhanced classifier such as Naïve Bayes struggles in treating the attribute correlation of the clinical trial dataset and resulted in poor classification performance. KNN performance depends on the initialization of k and perhaps, in this study, the ROC is the worst among other two classifiers due to improper parameter initialization.

We can also infer that the result contradicts to earlier hypothesis made in which standard classifiers perform better in clean dataset compared to dirty dataset because generally all of three weak classifiers outperform in dirty dataset. There might be two reasons that resulted to this; firstly, due to simple imputation method using mean imputation and secondly, we probably over cleaned the data. We realized that these two reasons are from data processing phase and a critical study of techniques on preprocessing the data should be done.

Future studies may be needed in comparing the classifier performance by evaluating robustness measure rather than only using Accuracy, ROC Area and Precision. In addition, improved imputation method and cleaning methods should be experimented for a better data quality. We cannot disregard the involvement of clinical data scientist in criticizing the data processing phase and performance evaluation as they will assist in a better assessment instead of depending on the result of the classifier as a whole.

Acknowledgement. This paper is a part of Master Dissertation Theses written in University of Manchester, UK. We would like to thank data scientists from Advance Analytics Centre, Astra Zeneca, Alderley Park, Chesire, UK for their review, support and suggestion on this study.

References

1. Rogers, S., Girolami, M.: A First Course in Machine Learning. CRC Press, Boca Raton (2015)
2. Simon, H.A.: Applications of Machine Learning and Rule Induction (1995)
3. Gamberger, D., Lavrač, N.: Noise detection and elimination applied to noise handling in KRK chess endgame. In: International Conference Inductive Logic Programming (1997)
4. Zhu, X., Wu, X.: Class noise vs. attribute noise: a quantitative study. Artif. Intell. Rev. **22**(3), 177–210 (2004)
5. Little, R.J., D'Agostino, R., Cohen, M.L., Dickersin, K., Emerson, S.S., Farrar, J.T., Frangakis, C., Hogan, J.W., Molenberghs, G., Murphy, S.A., Neaton, J.D., Rotnitzky, A., Scharfstein, D., Shih, W.J., Siegel, J.P., Stern, H.: The prevention and treatment of missing data in clinical trials. N. Engl. J. Med. **367**(14), 1355–1360 (2012)
6. Grubbs, F.E.: Procedures for detecting outlying observations in samples (1974)
7. Gamberger, D., Lavrač, N., Duzeroski, S.: Noise detection and elimination in data preprocessing: experiments in medical domains. Appl. Artif. Intell. **14**(2), 205–223 (2000)
8. Van Hulse, J., Khoshgoftaar, T.: Knowledge discovery from imbalanced and noisy data. Data Knowl. Eng. **68**(12), 1513–1542 (2009)
9. Zhu, X., Wu, X., Chen, Q.: Eliminating class noise in large datasets. In: ICML, pp. 920–927 (2003)
10. Hall, M.A.: Correlation-based feature selection for machine learning. Methodology **21i195–i20**, 1–5 (1999)
11. Quinlan, J.R.: Induction of decision trees. Mach. Learn. **1**(1), 81–106 (1986)
12. Zupan, B., Demšar, J., Kattan, M.W., Beck, J.R., Bratko, I.: Machine learning for survival analysis: a case study on recurrence of prostate cancer. Artif. Intell. Med. **20**(1), 59–75 (2000)
13. Kalapanidas, E., Avouris, N., Craciun, M., Neagu, D.: Machine learning algorithms: a study on noise sensitivity. In: Proceedings of the 1st Balcan Conference on Informatics, pp. 356–365, October 2003
14. Vannucci, M., Colla, V., Cateni, S.: An hybrid ensemble method based on data clustering and weak learners reliabilities estimated through neural networks. In: Rojas, I., Joya, G., Catala, A. (eds.) IWANN 2015. LNCS, vol. 9095, pp. 400–411. Springer, Heidelberg (2015)
15. Rokach, L.: Ensemble-based classifiers. Artif. Intell. Rev. **33**(1–2), 1–39 (2010)
16. Maclin, R., Opitz, D.: Popular Ensemble Methods: An Empirical Study, arXiv.org, vol. cs.AI, pp. 169–198 (2011)
17. Dietterich, T.G.: Ensemble methods in machine learning. In: Kittler, J., Roli, F. (eds.) MCS 2000. LNCS, vol. 1857, pp. 1–15. Springer, Heidelberg (2000)
18. Kohavi, R.: The power of decision tables. In: Machine Learning, ECML 1995, pp. 174–189 (1995)
19. Wets, G., Vanthienen, J., Timmermans, H.: Modelling decision tables from data. In: Wu, X., Kotagiri, R., Korb, K.B. (eds.) PAKDD 1998. LNCS, vol. 1394. Springer, Heidelberg (1998)
20. John, G.H.G., Langley, P.: Estimating continuous distributions in Bayesian classifiers. In: Proceedings of the Conference on Uncertainty in Artificial Intelligence, Montreal, Quebec, Canada, vol. 1, pp. 338–345 (1995)
21. Aha, D.W., Kibler, D., Albert, M.K.: Instance-based learning algorithms. Mach. Learn. **6**(1), 37–66 (1991)

22. Zweig, M.H., Campbell, G.: Receiver-operating characteristic (ROC) plots: a fundamental evaluation tool in clinical medicine. Clin. Chem. **39**(4), 561–577 (1993)
23. Witten, I.H., Frank, E., Hall, M.A.: Data Mining: Practical Machine Learning Tools and Techniques (Google eBook) (2011)
24. Li, M., Shang, C., Feng, S., Fan, J.: Quick attribute reduction in inconsistent decision tables. Inf. Sci. (Ny) **254**, 155–180 (2014)
25. Tomar, D., Agarwal, S.: A survey on data mining approaches for healthcare. Int. J. Bio-Sci. Bio-Technol. **5**(5), 241–266 (2013)
26. Everitt, B.S., Landau, S., Leese, M., Stahl, D.: Miscellaneous Clustering Methods (2011)

A Review of Feature Extraction Optimization in SMS Spam Messages Classification

Kamahazira Zainal[✉] and Mohd Zalisham Jali

Faculty of Science and Technology (FST), Universiti Sains Islam Malaysia (USIM),
71800 Nilai, Negeri Sembilan, Malaysia
kamahazira@raudah.usim.edu.my, zalisham@usim.edu.my

Abstract. Spam these days has become a definite nuisance to mobile users. Provision of Short Messages Services (SMS) has been intruded, in line with an advancement of mobile technology by the emergence of SMS spam. This issue has not only cause distressing situation but also other serious threats such as money loss, fraud, and false news. The focus of this study is to excavate the features extraction in classifying SMS spam messages at users' end. Its objective is to study the discriminatory control of the features and considering its informative or influence factor in classifying SMS spam messages. This study has been conducted by gathering research papers and journals from numerous sources on the subject of spam classification. The discovery offers a motivational effort for further execution in a wider perspective of combating spam such as measurement of spam's risk level.

Keywords: SMS spam · Spam classification · Spam filtering · Spam feature extraction · Feature extraction review

1 Introduction

Spam filtering in email system has been in a mature study that has improved with the use of available and advanced tools. However, this problem of spam is likely to be an unending conflict since it has evolved and rapidly developed into another medium such as SMS, social media and instant messaging. The lack of helpful tools to aid users in making a decision on how to respond to spam has elevated the records of losses in terms of financial, reputation and trustworthy [1–3].

The issue of spam has been pervasively spread in our daily communication channels. The mobile technology that revolutionizes the world of communication has revealed a new threat. As reported in media such in security bulletin [4–6], spam has its influence factor since its content might deceptive and lead users to severe effects. Despite many very good efforts, the amount of spam sent and received daily continues from rising. This rising amount also highly expected is occurring due to the advancement of mobile technology [1, 6].

Spam messages could be a potential expansion of internet crime. The evolution of typical mobile phone to a smartphone that comes with more intelligent and exciting

© Springer Nature Singapore Pte Ltd. 2016
M.W. Berry et al. (Eds.): SCDS 2016, CCIS 652, pp. 158–170, 2016.
DOI: 10.1007/978-981-10-2777-2_14

features such as mobile web browser and wireless fidelity or Wi-Fi support would bring the higher risk of spam's impact.

Spam used to be a platform to advertise a sale on products, but now it has become a tool to propagate lies, slander, and profanity and even has become a medium for criminal's activities. The potential loss and risk currently have been recorded as high and it is keeping increasing. Latest identified threat that derived from SMS spam known as SMiShing is a form of phishing that tends to deceive users into disclosing confidential information. A study by Yeboah-Boateng et al. [7] revealed that this kind of attack will cause a major and even catastrophic impact.

This paper essentially accumulates findings of SMS spam classification which include the most employed and accomplished feature extraction. This result then will be further enhanced as to improve the quality of suggested features. It is expected that this study will not only intend to differentiate spam and ham but also to predict the potential risk of a spam that later on might assist the user to decide on how to respond towards spam wisely.

The remaining of this paper is highlighted as follows. Section 2 addresses related works in SMS spam that include its evolution, type and spam management. Section 3 verifies a deeper detail of feature extraction employed in previous work. The following section then focuses on the type of available SMS messages corpus that can be deployed in this particular study. Section 5 elaborates the most selected features in depth and finally the conclusion and future works are discussed.

2 Literature Review

2.1 Evolution of Text Spam

Spam prevalently referred to something that is mostly unwanted and unsolicited bulk messages which commonly used to advertise a service or a product. Nevertheless, since last decade spam has been emerged as a tool to spread false news, or even an offer that has raised a high risk and hazardous situation as such money loss. Nowadays spam is no longer limited to emails and web pages and the increasing penetrations of spam in many forms have started becoming a nuisance and potential threat. This issue seems will likely be an unending conflict.

Testing has been done to prove that the same methodology and technique used in email system cannot be implemented effectively in combating spam for mobile application [8]. This is due to the different characteristic attributes in email spam and mobile spam. Further testing also has been executed by running SMS messages through email filtering tool as to distinguish spam from valid SMS messages. As done by Narayan et al. [9], they verified that the most outstanding spam filter used to detect email spam cannot be used in classifying SMS spam messages. In addition, via this testing also it is shown that some modification can be made to build email spam filters that are fit appropriately to be used in classifying SMS spam. As email spam has known as a conventional spam problem, Vural et al. [10] articulated that there is few more type of spam that has existed along with the technology advancement. This has to make spam as a co-evolutionary problem. These include SMS spam, VoIP spam, comment spam and messaging spam.

Comment spam usually imposed at the section of newspaper websites, where adverts are inserted in the comments section. While messaging spam, also known as SPam over Instant Messaging (SPIM) is the type that would be received over an instant messenger application. Voice Over Internet Protocol (VOIP) spam is the type of spam that could be received through automated voice messages over the VOIP phone. Mobile spam or also known as SMS spam is spam received on one's mobile device in the form of Short Messaging Services.

Studies also discussed on the short text messages which are not limited to mobile SMS, but also include instant messaging, blog comments, email summaries, and web social media spam. This study of short text messages can be referred to Mosquera et al. [11], Cormack et al. [12], Xia et al. [13] and Song et al. [14].

A study by Cormack et al. [15] suggested that the same feature of content-based extraction gave different empirical results for the different type of short messages. The findings somehow showed that a different extraction of features is needed in classifying spam for the different type of short messages. Hence, a study focus on every type of short messages is required. This paper will focus on one of the types; SMS spam because of its severe impact that has been recorded requires serious attention. Additionally, the dataset for the experiment is easily reached that make this research is even more expedient and practical.

2.2 Management of Spam

Technically, there is a solution available for users to be applied according to on how to react against spam. The most solution proposed to move SMS messages that are identified as spam to another folder besides Inbox, such as Spam box or merely delete it. Observing this issue that demands appropriate attention, a work in Zainal et al. [16] has elaborated that managing spam would include a few phases: classification, clustering, determination of its severity level and response towards spam.

In other works, as for solution and proof of performance, research has been implemented up to mobile devices level. Balubaid et al. [1] proved their proposed solution using Android platform. This application namely as SMS Controller has the feature of managing SMS which includes an auto-reply and spam filtering that can be activated at any time convenient to users. The simulation experiment result shows SMS Controller that has been developed is effective both in terms of efficiency and time. Sethi et al. [2] also have shown their proposed solution effectiveness in SMS spam filtering on Android platform. Simulation utilized a small data set of 100 SMS messages that has been employed during the training phase and another 100 SMS messages for testing purpose. Authors used specificity and sensitivity rate of spam detection to measure the effectiveness of spam filtering. Results showed that those 2 measurements returned a high value indicated that a well successful filtering method. In other work, a real-time mobile application for Android based mobile phones has been developed by Uysal et al. [3]

utilizing the proposed 2 different spam filtering schemes. SMS spam messages are silently filtered out without disturbing users.

The issue of combating spam is possible to be done technically. However, another strong factor that could bring spam to even more hazardous impact is the human's reaction. Even though the technology for spam detection is getting better, this issue still persists and cause relatively high adverse impact in the daily life of everyone globally.

A study done in 2005 by Sapronov [17] shows that people, who are a part of any system, are always going to be the weakest point in a security chain. They tend to open spam messages, suspicious attachments or even click on given unknown URL link because the factor of negligence or lacking of awareness, carelessness, and curiosity. In another study, Emm [18] suggested that there is a critical need to patch human vulnerabilities via education and raise awareness of security implications and potential online threats. Yeboah-Boateng *et al.* [7] examined that users found to be unwary or oblivious of the numerous cyber attacks against their mobile devices.

In some countries, laws enforcement has been seriously executed such as in China and United Kingdom. In early 2014, China authorities have arrested 1,530 suspects in SMS spam attack. China's Ministry of Public Security said that it had recognized 3,540 cases of suspected crimes; which include a case where a Liaoning Province gang is suspected of sending out over 200 million spam messages [19]. While in the United Kingdom [20], a number of mobile network operators have collaborated with the U.K. Information Commissioner's Office try to prevent and penalize companies that are sending spam messages.

2.3 Effect of Spam

Spam is changing and it has reduced as a traditional advertising medium but evolved into more fraud, malware, and phishing that involved criminal activities. A study done by Kaspersky [4] showed that at this time, phishing attacks are moving from an illegal access of banking accounts to email and social networking.

Mobile networks are now well integrated with the Internet; therefore it is facing the equal threats as Internet. The serious needs of SMS spam filtering are mentioned by Delany *et al.* [21] and it is relatively acquired to the state of mobile phone technology.

Zalpuri *et al.* [22] described that SMS spam could lead to serious threats such as message disclosure, man-in-the-middle-attack and also viruses that potentially caused a critical impact. Other than that, Nuruzzaman *et al.* [23] defined SMS spam as annoying, wasting time and abused personal privacy.

Furthermore, a study by Nikiforakis *et al.* [24] shown that URL link usually is partial of spam content used by malicious advertisers to launch drive-by downloads, convince users to install malicious software like ransomware or redirect them to another different page consists of inappropriate ads such as adults advertisement. These ad-based URL links commonly use shortening services.

3 Feature Extractions: Related Works

Spam has evolved in many aspects such as its platform diverges from email to website spam; the impact of spam also has been more hostile which extended to criminal activities. Since past decade, SMS spam has shown a tremendous increment together with an advancement of the mobile phone. The ability of nowadays mobile phone or well known as a smartphone which its capability to integrate with the Internet network once connected, increase the risk of SMS spam even higher.

SMS are usually shorter than email messages. Only 160 characters (which 1 character is 7 bits) or also equal to 140 bytes (which 1 byte is 8 bits); are allowed in a standard text SMS. This fewer words could be a problem in analyzing messages because there is less information to work with. In addition to that, users tend to use acronyms when writing SMS and abbreviations used by SMS users are not standard for a language and based on colloquial of those users communities. A research was done by Almeida et al. [25] suggested that these could affect the filters accuracy.

Narayan et al. [9] pointed out that SMS spam filtering can be applied at two points; in a telecommunication network or also known as SMS Service Centre (SMSC) and at the end systems or mobile devices. Most SMSC using the non-content features technique to recognize spam, prior sending the SMS messages to users. This type of service-side solution is discussed in Xu et al. [26] whereby elements such as a number of messages, message size, and time of day are the partial consideration in determining SMS messages as a spam. However, an employment of content features technique also showed promises solution, as validated in Belém et al. [27]. Authors of this study have combined the features of Bayesian classifier and word grouping. The results demonstrated that the proposed features reached true spam index detection at a rate of 99.9 %.

As described in the same paper of Narayan et al. [9], they gathered 12 SMS spam filtering apps from Google Play and tested its effectiveness. Results from their empirical study showed that these apps ineffectively able to detect spam in SMS messages since the highest accuracy were only 56 %. Therefore, a further study is needed to design and develop a tool for the implementation of SMS spam filtering.

A work done by Mathew et al. [28] compared 32 of algorithms in classifying SMS messages by using Weka.[1] Their empirical results showed that Naïve Bayesian has detected spam in SMS messages with the highest accuracy among these algorithms. A similar experiment also has been done by Zainal et al. [29] using both Weka and Rapid-Miner[2] as a tool. Surprisingly, their results via these both tools suggested that SVM was the best algorithm in recognizing spam instead of Naïve Bayesian, although both papers stated using the same collection of SMS messages corpus.

The same findings also resulted in a work done by Almeida et al. [25]. In this experiment, SVM outperforms other 14 algorithms with the highest accuracy rate at 97.5 % with combining of 2 additional features (token1 and token2). A year later, authors produced another paper with similar technique but with an additional 3 more algorithms

---

[1] http://www.cs.waikato.ac.nz/ml/weka/.
[2] https://rapidminer.com/.

tested in the experiment [30]. As shown in their previous work, SVM still demonstrated the best performance and remarkable accuracy rate in spam filtering.

A review of SMS spam detection techniques by Warade *et al.* [31] has simplified 12 type of research that cover up almost all available techniques in spam filtering that include the content feature, lexical semantics, stylistic feature and bioinformatics pattern matching algorithm. All possible techniques that have been verified for its effectiveness in spam filtering have been discussed. A year later, a survey done by Zalpuri *et al.* [22], not only treat spam as one of the SMS threat but it is also identified prone to other several attacks such as message disclosure, man-in-the-middle attack and viruses. In this review, SVM has been suggested for classifying spam messages.

A content feature usually consists of spam keywords, URL links, monetary value, special characters, emotion symbols and function words. Non-content features consider message metadata such as length, the number of characters, white spaces, the number of terms, date, time and location wise. The summary review for utilization of features extraction and tools that have been studied in SMS spam filtering is tabulated in Fig. 1. This figure is not intended to be all-inclusive but rather to illustrate the most widely used of feature extraction and tools in SMS spam filtering.

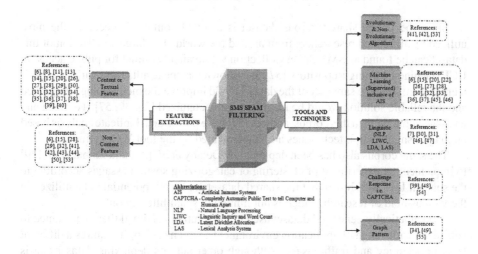

Fig. 1. Simplification of SMS spam filtering methodology which includes type of feature extractions and tools.

4   Data Set of SMS Messages Corpus

The dataset is a mandatory requirement for any research as to be deployed with further experiment. Depending on the nature of research problem, the reliable dataset is crucial for any scientific research. As described in Oda [56], a good corpus of the dataset for spam experiment definitely required in order to verify its accuracy detection empirically.

With respect to the research related to SMS messages, there is a few type of dataset sources that can be categorized into 3 groups; public shared dataset, synthetic dataset and own collection by researchers. This categorization is identified through reviewing of many research papers related to SMS spam classification. This can be simplified in Fig. 2.

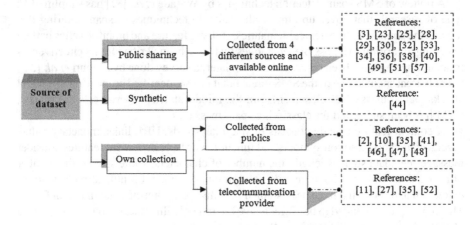

Fig. 2. Available sources of SMS messages corpus.

Since the public shared corpus dataset is available online, it becomes the most utilized by all related researchers from around the world. The largest collection of this dataset can be found as SMS Spam Collection v.1, available online for public access at UCI Machine Learning Repository [57]. It is shown that the results from various experiments by utilizing the same set of the dataset could improve the methodology employed and the reliable solution would be identified and recognized [25, 30, 57]. To avoid any message duplication in this corpus, authors [57] have performed duplicate analysis based on plagiarism detection techniques and implemented by the tool WCopyfind.[3]

The same corpus also has been deployed by Delany *et al.* [21] and Zainal *et al.* [29] to implement an experiment of clustering or categorizing spam messages according to the subject. It is also shown that the shared dataset are highly potential to be utilized in the various field of research, which is not limited to spam filtering only.

As synthetically generated dataset used in [44], authors claimed that they wanted to consider a wide range of application environments, which each of it requires a different level of accuracy and traffic usage. Although other authors deploying dataset that is gathered from public or telecommunication provider is to coordinate the changing nature of SMS spam. Consequently, this also expected to create a robust model by utilizing a bigger size of the dataset.

5   In-depth of Most Selected Features: Content-Based a.k.a Textual Feature

A proof of performance study is conducted to understand how this content-based feature or also known as textual feature assisting in classifying spam messages. Implemented

---

[3] http://plagiarism.bloomfieldmedia.com/wordpress/software/wcopyfind/.

experiment is intentional to investigate the effect of particular term or spam word in recognizing spam messages, with and without pre-processing phase. RapidMiner tool with version 5.3.015 is used to run the message corpus and finally derived the frequency of occurrence for every term in the corpus.

It is apparent that as illustrated in Fig. 1, the content-based feature is the most widely used and giving remarkable result while the utilization of machine learning is the most supportive tool in this work of spam filtering. This experiment is solely utilizing the content of the message.

One of the most important requirements for the proper evaluation of behavioral algorithms for the classification of spam is a realistic data set. The same collection of SMS corpus that is publicly shared for researchers has been deployed in this clear-cut experiment. As aforementioned in the previous section, this dataset can be accessed at UCI Machine Learning Repository [57]. This set of corpus consists of 5,574 SMS messages which include labeled messages with 4,827 for ham messages and 747 for spam messages. These bulks of messages then run in RapidMiner in order to count its term occurrences. Initially, as common practice in text mining, all messages have been cleaned in text pre-processing phase. This particular phase is also known as a pre-treatment process; whereby this phase requires an extraction of unique characteristic to distinguish between a spam and ham. As elaborated in Zhang et al. [53], this phase is extremely important as it has direct relation and effect with the quality of classification outcome.

However, when the same dataset is executed without the pre-processing phase, the result for word's spam probabilities become dissimilar from the previous testing.

Table 1. Comparative analysis of spam probability value for the dataset, at the initial stage, has been running with pre-processing and without pre-processing phase.

Stem word	Term in SMS message	With pre-processing			Without pre-processing		
		Ham	Spam	Spam probability	Ham	Spam	Spam probability
free	FREE	–	–	–	0	87	1.0000
	Free	–	–	–	4	34	0.8947
	free	40	163	0.8029	36	42	0.5384
Total occurrences		203			203		
claim	CLAIM	–	–	–	0	1	1.0000
	Claim	–	–	–	0	54	1.0000
	claim	0	82	1.0000	0	26	1.0000
	claims	–	–	–	0	1	1.0000
Total occurrences		82			82		
call	CALL	–	–	–	7	14	0.6666
	Call	–	–	–	13	95	0.8796
	Calls	–	–	–	0	9	1.0000
	call	213	264	0.5534	149	137	0.4790
	calls	–	–	–	10	6	0.3750
	called	–	–	–	22	0	0
	calling	–	–	–	12	3	0.2
Total occurrences		477			477		

Surprisingly, the result has returned even a significant value. For example, word **FREE** giving value 1.0 for spam probability, but **Free** resulted as 0.89 in value and **free** as 0.54. It has been discussed earlier in Almeida *et al.* [25] and Almeida *et al.* [30] that pre-processing phase could weaken the value of accuracy for spam classification. That particular claim could be confirmed with the outcome of this testing. A further quantitative finding is illustrated in Table 1.

With reference to Sethi *et al.* [2] and Zhang *et al.* [53], probability of a word that possibly present in spam messages can be measured as follows:

$$S_p = f_1 / (f_1 + f_2) \tag{1}$$

where,

S_p = spam probabilities
f_1 = word of frequency in spam messages
f_2 = word of frequency in ham messages

Every term is measured using Eq. (1) of spam probability value which does mean as follows:

- 1.00, this value means that term uniquely only exist in spam messages and can be considered as spam words;
- 0.99–0.51, the term exists both in spam and ham messages but the frequency of occurrence in spam still higher than ham messages;
- 0.50 considering the term exist with the same frequency in both ham and spam messages;
- 0.01–0.49 indicate that the term occurred more in ham messages than spam messages; and
- 0.00 explained that none of the term exists in spam messages.

The present findings also suggest that the precise combination of spam words in a spam message with the right value of weightage will influence the tendency of a message to be identified as spam. The accuracy rate of spam detection then will be determined by the employed classifier.

Noticeably from Table 1, a message that has been running with pre-processing and without pre-processing will give a different spam probability value, which this can be recognized as a significant different. As for example, terms **free** contributing 3 significant values (without pre-processing) as to assess the spam's probability of a message, compared to the value of spam probability for the same term but has been run with pre-processing at the initial phase.

This discovery also potentially to be use as feature construction for determination of spam risk level that needs to be further analyzed for proof of concept. It is expected that further research of this paper will be an extension part from previous related works in [16, 58], as to complete the cycle of spam management that involve 3 stages; classification of spam, clustering or categorize the spam and determination of risk level in SMS spam messages. Instead of using classifier such as SVM and NB, a Dendritic Cell

Algorithm (DCA) from the Danger Theory of Artificial Immune Systems (AIS) will be applied to assess the severity of identified risk in a spam message.

6 Conclusion

There are a variety of works that has been done in reviewing the SMS spam detection efficiency. Identifying distinctive features are most likely to be crucial in SMS spam classification. This review has encountered a number of feature types in SMS spam filtering which have resulted at a different level of accuracy. It is also found that the spam filter for email is less practical for SMS messages. Even though there are already available researches on this subject, a comparative study would be beneficial to the real world application and substantiate the earlier findings. A proof of performance testing has been conducted to have a better perspective of spam words influence in recognizing spam messages.

As for the content-based filtering, it is believed that relevant feature able to increase the classification confidence. The contribution of this study is obvious as the resulting outcomes can be capitalized as guidelines to determine the level of spam's risk or impact. We are fully aware that more conclusive results are required to determine the effectiveness of this claim. This proposed feature seems very appealing and will be proven via further experiments in our future works. Other than reliable feature construction, it is also notable that usable dataset or SMS messages corpus are definitely crucial in this research.

Acknowledgement. This research is fully funded by the Ministry of Higher Education of Malaysia and Research Management Centre of USIM via grant research with code USIM/FRGS/FST/32/50315.

References

1. Balubaid, M.A., Manzoor, U., Zafar, B., Qureshi, A., Ghani, N.: Ontology based SMS controller for smart phones. Int. J. Adv. Comput. Sci. Appl. **6**(1), 133–139 (2015)
2. Sethi, G., Bhootna, V.: SMS spam filtering application using Android. Int. J. Comput. Sci. Inf. Technol. (IJCSIT) **5**(3), 4624–4626 (2014)
3. Uysal, A.K., Gunal, S., Ergin, S., Gunal, E.S.: The impact of feature extraction and selection on SMS spam filtering. IEEE **19**(5), 67–72 (2013)
4. Gudkova, D.: Kaspersky Security Bulletin-Spam Evolution 2013 (2013)
5. Gohring, N.: Handful of Spam. Newsfront (2001)
6. SophosLabs: Android: Today's biggest target. Sophos Security Threat report 2013 (2013)
7. Yeboah-Boateng, E.O., Amanor, P.M.: Phishing, SMiShing & Vishing: an assessment of threats against mobile devices. J. Emerg. Trends Comput. Inf. Sci. **5**(4), 297–307 (2014)
8. Hidalgo, J.M.G., Bringas, G.C., Sánz, E.P.: Content based SMS spam filtering (2003)
9. Narayan, A., Saxena, P.: The curse of 140 characters: evaluating the efficacy of SMS spam detection on Android. ACM (2013)
10. Vural, I., Venter, H.S.: Combating mobile spam through Botnet detection using artificial immune systems. J. Univ. Comput. Sci. **18**(6), 750–774 (2012)

11. Mosquera, A., Aouad, L., Grzonkowski, S., Morss, D.: On detecting messaging abuse in short text messages using linguistic and behavioral patterns. Arxiv - Social Media Intelligence (2014). http://arxiv.org/abs/1408.3934

12. Cormack, G.V., Gómez Hidalgo, J.M., Sánz, E.P.: Spam filtering for short messages. In: Proceedings of the 16th ACM - Conference on Information and Knowledge Management - CIKM 2007 (2007). doi:10.1145/1321440.1321486

13. Xia, H., Fu, Y., Zhou, J.: Intelligent spam filtering for massive short message stream. Int. J. Comput. Math. Electr. Electron. Eng. **32**(2), 586–596 (2013). doi: 10.1108/03321641311296963

14. Song, G., Ye, Y., Du, X., Huang, X., Bie, S.: Short text classification: a survey. J. Multimedia **9**(5), 635–643 (2014). doi:10.4304/jmm.9.5

15. Cormack, G.V, Hidalgo, J.M.G., Sánz, E.P.: Feature engineering for mobile (SMS) spam filtering. In: SIGIR, pp. 1–2 (2007)

16. Zainal, K., Zalisham, M.: A perception model of spam risk assessment inspired by danger theory of artificial immune systems. In: International Conference on Computer Science and Computational Intelligence (ICCSCI), vol. 59, pp. 152–161 (2015). doi:10.1016/j.procs.2015.07.530

17. Sapronov, K.: The human factor and information security. Securelist (2005)

18. Emm, D.: Patching human vulnerabilities. Securelist (2010)

19. Farrell, N.: China arrests mobile spammers (2014)

20. Ricknas, M.: UK mobile operators join forces to combat text spam (2014)

21. Delany, S.J., Buckley, M., Greene, D.: SMS spam filtering: methods and data. Expert Syst. Appl. **39**(10), 9899–9908 (2012). Elsevier

22. Zalpuri, N., Arora, M.: An efficient model for S.M.S security and SPAM detection: a review. Int. J. Comput. Sci. Eng. (IJCSE) **3**(12), 1–6 (2015)

23. Nuruzzaman, M.T., Lee, C., Choi, D.: Independent and personal SMS spam filtering. In: IEEE International Conference on Computer and Information Technology, pp. 429–435 (2011). doi:10.1109/CIT.2011.23

24. Nikiforakis, N., Maggi, F., Stringhini, G., Rafique, M.Z.: Stranger danger: exploring the ecosystem of Ad-based URL shortening services. ACM, April 2014. doi: 10.1145/2566486.2567983

25. Almeida, T.A., Gomez, J.M., Yamakami, A.: Contributions to the study of SMS spam filtering: new collection and results. ACM (2011)

26. Xu, Q., Xiang, E.W., Yang, Q., Du, J., Zhong, J.: SMS spam detection using noncontent features. IEEE Intell. Syst. **27**(6), 44–51 (2012)

27. Belém, D., Duarte-Figuriredo, F.: Content filtering for SMS systems based on Bayesian classifier and word grouping. IEEE (2011)

28. Mathew, K., Issac, B.: Intelligent spam classification for mobile text message. In: International Conference on Computer Science and Network Technology, pp. 101–105 (2011)

29. Zainal, K., Sulaiman, N.F., Jali, M.Z.: An analysis of various algorithms for text spam classification and clustering using RapidMiner and Weka. Int. J. Comput. Sci. Inf. Secur. (IJCSIS) **13**(3), 66–74 (2015). http://sites.google.com/site/ijcsis/

30. Almeida, T.A., María, J., Hidalgo, G., Silva, T.P.: Towards SMS spam filtering: results under a new dataset. Int. J. Inf. Secur. Sci. **2**(1), 1–18 (2012)

31. Warade, S.J., Tijare, P.A., Sawalkar, S.N.: An approach for SMS spam detection. Int. J. Res. Advent Technol. **2**(12), 8–11 (2014)

32. Al-Hassan, A.A., EL-Alfy, E.-S.M.: Dendritic cell algorithm for mobile phone spam filtering. In: International Conference on Ambient Systems, Networks and Technologies (ANT), vol. 52, pp. 244–251 (2015)
33. Tan, H., Goharian, N., Sherr, M.: $100,000 Prize Jackpot. Call Now! Identifying the pertinent features of SMS spam. In: SIGIR, pp. 1–2 (2012)
34. Sulaiman, N.F., Jali, M.Z.: A new SMS spam detection method using both content-based and non content-based features. In: Sulaiman, H.A., Othman, M.A., Othman, M.F.I., Rahim, Y.A., Pee, N.C. (eds.) Advanced Computer and Communication Engineering Technology, LNEE, vol. 362, pp. 505–514. Springer, Switzerland (2015)
35. Mujtaba, G., Yasin, M.: SMS spam detection using simple message content features. J. Basic Appl. Sci. Res. 4(4), 275–279 (2014)
36. Karami, A., Zhou, L.: Exploiting latent content based features for the detection of static SMS spams. In: Proceedings of the American Society for Information Science and Technology, March 2014. doi:10.1002/meet.2014.14505101157
37. Karami, A., Zhou, L.: Improving static SMS spam detection by using new content-based features. In: 20th Americas Conference on Information Systems, pp. 1–9 (2014)
38. Mahmoud, T.M., Mahfouz, A.M.: SMS spam filtering technique based on artificial immune system. Int. J. Comput. Sci. Issues (IJCSI) 9(2), 589–597 (2012)
39. Khemapatapan, C.: Thai-English spam SMS filtering. In: IEEE - Asia Pacific Conference on Communications (APCC), pp. 226–230 (2010)
40. Rafique, M.Z.: Graph-based learning model for detection of SMS spam on smart phones. IEEE 53(9), 1689–1699 (2012). doi:10.1017/CBO9781107415324.004
41. Yadav, K., Kumaraguru, P., Goyal, A., Gupta, A., Naik, V.: SMSAssassin: crowdsourcing driven mobile-based system for SMS spam filtering. In: Proceedings of the 12th ACM Workshop on Mobile Computing Systems and Applications - HotMobile, USA, pp. 1–6 (2011). doi:10.1145/2184489.2184491
42. Yadav, K., Saha, S.K., Kumaraguru, P., Kumra, R.: Take control of your SMSes: designing an usable spam SMS filtering system. In: IEEE International Conference on Mobile Data Management, pp. 352–355 (2012). doi:10.1109/MDM.2012.54
43. Uysal, A.K., Gunal, S., Ergin, S., Gunal, E.S.: A novel framework for SMS spam filtering. IEEE (2012)
44. Yoon, J.W., Kim, H., Huh, J.H.: Hybrid spam filtering for mobile communication. Comput. Secur. 29(4), 446–459 (2010). doi:10.1016/j.cose.2009.11.003
45. Cao, L., Nie, G.H., Liu, P.F.: Ontology-based spam detection filtering system. IEEE, pp. 282–284 (2011)
46. Junaid, M.B., Farooq, M.: Using evolutionary learning classifiers to do mobile spam (SMS) filtering. In: GECCO, pp. 1795–1801 (2011)
47. Rafique, M.Z., Alrayes, N., Khan, M.K.: Application of evolutionary algorithms in detecting SMS spam at access layer. In: GECCO-2011: Proceedings of the 13th Annual Genetic and Evolutionary Computation Conference, pp. 1787–1794 (2011). doi:10.1145/2001576.2001816
48. Androulidakis, I., Stefan, M.P.S.J., Vlachos, V., Papanikolaou, A.: Spam goes mobile: filtering unsolicited SMS traffic. In: 2012 20th Telecommunications Forum, TELFOR 2012 - Proceedings, vol. 7, pp. 1452–1455 (2012). doi:10.1109/TELFOR.2012.6419492
49. Iyer, K.B.P., Shanthi, V.: Privacy preferences for Geo-Calendar based SMS using intelligent text configurator. Int. J. Comput. Commun. Eng. 2(1), 82–85 (2013)
50. Rafique, M.Z., Farooq, M.: SMS spam detection by operating on byte-level distributions using Hidden Markov Models (HMMS). In: Virus Bulletin Conference, pp. 1–7 (2010)

51. El-Alfy, E.-S.M., Alhasan, A.A.: Spam filtering framework for multimodal mobile communication based on dendritic cell algorithm. In: Future Generation Computer Systems. Elsevier B.V (2016). doi:10.1016/j.future.2016.02.018
52. Zhang, J., Li, X., Xu, W., Li, C.: Filtering algorithm of spam short messages based on artificial immune system. IEEE, pp. 195–198 (2011)
53. Zhang, H., Wang, W.: Application of Bayesian method to spam SMS filtering. IEEE, pp. 1–3 (2009)
54. Shirali-Shahreza, M.H., Shirali-Shahreza, M.: An anti-SMS-spam using CAPTCHA. In: International Colloquium on Computing, Communication, Control and Management, pp. 318–321 (2008). doi:10.1109/CCCM.2008.247
55. Zhao, Y., Zhang, Z., Wang, Y., Liu, J.: Robust mobile spamming detection via graph patterns. In: 21st International Conference on Pattern Recognition (ICPR 2012), pp. 983–986 (2012)
56. Oda, T.: A Spam-Detecting Artificial Immune System (2005)
57. Almeida, T.A., Hidalgo, J.M.G.: UCI Machine Learning Repository (2012). http://archive.ics.uci.edu/ml/datasets/SMS+Spam+Collection#
58. Sulaiman, N.F., Jali, M.Z.: Integrated mobile spam model using artificial immune system algorithms. In: Knowledge Management International Conference (KMICe) (2014)

Assigning Different Weights to Feature Values in Naive Bayes

Chang-Hwan Lee[✉]

Department of Information and Communications, DongGuk University, Seoul, Korea
chlee@dgu.ac.kr

Abstract. Assigning weights in features has been an important topic in some classification learning algorithms. While the current weighting methods assign a weight to each feature, in this paper, we assign a different weight to the values of each feature. The performance of naive Bayes learning with value-based weighting method is compared with that of some other traditional methods for a number of datasets.

Keywords: Feature weighting · Feature selection · Naive Bayes · Kullback-Leibler

1 Introduction

In some classifiers, the algorithms operate under the implicit assumption that all features are of equal value as far as the classification problem is concerned. However, when irrelevant and noisy features influence the learning task to the same degree as highly relevant features, the accuracy of the model is likely to deteriorate. Since the assumption that all features are equally important hardly holds true in real world application, there have been some attempts to relax this assumption in classification. Zheng and Webb [1] provide a comprehensive overview of work in this area.

The first approach for relaxing this assumption is to combine feature subset selection with classification learning. It is to combine a learning method with a preprocessing step that eliminates redundant features from the data. Feature selection methods usually adopt a heuristic search in the space of feature subsets. Since the number of distinct feature subsets grows exponentially, it is not reasonable to do an exhaustive search to find optimal feature subsets.

Another major way to help mitigate this weakness, feature independence assumption, is to assign weights to important features in classification. Since features do not play the same role in many real world applications, some of them are more important than others. Therefore, a natural way to extend classification learning is to assign each feature different weight to relax the conditional independence assumption. Feature weighting is a technique used to approximate the optimal degree of influence of individual features using a training set. While feature selection methods assign 0/1 values as the weights of features, feature weighting is more flexible than feature subset selection by assigning continuous weights.

© Springer Nature Singapore Pte Ltd. 2016
M.W. Berry et al. (Eds.): SCDS 2016, CCIS 652, pp. 171–179, 2016.
DOI: 10.1007/978-981-10-2777-2_15

When successfully applied, important features are attributed a high weight value, whereas unimportant features are given a weight value close to zero. There have been many feature weighting methods proposed in the machine learning literature, mostly in the domain of nearest neighbor algorithms [2]. They have significantly improved the performance of classification algorithms.

In this paper, we propose a new paradigm of weighting method, called *value weighting* method. While the current weighting methods assign a weight to each *feature*, we assign a weight to each *feature value*. Therefore, the value weighting method is a more fine-grained weighting method than the feature weighting method. The new value weighting method provides a potential of expanding the expressive power of classification learning, and possibly improves its performance. Since features weighting methods improve the performance of the classification learning, we will investigate whether assigning weights to feature *values* can improve the performance even further.

The main contribution of this paper is to provide a new paradigm of weighting method in classification learning. While there have been a few work focused on feature weighting in the literature, to our best knowledge, there has been no work which assigns different weight to each feature value in classification leaning. We extended the current hypothesis space of classification learning into next level by introducing a new set of weight space to the problem.

In this paper, we study the value weighting method in the context of naive Bayesian algorithm. We have chosen naive Bayesian algorithm as the template algorithm since it is one of the most common classification algorithms, and many researchers have studied the theoretical and empirical results of this approach. It has been widely used in many data mining applications, and performs surprisingly well on many applications [3].

There have been only a few methods for combining feature weighting with naive Bayesian learning [4–6]. The feature weighting methods in naive Bayes are known to improve the performance of classification learning. We will investigate, in this paper, whether the value weighting method provides enhanced performance in the context of naive Bayes.

2   Related Work

A number of approaches have been proposed in the literature of feature selection [7–10]. While feature selection methods assign 0/1 value to each feature, feature weighting assigns a continuous value weight to each feature. Feature weighting can be viewed as learning bias, and many feature weighting methods have been applied mostly to nearest neighbor algorithms [2]. While there have been many research for assigning feature weights in the context of nearest neighbor algorithms, very little work of weighting features is done in naive Bayesian learning.

In this section, we focus on feature weighting methods in naive Bayesian learning. The methods for calculating feature weights can be roughly divided into two categories: filter methods and wrapper methods [11]. These methods are distinguished based on the interaction between the feature weighting and classification. The class of filter-based methods contains algorithms that use no input other than the training data itself to calculate the feature weights, whereas wrapper-based algorithms use feedback from a classifier to guide the search.

Hall [5] proposed a feature weighting algorithm for naive Bayes using decision tree, called DTNB. This method estimates the degree of feature dependency by constructing unpruned decision trees and looking at the depth at which features are tested in the tree. A bagging procedure is used to stabilize the estimates. Features that do not appear in the decision trees receive a weight of zero.

Lee et al. [12] proposed a method for calculating feature weights in naive Bayes. They calculated the feature weights of naive Bayes using Kullback-Leibler measure. The averaged amount of information for a feature is calculated, and it showed improvement over regular naive Bayes and other supervised learning methods.

In case of wrapper (feedback) method, the performance feedback from the classification algorithm is incorporated in determining feature weights. In wrapper methods, the weights of features are determined by how well the specific feature settings perform in classification learning.

Zhang and Sheng [6] investigated a weighted naive Bayes with accurate ranking from data. They used various weighting methods including the gain ratio method, the hill climbing method, the Markov Chain Monte Carlo method, and combinations of these methods. The performance of the methods is measured using AUC, and a weighted naive Bayes is claimed to produce accurate ranking outperforms naive Bayes.

Gartner [4] employs feature weighting performed by SVM. The algorithm looks for an optimal hyperplane that separates two classes in given space, and the weights determining the hyperplane can be interpreted as feature weights in the naive Bayes classifier. The weights are optimized such that the danger of overfitting is reduced. It can solve the binary classification problems and the feature weight is based on conditional independence. They showed that the method compares favorably to state-of-the-art machine learning approaches.

3   Background

The naive Bayesian classifier is a straightforward and widely used method for supervised learning. It is one of the fastest learning algorithms, and can deal with any number of features or classes. Despite of its simplicity in model, naive Bayesian performs surprisingly well in a variety of problems. Furthermore, naive Bayesian learning is robust enough that small amount of noise does not perturb the results. In this paper we implement the value weighting method in the context of naive Bayes using filter method.

The naive Bayesian learning uses Bayes theorem to calculate the most likely class label of the new instance. Since all features are considered to be independent given the class value, the classification on d is defined as follows

$$\mathcal{V}_{NB}(d) = \text{argmax}_c \, P(c) \prod_{a_{ij} \in d} P(a_{ij}|c)$$

where a_{ij} represents the j-th value of the i-th feature.

The naive Bayesian classification with feature weighting is now represented as follows.

$$\mathcal{V}_{FWNB}(d) = \text{argmax}_c \, P(c) \prod_{a_{ij} \in d} P(a_{ij}|c)^{w_i} \qquad (1)$$

where $w_i \in \mathbb{R}$ represents the weight of feature. In this formula, unlike traditional naive Bayesian approach, each feature i has its own weight w_i. Since feature weighting is a generalization of feature selection, it involves a much larger search space than feature selection.

Current feature weighting methods have some limitations that each feature value has the same significance with respect to the target concept. A feature takes on a number of discrete values, and each value might have different importance with respect to the target values. Since the current feature weighting methods assign a weight to each feature, all values of the given feature are given the same weight. As an illustrative example of value weighting paradigm, let us consider the following example.

Example 1: Suppose the *Gender* feature has values of *male* and *female*, and the target feature has the value of y and n. Suppose their corresponding probabilities are given as

$$p(y) = 0.9, \ p(n) = 0.1, \ p(y|female) = 0.1$$

$$p(n|female) = 0.9, \ p(y|male) = 0.99, \ p(n|male) = 0.01.$$

Traditional feature weighting methods give the same weight to both cases whether the feature value is either *male* or *female*. When *female* value is observed, it significantly impacts the probability distribution of the target feature. Meanwhile, the observation of *male* value does not change much of the target distribution. Majority of the significance of *Gender* feature takes place when its value is *female*.

However, by assigning the same weight to each feature value, we are not able to discriminate the embedded significance among the feature values.   □

In this paper, we think of a problem in which weights are assigned in a more fine-grained way. Unlike feature weighting methods, we assign different weight for each feature *value*. We call this method as *value weighting* method.

The proposed value weighting method is more fine-grained and provides a new weighting paradigm in classification learning. The basic assumption behind the value weighting method is that each feature value has different significance with respect to class value. When we say a certain feature is important/significant, we think the importance of the feature can be decomposed. As we have seen in Example 1, feature values are not equally important.

Some feature values are more important than other feature values. In the case of Example 1, we can see that the value of *female* is much more important than that of *male* with respect to the target feature (pregnancy-related disease). If we assign the same weight to each feature value, we will lack the capability to discriminate the predicting power residing across feature values.

The formula of value weighting method is defined as follows.

$$\mathcal{V}_{VWNB}(d) \;=\; \mathrm{argmax}_c P(c) \prod_{a_{ij} \in d} P(a_{ij}|c)^{w_{ij}} \qquad (2)$$

where $w_{ij} \in \mathbb{R}$ represents the weight of feature value a_{ij}. You can easily see that each feature value is assigned a different weight. The w_{ij} can be any real value, representing the significance of feature value a_{ij}.

By providing more parameters in naive Bayes learning, it expands the dimension of hypothesis space of naive Bayes learning. Introducing a new dimension in naive Bayes increases the expressive power of naive Bayes, which in turn can possibly improve the performance of naive Bayesian learning. In this paper, we will investigate whether the proposed *value weighting* method can improve the performance of naive Bayesian learning even further, and its performance will be compared with that of other methods.

4   Value Weighted Learning

A feature takes on a number of discrete values, and each feature value has different importance with respect to the target value. In this paper, we propose a new method for calculating the weights for feature values using filter approach. We will use an information-theoretic filter method for assigning weights to feature values. We choose the information theoretic method because it has strong theoretical background for deciding and calculating weight values.

The basic assumption of the value weighting method is that when a certain feature value is observed, it gives a certain amount of information to the target feature. The more information a feature value provides to the target feature, the more important the feature value becomes. The critical part now is how to define or select a proper measure which can correctly measure the amount of information.

The first candidate we can use for this purpose is information gain. Information gain is a widely used method for calculating the importance of features, including in decision tree algorithms. It is quite an intuitive argument that a feature with higher information gain deserves higher weight. Quinlan proposed a classification algorithm C4.5 [13], which introduces the concept of information gain. C4.5 uses the information gain (or gain ratio) that underpins the criterion to construct the decision tree for classifying objects. It calculates the difference between the entropy of a priori distribution and that of a posteriori distribution of class, and uses the value as the metric for deciding branching node. The information gain used in C4.5 is defined as follows.

$$H(C) - H(C|A) =$$
$$\sum_a P(a) \sum_c P(c|a)\log P(c|a) - \sum_c P(c)\log P(c) \qquad (3)$$

Equation (3) represents the discriminative power of a feature and this can be regarded as the weight of a feature. Since we need the discriminative power of a feature *value*, we cannot directly use Eq. (3) as the measure of discriminative power of a feature value.

In this paper, let us define $\mathcal{IG}(C|a)$ as the instantaneous information that the event $A = a$ provides about C, i.e., the information gain that we receive about C given that $A = a$ is observed. The $\mathcal{IG}(C|a)$ is the difference between a priori and a posteriori entropies of C given the observation a, and is defined as

$$\mathcal{IG}(C|a) = H(C) - H(C|a)$$
$$= \sum_c P(c|a)\log P(c|a) - \sum_c P(c)\log P(c) \qquad (4)$$

While the information gain used in C4.5 is the information content of a specific feature, the information gain defined in Eq. (4) is that of a specific observed value. Therefore, Eq. (4) can be a candidate for the weight measure of a feature value ($A = a$).

However, although $\mathcal{IG}(C|a)$ is a well known formula, there is a fundamental problem with using $\mathcal{IG}(C|a)$ as the measure of value weight. The first problem is that $\mathcal{IG}(C|a)$ can be zero even if $P(c|a) \neq P(c)$ for some c. For instance, consider the case of an n-valued feature where a particular value of $C = c$ is particularly likely a priori ($p(c) = 1 - \epsilon$), while all other values in C are equally unlikely with probability $\epsilon/(n-1)$. As for $\mathcal{IG}(C|a)$, it can not distinguish the permutation of these probabilities, i.e., an observation which predicts the relatively rare event $C = c$. Since it cannot distinguish between particular events, $\mathcal{IG}(C|a)$ would yield zero information for such events.

Another problem of using $\mathcal{IG}(C|a)$ as the weight of feature value is that this formula can give negative value. It is very unnatural for the weight of a feature value to be negative. Due to these problems described so far, we do not use the $\mathcal{IG}(C|male)$ form as the measure of value weight.

Instead of information gain, in this paper, we employ Kullback-Leibler measure. This measure has been widely used in many learning domains since it originally was proposed in [14]. The Kullback-Leibler measure (denoted as $\mathcal{KL}$) for a feature value a_{ij} is defined as

$$\mathcal{KL}(C|a_{ij}) = \sum_c P(c|a_{ij})\log\left(\frac{P(c|a_{ij})}{P(c)}\right) \qquad (5)$$

where a_{ij} means the j-th value of i-th feature. The formula $\mathcal{KL}(C|a_{ij})$ is the average mutual information between the events c and a_{ij} with the expectation taken with respect to a posteriori probability distribution of C. (The original notation should be $\mathcal{KL}(a_{ij}|C)$. However, we use $\mathcal{KL}(C|a_{ij})$ because it is more meaningful in this paper.) The difference is subtle, yet significant enough that the $\mathcal{KL}(C|a_{ij})$ is always non-negative, while $\mathcal{IG}(C|a_{ij})$ may be either negative or positive.

$\mathcal{KL}(C|a_{ij})$ appears in information theoretic literature under various guises. For instance, it can be viewed as a special case of the cross-entropy or the discrimination, a measure which defines the information theoretic similarity between two probability distributions. In this sense, the $\mathcal{KL}(C|a_{ij})$ is a measure of how dissimilar our a priori and a posteriori beliefs are about C–a useful feature value should have a high degree of dissimilarity. Therefore, we employ the $\mathcal{KL}$ measure in Eq. 5 as a measure of divergence, and the information content of a feature value a_{ij} is calculated with the use of the $\mathcal{KL}$ measure.

The final weight of feature value is represented as the multiplication form of $P(a_{ij})$ with the $\mathcal{KL}$ measure of the feature value. Therefore, the final formula (w_{ij}) for the feature value a_{ij} Eq. (6) is given as

$$w_{ij} = r_{ij}\mathcal{KL}(C|a_{ij}) = P(a_{ij})\mathcal{KL}(C|a_{ij}) \tag{6}$$

4.1 Smoothing

Finally, when we calculate the value weights, we need an approximation method to avoid the problem that the denominator of equations (i.e., $P(c)$) being zero. We use Laplace smoothing for calculating probability values, and the Laplace smoothing methods used in this paper are defined as

$$P(a_{ij}) = \frac{\#(a_{ij}) + 1}{N + |a_i|}, \ \ P(c) = \frac{\#(c) + 1}{N + L} \ \text{and} \tag{7}$$

$$P(c|a_{ij}) = \frac{\#(a_{ij} \wedge c) + 1}{\#(a_{ij}) + L} \tag{8}$$

where $|a_i|$ and L mean the number of values in feature a_i and the number of class values, respectively.

5 Experimental Evaluation

In this section, we provide empirical results on several benchmark datasets and compare the value weighted naive Bayes (VWNB) algorithm against other state-of-the-art supervised algorithms.

We selected 10 datasets from the UCI repository [15]. The continuous features in datasets were discretized using the method in [16]. The characteristics of the datasets we used are omitted due to space limitations. To evaluate the performance, we used 10-fold cross validation method.

Firstly, since the NB-related algorithm is the gold standard algorithm in this paper, we compare the performance of VWNB with that of other naive Bayes with feature weighting. These include (1) naive Bayes (NB), (2) DTNB [5] and (3) naive Bayes with feature weighting (FWNB) [12]. VWNB is also compared with Logistic Regression. We used Weka software [17] to run these programs.

Table 1. Accuracies of the methods

Dataset	VWNB	DTNB	FWNB	NB	Logistic
balance	†71.28	70.72	*71.44	70.72	69.76
cmc	*56.17	54.04	52.25	52.82	†55.80
credit	75.14	69.90	75.35	†75.90	*76.10
crx	*86.00	85.90	†85.93	85.91	85.60
dermatology	*98.05	96.93	97.91	†98.04	97.49
diabetes	†77.96	77.60	†77.96	77.86	*78.65
flare	74.29	*75.14	74.57	74.39	†75.05
glass	*76.25	†76.17	74.22	74.30	75.23
haberman	72.41	*73.69	72.11	72.71	†73.37
ionosphere	88.92	*90.31	†88.99	88.89	86.89

Table 1 shows the results of the accuracies of these methods. The numbers with * mark mean the best accuracies among the methods, and the † mark means the second best accuracy.

As we can see in Table 1, the performance of the value weighting method is quite impressive. The proposed method (VWNB) shows the top or the second performance in 6 cases out of 10 datasets. Its performance is quite competitive with that of other algorithms such as Logistic Regression. These results indicate that assigning weights to each feature value can improve the performance of the classification task for the naive Bayesian method in many cases.

6 Conclusions

In this paper, a new paradigm of weighting method, called *value weighting* method, is proposed. Unlike current feature weighting methods, we propose a more fine-grained weighting methods, where we assign a weight to each feature value. An information-theoretic filter method for calculating value weights was developed.

The proposed weighting method is implemented and tested in the context of naive Bayes. The experimental results show that the value weighting method is successful and shows better performance in most cases than its counterpart algorithms. In light of these evidences, we could improve the performance of classification learning even further by using a value weighting approach.

As future work, we will combine the value weighting method with other classification algorithm, and study whether the value weighting method could show the better performance regardless of algorithms.

Acknowledgements. This work was supported by the Korea Research Foundation (KRF) grant funded by the Korea government (MEST) (No. 2014-R1A2A1A11051011).

References

1. Zheng, Z., Webb, G.I.: Lazy learning of Bayesian rules. Mach. Learn. **41**, 53–84 (2000). Kluwer Academic Publishers
2. Wettschereck, D., Aha, D.W., Mohri, T.: A review and empirical evaluation of feature weighting methods for a class of lazy learning algorithms. AI Rev. **11**, 273–314 (1997)
3. Domingos, P., Pazzani, M.: On the optimality of the simple Bayesian classifier under zero-one loss. Mach. Learn. **29**(2–3), 103–130 (1997)
4. Gärtner, T., Flach, P.A.: Wbcsvm: weighted Bayesian classification based on support vector machines. In: The Eighteenth International Conference on Machine Learning (2001)
5. Hall, M.: A decision tree-based attribute weighting filter for naive Bayes. Knowl. Based Syst. **20**(2), 120–126 (2007)
6. Zhang, H., Sheng, S.: Learning weighted naive Bayes with accurate ranking. In: ICDM 2004: Proceedings of the Fourth IEEE International Conference on Data Mining (2004)
7. Ratanamahatana, C.A., Gunopulos, D.: Feature selection for the naive Bayesian classifier using decision trees. Appl. Artif. Intell. **17**(5–6), 475–487 (2003)
8. Kohavi, R.: Scaling up the accuracy of naive-Bayes classifiers: a decision-tree hybrid. In: Second International Conference on Knowledge Discovery and Data Mining (1996)
9. Langley, P., Sage, S.: Induction of selective Bayesian classifiers. In: Proceedings of the Tenth Conference on Uncertainty in Artificial Intelligence, pp. 399–406 (1994)
10. Friedman, N., Geiger, D., Goldszmidt, M., Provan, G., Langley, P., Smyth, P.: Bayesian network classifiers. Mach. Learn. **29**, 131–163 (1997)
11. Kohavi, R., John, G.H.: Wrappers for feature subset selection. Artif. Intell. **97** (1–2), 273–324 (1997)
12. Lee, C.-H., Gutierrez, F., Dou, D.: Calculating feature weights in naive Bayes with Kullback-Leibler measure. In: 11th IEEE International Conference on Data Mining (2011)
13. Quinlan, J.R.: C4.5: Programs for Machine Learning. Morgan Kaufmann Publishers Inc., San Francisco (1993)
14. Kullback, S., Leibler, R.A.: On information and sufficiency. Ann. Math. Stat. **22**(1), 79–86 (1951)
15. Frank, A., Asuncion, A.: UCI machine learning repository (2010). http://archive.ics.uci.edu/ml
16. Fayyad, U.M., Irani, K.B.: Multi-interval discretization of continuous-valued attributes for classification learning. In: International Joint Conference on Artificial Intelligence, pp. 1022–1029 (1993)
17. Hall, M., Frank, E., Holmes, G., Pfahringer, B., Reutemann, P., Witten, I.H.: The weka data mining software: an update. SIGKDD Explor. **11**, 10–18 (2009)

An Automatic Construction of Malay Stop Words Based on Aggregation Method

Khalifa Chekima and Rayner Alfred[(✉)]

Faculty of Computing and Informatics, Universiti Malaysia Sabah, Kota Kinabalu, Malaysia
k.chekima@gmail.com, ralfred@ums.edu.my

Abstract. In information retrieval, the key to an effective indexing can be achieved through the removal of stop words. Despite having many theories and algorithms related to the construction of stop words in many languages, yet, most of the Malay stop words used are either utilized/borrowed from English stop words, or constructed manually by different researchers which happen to be costly, time consuming and susceptible to error. In other words, no standard stop word list has been constructed for Malay language yet. In this study, we propose an aggregation technique using three different approaches for an automatic construction of general Malay Stop words. The first approach based on statistical method, by considering words' frequencies (highest and lowest) against their ranks, this method inspired by zipf's law. The second approach by considering words' distribution against documents using variance measure. The third approach by computing how informative a word is by using Entropy measure. As a result, a total of 339 Malay stop words were produced. The discussion and implication of these findings are further elaborated.

Keywords: Information retrieval · Malay stop-words · Natural language processing

1 Introduction

Stop words, by definition, are meaningless words that have low discrimination power [1]. In natural language processing, elimination of stop words is a common procedure during data preprocessing stage. According to [2] Stop words can be viewed as a homogenously distributed noise signal that should be filtered from a text, as these words have no discriminative values. On the other hand, [3] reported that, the ten most commonly used words in English normally account 20 to 30 percent of the words in a document. It is usually worth to eliminate all stop words terms when indexing a document or processing queries. As reported by [4], elimination of stop words improve the accuracy of text processing for categorization.

Several stop words list have been constructed for English language in the past, which mostly are based on statistical approach (frequencies) [5–7]. Malay language like any other language comprises of stop words. However, to recent date, no commonly accepted stop word list has been developed for Malay language. Most of researches [22–26] dealing with Malay natural language processing have used one of the two techniques,

© Springer Nature Singapore Pte Ltd. 2016
M.W. Berry et al. (Eds.): SCDS 2016, CCIS 652, pp. 180–189, 2016.
DOI: 10.1007/978-981-10-2777-2_16

manual construction of stop words, or use/borrow from general English stop words. Utilizing English general stop words list have an obvious drawback of being too general, it does not fit well when dealing with domain based natural language processing. On the other hand, manual construction of stop words list tend to be costly, time consuming and vulnerable to errors.

The rest of the paper is organized as follow. Section 2 discusses related work. Section 3 discusses our approach, this include an explanation on data collection and preparation as well as methods adopted. Section 4 discusses the evaluation and result. Finally, we conclude our paper in Sect. 5.

2 Related Work

The first work on stop words removal is credited to Hans Peter Luhan (1957), who suggested that words in natural language texts can be grouped into keyword terms and non-keyword terms. Inspired by Luhan's work, few stop words have been constructed such as the well-known Brown stop words list [8] which consists of 421 stop words and the Van stop word list [5], which consists of 250 stop words. These lists known as the standard or the classic stop lists. Despite their popularity, the classic stop words encounter two major limitations. First, by being outdated, they do not cover new emerging stop words on the web [1, 9, 10]. Secondly, they are too generic and provide off-topic and domain independent stop words.

To overcome these limitations, researchers proposed several methods to automatically construct stop words list. These methods can be grouped into two categories, methods based on the information gain standards [1, 11], and methods based on zipf's law [12–14]. Zipf (1949) perceived that in a data, the frequency of a given word is inversely proportional to its rank. Inspired by this Zip'f law, several methods assume that stop words match those of top ranks [1, 12], while others take into account that both of top and low-ranked words as stop words [13].

Information gain methods on the other hand, rely on the amount of information a term conveys. The idea behind this is that, stop words are words with low explanatory values. Few methods have been applied to compute how informative a word is, these includes term entropy measure [9, 10], the Maximum Likelihood Estimation measure [11], the Kullback-Leibler (KL) and divergence measure [1].

A considerable number of researchers have shown interest in constructing stopwords for non-English language by deploying above mentioned methods. Few non-English stop words will be highlighted in this section such as French stop words, Arabic stop words, Chinese stop words, etc.

Yoa and zen [15] construct a customized Chinese stop wordlist containing a total of 1289 words. Their idea was to combine domain based stop word list with the classical stop word list. Feng Zou et al. [16] on the other hand, proposed construction of Chinsese stop words based on an aggregated methodology. Statistical and information model utilized in their methodology. As a result, a much general stop word constructed compare to the existing Chinese stop words.

Savoy [17] constructed French general stop word list. In establishing their work, they followed the guidelines described in [18]. First, all distinct words appeared in French corpora are sorted in a descending order from highest to lowest. The first 200 words with heights frequencies extracted. Secondly, the list is then inspected to remove all numbers (e.g., "1992", "1"), plus all nouns and adjectives more or less directly related to with the main subjects of the underlying collections. Thirdly, they included some non-information-bearing words. As a result, total of 215 words have been identified as French general stop words.

Alhadidi and Alwedyan [19] applied a hybrid approach to construct Arabic stop-word. They used dictionary based approach and a set of rules (algorithm) based approach. On the other hand, A. Alajmi et al. [20] proposed a statistical approach to extract Arabic stop words. The presented Arabic stop words outperform the general Arabic stop words in term of text categorization task.

Zheng and Gaowa [21] proposed to improve Mongolian language document retrieval by proposing the construction of stop word list by combining statistical method and part of speech method. Using stop words generated from presented method shows a tremendous result compared to those stop words obtained using statistical method.

Kwee et al. [27] published a list of 216 Malay stop words. Some of the stop words generated were directly translated from English stop words, such as 'setiap' which means 'each', 'sesudah' means 'after' and 'maka' which means 'therefore'. Some others were gathered from Malay documents, such as 'ayuh', 'alamak', and 'amboi'. From an observation on this list, we discovered that there are many Malay stopwords that have not been listed, furthermore, some of the stopwords present in the list are less likely to occur in formal Malay text such as "ee", "ehem", "auh", "au", "oh","ah","wah", "nun", "ha", "haah", "ai", etc.

3   Our Approach

3.1   Data Collection and Preparation

To ensure the quality of the corpus, data were collected from Dewan Bahasa dan Pustaka (DBP), The Institute of Language and Literature, a government body responsible for coordinating the use of the Malay language in Malaysia. Two features worth to mention about data collected from DBP, first, data from DBP less likely to have grammatical nor typo error, as it is handled/checked manually by linguistic professionals. Secondly, DBP offers data collection from variety of domains (more than 50 domains) as well as variety of data sources (7 different sources).

Next, all of data collected are merged to form one corpus called general corpus words (GCW). To normalize the data, all non-alphabet characters (noise) have been removed this includes periods, digits and symbols. Next, corpus data converted to lowercase to ensure same words with upper or lowercases are identical. As a result, a total of 7,363,578 tokens consist of 27,332 distinct words as size of our corpus, covering nine different domains, namely religion, entertainment, education, sports, business, science, politics, technology and economy. GCW details are illustrated under Table 1.

Table 1. Corpus details

Item	Domain	No. of words	Source DBP
1	Religion	1,823,204	Books, Magazines, Newspapers
2	Entertainment	1,718,455	Books, Magazines, Newspapers
3	Education	523,032	Books, Magazines, Newspapers
4	Sports	139,100	Books, Magazines, Newspapers
5	Business	23,683	Books, Magazines, Newspapers
6	Science	856,470	Books, Magazines, Newspapers
7	Politics	1,645,326	Books, Magazines, Newspapers
8	Technology	360,559	Books, Magazines, Newspapers
9	Economy	273,749	Books, Magazines, Newspapers

3.2 Statistical Method

Stop words are known to have two common characteristics, one having high frequency of occurrence, second, having a stable/high distribution [2]. From there we decided to utilize the first feature for the statistical method by taking into account the terms and their frequencies, similar to those researches [1, 12, 13] inspired by zipf's law. In the proposed method, Stop words match those of top and low-ranked words.

To identify words and their corresponding frequencies, distinct words from document extracted with their corresponding frequencies (occurrence), word with highest frequency ranked on descending order (height to lowest). Instead of using word's frequencies $f(w_j)$, we normalize our results by presenting our data based on words probability $Pr(w_j)$ against a document, as frequencies gap between one word and another tend to be big. This can be achieved by summing up word's frequencies in a particular document D_i divided by the number of tokens n in that document. The formula for words probability as following:

$$\Pr\left(w_j\right) = \frac{\sum f\left(w_j\right)}{\sum n} \tag{1}$$

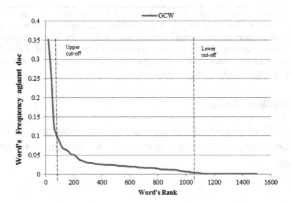

Fig. 1. Word's rank against probability frequencies in general corpus.

Next, Words' rank frequency graph plotted as shown in Fig. 1. The size of the Stop word list matches the elbow shown in Fig. 1. List of stop words with high frequencies are those terms that appears above the upper cut-off, as for the stop word list with low frequencies (frequency 1) are those appear below the lower cut-off. Result of top and lowest twenty words are shown in Table 2 and Table 3 respectively.

Table 2. Top 20 words with highest frequency

Word	Translation	Frequency	Rank	Pr (%)
dan	and	25,891	1	0.351609
yang	the	19,129	2	0.259779
di	at	11,485	3	0.15597
dengan	with/by	6,561	4	0.089101
tidak	not	6,208	5	0.084307
itu	that	6,107	6	0.082935
untuk	to	5,977	7	0.08117
ini	this	5,504	8	0.074746
kepada	to/for	4,290	9	0.05826
mereka	they	3,840	10	0.052149
sebagai	as	2,951	11	0.040076
akan	will	2,806	12	0.038106
juga	also	2,613	13	0.035485
pada	on	2,597	14	0.035268
oleh	by	2,523	15	0.034263
atau	or	2,511	16	0.0341
kerana	because	2,481	17	0.033693
bagi	for	2,476	18	0.033625
perlu	need	2,449	19	0.033258
baru	new	2,400	20	0.032593

Table 3. Lowest 20 Words with Frequency of 1

dedehan (exposure), mukadimah (preamble), kiriman (shipment), Kerjasama (cooperation), berkuasanya (the reign), mengiktirafkan (recognize), memintanya (asked), ketidakprihatinan (indifference), mengekori (follow), beriltizam(committed), imbangannya (counterpart), Merusak (ruin), mertuanya(in-law), diragukan (doubt), fakulti (faculty), sunda (tone), menukarnya (change), imbangannya (counterpart), berempat (foursome), dukungan (endorsement)

3.3   Data Distribution Method

Another characteristic of stop word is being homogenously distributed. The more spread the word the higher chances of it being stop word. We measure data spread using the following formula:

$$s^2 = \frac{\sum (f(w_j) - \overline{f(w_j)})^2}{n-1} \tag{2}$$

Where $f(w_j)$ represents a word's frequency in document D, and n represents the number of distinct word in document D. The $\overline{w}$ refers to mean value, it can be donated by the following formula:

$$\overline{w} = \frac{\sum f(w_j)}{n} \tag{3}$$

The result of the top twenty words' distribution/spread based on variance values is illustrated in Table 4. As can be observed from Tables 2 and 4, the top twenty words based frequency tend to have high distribution (variance value). At least for the first twenty stop words produces by the first two methods, the overlap between the two lists is high. We can make an early observation that the top frequent words tend to have high distribution/spread.

Table 4. Top 20 words with highest spread/distribution

Word	Translation	Variance
dan	and	24018.22
yang	the	13013.47
di	at	4602.28
dengan	with/by	1448.27
tidak	not	1290.31
itu	that	1246.80
untuk	to	1191.88
ini	this	1002.52
kepada	to/for	591.47
mereka	they	466.45
sebagai	as	263.10
akan	will	235.41
juga	also	200.95
pada	on	198.28
oleh	by	185.81
atau	or	183.84
kerana	because	178.95
bagi	for	178.14
perlu	need	173.81
baru	new	166.084

3.4   Data Entropy Measures Method

Based on information theory, stop words are those words that carry little information. To measure the informativeness of a word, Entropy measure method is deployed. Based on Entropy measures, the higher the entropy value of a particular word, the lower the information carried by that word, which made it a potential stop word. To measure word's entropy, we used the following formula:

$$H(w_i) = \sum_{i=1}^{n} P(w_i) \log(\frac{1}{P(w_i)}) \tag{4}$$

The result of top twenty words based on entropy measures are presented in Table 5. As can be observed from Table 5, the top twenty words using entropy measures tend to be slightly different from those produced by the other two methods. Yet, the overlap between the top 20 stop words of all three lists is recorded to be high too.

Table 5.   Top 20 words with highest entropy

Word	Frequency	Entropy
dan	and	0.159609
yang	the	0.152074
di	at	0.125861
dengan	with/by	0.093567
tidak	not	0.090557
itu	that	0.089674
untuk	to	0.088524
ini	this	0.084195
kepada	to/for	0.07193
menjadi	become	0.066894
mereka	they	0.055991
orang	person	0.054073
sebagai	as	0.051452
akan	will	0.051231
juga	also	0.050201
pada	on	0.050033
oleh	by	0.049612
atau	or	0.049541
kerana	because	0.049159
bagi	for	0.048462

3.5   Aggregation Method

The three stop word lists were aggregated to generate the final Malay stop word list. The combination of the three lists redefined the stop words as those with high occurrences, stable distribution and less informative. Since there is no clue on the ideal number of expected stop words for the Malay language, we examined the range of stop words in other languages. We learnt that in most cases, the existing stop word lists in other languages range from 200 to 450 words. Due to that, we choose the average of that range by extracting the highest 300 words from each of the three stop word list as potential final stop words. To rank stop words, we use intersection and union techniques. Using these techniques, we managed to rank stop words into three ranks.

Denote TG as top group with highest rank, SG as second top group and LG as lowest group with lowest rank. On the other hand, denote SWL as top 300 stop word list from statistical method, DWL as top 300 word list from distribution method and EWL as top 300 word list from entropy method. Figure 2 shows an overall picture of the three stop word list generated earlier and the three levels of potential stop words. TG denoted by vertical lines, SG denoted by horizontal line and LG denoted by dots.

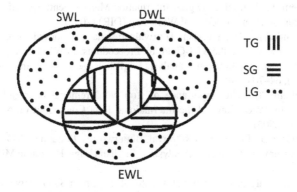

Fig. 2. Illustrate the three levels of the potential stop words list based on the stop word list generated earlier from statistical, entropy and data distribution methods.

Based on the aggregation method mentioned above, a total of 502 words been identified as potential stop words. TG group recorded total of 137 words, SG group recorded total of 126 words, and LG group recorded total of 239 words.

4   Evaluation and Results

Since there is no standard stop words for Malay language that can be used as comparison to our stop word list, a manual evaluation is carried out by language expert. A small number of words have been removed from group TG and group SG such as "orang" (person), "sekolah" (school), "Malaysia", etc. As for group LG, a considerable amount of words have been removed from its potential stop words list compared to the other two groups. This shows that aggregation method tend to produce a better potential stop-word list compared to using a single approach. After removing all non-potential stop words, the stop word list is brought to its final with the total of 339.

5 Conclusion

In this study, first, we demonstrated the steps involved in generating stop word list by considering three common characteristics of stop words, one having high occurrences, second having a stable distribution and third carrying little information. Next, an aggregation method is introduced as final step to construct Malay stopword list. Aggregation method tends to produce top potential stop words like in TG and SG compared to a single method like in LG.

References

1. Lo, R.T.-W., He, B., Ounis, I.: Automatically building a stopword list for an information retrieval system. In: Journal on Digital Information Management: Special Issue on the 5th Dutch- Belgian Information Retrieval Workshop (DIR) (2005)
2. Saif, H.; Fernandez, M.; He, Y.; Alani, H.: Evaluation datasets for twitter sentiment analysis a survey and a new dataset, the STS-Gold. In: Proceedings, 1st Workshop (2013)
3. Burchfield, R.: Review of Frequency analysis English usage: Lexicon and grammar. J. Engl. Linguist. **18**(1), 64–70 (1985)
4. Silva, C., Ribeiro, B.: The importance of stop word removal on recall values in text categorization. In: Proceedings of the International Joint Conference on Neural Networks, pp. 1661–1666 (2003)
5. Van Rijisbergen, C.J.: Information Retrieval, Butterworths, London (1975)
6. Francis, W., Kucera, H.: Frequency Analysis of English Usage. Houghton Mifflin, New York (1982)
7. Fox, C.: Lexical Analysis and Stop List. Prentice-Hall, Upper Saddle River (1990)
8. Fox, C.: Information retrieval data structures and algorithms. Lexical Analysis and Stoplists, pp. 102– 130 (1992)
9. Sinka, M. P., Corne, D.: Evolving better stoplists for document clustering and web intelligence. In: HIS, pp. 1015–1023 (2003a)
10. Sinka, M.P., Corne, D.W.:Towards modernized and web-specific stoplists for web document analysis. In: Proceedings of the IEEE/WIC International Conference on Web Intelligence, WI 2003, IEEE (2003b)
11. Ayral, H., Yavuz, S.: An automated domain specific stop word generation method for natural language text classification. In: 2011 International Symposium on Innovations in Intelligent Systems and Applications (INISTA), pp. 500–503. IEEE (2011)
12. Trumbach, C.C., Payne, D.: Identifying synonymous concepts in preparation for technology mining. J. Inf. Sci. **33**(6), 660–677 (2007)
13. Makrehchi, M., Kamel, M.S.: Automatic extraction of domain-specific stopwords from labeled documents. In: Macdonald, C., Ounis, I., Plachouras, V., Ruthven, I., White, R.W. (eds.) ECIR 2008. LNCS, vol. 4956, pp. 222–233. Springer, Heidelberg (2008)
14. Forman, G.: An extensive empirical study of feature selection metrics for text classification. J. Mach. Learn. Res. **3**, 1289–1305 (2003)
15. Yao, Z., Ze-wen, C.: Research on the construction and filter method of stop-word list in text Preprocessing. In: Fourth International Conference on Intelligent Computation Technology and Automation (2011)
16. Zou, F., Wang, F.L., Deng, X., Han, S., Wang, L.S.: Automatic construction of chinese stop word list. In: Proceedings of the 5th WSEAS International Conference on Applied Computer Science, Hangzhou, China, pp. 1010–1015, 16–18 April 2006

17. Savoy, J.: A stemming procedure and stopword list for general French corpora. J. Am. Soc. Inf. Sci. **50**(10), 944–952 (1999)
18. Fox, C.: A stop list for general text. ACM-SIGIR Forum **24**, 19–35 (1990)
19. Alhadidi, B., Alwedyan, M.: Hybrid stop-word removal technique for Arabic language. Egypt. Comput. Sci. J. **30**(1), 35–38 (2008)
20. Alajmi, A., Saad, E.M., Darwish, R.R.: Toward an ARABIC stop-words list generation. Int. J. Comput. Appl. **46**(8), 8–13 (2012)
21. Zheng, G., Gaowa,G.: The selection of Mongolian stop words. In: IEEE International Conference on Intelligent Computing and Intelligent Systems (ICIS) (2010)
22. Alsaffar, A., Omar, N.: Study of feature selection and machine learning algorithm for Malay sentiment classification. In: International Conference on Information Technology and Multimedia (2014)
23. Samasudin, N., Puteh, M., Hamdan, A.R., Nazri, M.Z.N.: Is artificial immune system suitable for opinion mining? In: 4th Conference on Data Mining and Optimization (DMO) (2012)
24. Shamasuddin, N., Puteh, M.: Bess or xbest: mining the Malaysian online reviews. In: 3rd Conference on Data Mining and Optimization (2011)
25. Darwich, M., Noah, S.A.M., Omar, N.: Inducing a domain-independent sentiment lexicon in Malay. In: Jaist Symposium on Advance Science And Technology (2015)
26. Isa, N., Puteh, M., Kamarudin, R.M.H.R.: Sentiment classification of Malay newspapers using immune network (SCIN). In: Proceeding of the World Congress on Engineering, UK (2013)
27. Kwee, A.T., Tsai, F.S., Tang, W.: Sentence-level novelty detection in English and Malay. In: Theeramunkong, T., Kijsirikul, B., Cercone, N., Ho, T.-B. (eds.) PAKDD 2009. LNCS, vol. 5476, pp. 40–51. Springer, Heidelberg (2009)

Modeling and Forecasting *Mudharabah* Investment with Risk by Using Geometric Brownian Motion

Nurizzati Azhari and Maheran Mohd Jaffar[✉]

Department of Mathematics, Faculty of Computer and Mathematical Sciences, University Teknologi Mara, 40450 Shah Alam, Selangor Darul Ehsan, Malaysia
nurizzati_azhari@yahoo.com.my,
maheran@tmsk.uitm.edu.my

Abstract. This study developed *mudharabah* investment with risk model by considering the rate of return as a total of deterministic profit rate and a function of white noise that is geometric Brownian motion. The result shows that the investment is considered as accurately forecast when using this developed model. The profit from *mudharabah* investment is compared with single party investment. The result obtained shows that the profit difference between *mudharabah* investment and single party investment is very small. It is verified that the developed model can be used in forecasting the investment and profit for two parties.

Keywords: *Mudharabah* · Geometric Brownian motion · Investment

1 Introduction

Islamic finance confident that one of the aspects which help economic growth is a system that can respond positively to economic problems. According to Muhammad [9], the profit sharing concepts is one of Islamic concepts that are required for Islamic finance. He believes that in order for Islamic finance to be successful, the profit sharing concept such as *mudharabah* should be involved aggressively. This is because this concept can replace entirely the institution of interest (*riba*) which is totally forbidden in Islam. Besides that, Noraziah and Abdul (2011) say that in order to ensure justice and improved efficiency, the institution of interest should be replaced to profit sharing concept in order to avoid transactions involve injustice and inefficiency.

The two broad categories of profit sharing concept that are widely used are *mudharabah* and *musyarakah* concepts. *Mudharabah* is a form of partnership in which one partner provides all of the capital provider and other party responsible for the management of the project. Profits from the investment are distributed according to a fixed or predetermined ratio. Meanwhile, *musyarakah* operates when partners contribute capital to a project and they share the risks and rewards. In term of profits, they share according to preagreed ratio but losses are shared in exact proportion to the capital invested by each party. Nevertheless in *musyarakah*, all partners have the right but not the obligation to participate in the management of the project which explains why the

© Springer Nature Singapore Pte Ltd. 2016
M.W. Berry et al. (Eds.): SCDS 2016, CCIS 652, pp. 190–199, 2016.
DOI: 10.1007/978-981-10-2777-2_17

profit sharing ratio is mutually agreed upon and may be different from the investment in the total capital [1].

The profit sharing concept also can be applied in investment such as purchase of bonds, stocks (equity) or real estate property [4]. Since the profit sharing concept operates based on Islamic principle, therefore it is known as Islamic investments. Islamic investments function the same way as conventional investments but the different is the investment activities must follow Islamic law. According to Faleel [3], most Islamic equity funds are based on *mudharabah* concept where the fund management companies invest in *syariah* compliance stock and after they sell the stock, the profit gained in investment is distributed to the investors. However, the Islamic equity funds also can use other contracts such as *wakalah* where investor gives capital to an agent that they pay some fees to them but this contract is not often used.

Since the profit sharing concept can widely use in investment of equity, thus the model on profit sharing concept is needed. There are some researches study on profit sharing concept model which are *mudharabah* and *musyarakah*. Maheran [6] discusses the *mudharabah* and *musyarakah* investment without risk. According to Mohamed [8] investment of stock involve element of risk. He also say that greater degree of risk cause reluctant of people to invest unless a greater amount of return is expected. The development model on profit sharing investment that involve risk is needed to help in reduce possibility of exposure to risk. There had been research on *mudharabah* and *musyarakah* with risk models done by Aslina [1] and Azizah [2] respectively. However, both of the researchers used different way in order to develop the model. Aslina [1] develops the *mudharabah* model with risk by using stochastic integration while Azizah [2] develops the model with risk by using geometric Brownian motion. From these studies, it shows that the method used by Azizah [2] is easier to understand as compared to Aslina (2015) that uses stochastic calculus. Besides that, stochastic calculus with Ito's lemma requires investors to have deeper knowledge and good background in mathematics to understand. Since there is no model on *mudharabah* with risk using geometric Brownian motion, thus this study attempts to develop the *mudharabah* investment with risk using geometric Brownian motion. This model can be used by investors in planning their investment.

Geometric Brownian motion is a lognormal, continuous time stochastic process where the movement of variable such as stock price is random and continuous. This can be explained that the stock prices changing continually over very small intervals of time and the position where the change of state on the assets is being altered by random amounts [11]. In this study, the geometric Brownian motion is considered as the foundation in development of *mudharabah* investment with risk model in forecasting stock investment.

2   The Development of New *Mudharabah* Investment with Risk Model for Two Parties

In order to understand the concept of *mudharabah* investment, assume that the investment involves two parties which are capital provider and entrepreneur. The joint venture begins when capital provider who provides capital, invest in the stock market while the entrepreneur manages the investment. This can be explained at initial time t_0, capital that is equity of capital provider is invested in the stock market but there is no initial capital made by entrepreneur in this time. However equity of entrepreneur will exist by the time profit is gained. Then the profit will be shared between the capital provider and entrepreneur with a ratio of $k : (1 - k)$. The assumptions for the *mudharabah* investment model are as follows [2]:

a. The investment is between two parties which are capital provider and entrepreneur.
b. One drift and one volatility are used in investing one stock market.
c. The investment only in *syariah* counters.
d. The parameter $k : (1 - k)$ are between 0 and 1. It can be in the form of decimal or percentage.
e. It is assumed there is no external factor such as natural disaster, seasonal factors, political issues, announcement from the company and etc.
f. It is assumed the *syariah* stock price and *mudharabah* investment follows Brownian motion.

From Eqs. (2.1) and (2.2) discussed by Maheran [6, 7], Aslina [1] proposed the equity of two parties in *mudharabah* investment which are capital provider and entrepreneur can be defined as Definitions 2.1 and 2.2.

$$E_t = E_{t-1} + r_t k E_{t-1} \tag{2.1}$$

$$Q_t = Q_{t-1} + r_t(1 - k) E_{t-1} \tag{2.2}$$

Definition 1. *Mudharabah* equity of capital provider at a certain time is equal to the equity of capital provider per unit of time of the previous period plus the profit per unit of time of the equity of the capital provider from previous period.

Definition 2. *Mudharabah* equity of entrepreneur at a certain time is equal to the equity of entrepreneur per unit of time of the previous period plus the profit per unit of time of the equity of capital provider prior to entrepreneur from previous period.

Equations (2.1) and (2.2) may be written in a matrix form as follows [9]:

$$\begin{bmatrix} E_t \\ Q_t \end{bmatrix} = \begin{bmatrix} E_{t-1} \\ Q_{t-1} \end{bmatrix} + \begin{bmatrix} r_t k E_{t-1} \\ r_t(1 - k) E_{t-1} \end{bmatrix}$$

$$\begin{bmatrix} E_t \\ Q_t \end{bmatrix} = \begin{bmatrix} E_{t-1} \\ Q_{t-1} \end{bmatrix} + \begin{bmatrix} r_t k & 0 \\ r_t(1 - k) & 0 \end{bmatrix} \begin{bmatrix} E_{t-1} \\ Q_{t-1} \end{bmatrix} \tag{2.3}$$

$$\begin{bmatrix} E_t \\ Q_t \end{bmatrix} - \begin{bmatrix} E_{t-1} \\ Q_{t-1} \end{bmatrix} = \begin{bmatrix} r_t k & 0 \\ r_t(1 - k) & 0 \end{bmatrix} \begin{bmatrix} E_{t-1} \\ Q_{t-1} \end{bmatrix}.$$

Equation (2.3) can be further enhanced to be

$$\mathbf{X}_t - \mathbf{X}_{t-1} = \mathbf{A}_t \mathbf{X}_{t-1} \qquad\qquad t = 1, 2, 3, \ldots \tag{2.4}$$

where $\mathbf{X}_t = \begin{pmatrix} E_t \\ Q_t \end{pmatrix}$ and $\mathbf{A}_t = \begin{pmatrix} r_t k & 0 \\ r_t(1 - k) & 0 \end{pmatrix}$.

Equation (2.4) is a discrete model for the *mudharabah* investment. Aslina [1] extends Eq. (2.4) to be continuous model for *mudharabah* investment as in Eq. (2.5). The time unit is small enough such that the nano-seconds $t \to \infty$ until $\mathbf{X}_t - \mathbf{X}_{t-1} \to \dot{\mathbf{X}}(t)$ which gives

$$\dot{\mathbf{X}}(t) = \mathbf{A}(t)\mathbf{X}(t) \tag{2.5}$$

where $\dot{\mathbf{X}}(t) = \begin{pmatrix} \dot{E}(t) \\ \dot{Q}(t) \end{pmatrix}$, $\mathbf{X}(t) = \begin{pmatrix} E(t) \\ Q(t) \end{pmatrix}$ and $\mathbf{A}(t) = \begin{pmatrix} r(t)k & 0 \\ r(t)(1-k) & 0 \end{pmatrix}$.

In Eq. (2.5), $\mathbf{X}(t)$ is the investment or equity of capital provider and entrepreneur while $\mathbf{A}(t)$ is the profit rate at time t. The difference is Eq. (2.4) is in discrete form but Eq. (2.5) is in continuous form.

If the rate of profit $r(t)$ is fixed that is r, therefore matrix $\mathbf{A}(t)$ is scalar $\mathbf{N}$. Then Eq. (2.5) becomes as follows:

$$\dot{\mathbf{X}}(t) = \mathbf{N}\mathbf{X}(t). \tag{2.6}$$

Solving linear equation of (2.6) which obtains

$$\mathbf{X}(t) = \mathbf{X}(0)e^{\mathbf{N}t} \tag{2.7}$$

where $\mathbf{N}$ is a matrix $\begin{bmatrix} rk & 0 \\ r(1-r) & 0 \end{bmatrix}$.

In order to get values of $e^{\mathbf{N}t}$ as detail, the eigenvalues of matrix $\mathbf{N}$ must be obtained first. The eigenvalues that have been obtained are $\lambda_1 = 0$ and $\lambda_2 = rk$.

Since the matrix $\mathbf{N}$ have different eigenvalues $\lambda_1 \neq \lambda_2$, the solution of $e^{\mathbf{N}t}$ is obtained and it is different from the result obtained by Aslina [1]. The $e^{\mathbf{N}t}$ is solved as follows: Since matrix $\mathbf{N}$ is 2×2 matrix, then we assume that $R(x)$ is degree one.

$$R(x) = a_0 + a_1 x$$

$$\begin{aligned}
f(\mathbf{N}) &= a_0 I + a_1 \mathbf{N}t \\
&= a_0 \begin{pmatrix} 1 & 0 \\ 0 & 1 \end{pmatrix} + a_1 \begin{pmatrix} rk & 0 \\ r(1-r) & 0 \end{pmatrix} t \\
&= a_0 \begin{pmatrix} 1 & 0 \\ 0 & 1 \end{pmatrix} + a_1 \begin{pmatrix} rkt & 0 \\ r(1-r)t & 0 \end{pmatrix} \\
&= \begin{pmatrix} a_0 + a_1 rkt & 0 \\ a_1 r(1-r)t & a_0 \end{pmatrix} \\
&= \begin{pmatrix} A_{11} & A_{12} \\ A_{21} & A_{22} \end{pmatrix}.
\end{aligned}$$

The coefficient a_0 and a_1 can be obtained from

$$e^{\lambda_1 t} = R(\lambda_1 t)$$
$$e^{\lambda_2 t} = R(\lambda_2 t)$$
$$e^{\lambda_1 t} = R(0t)$$
$$= a_0$$

(2.8)

$$e^{\lambda_2 t} = R(\lambda_2 t)$$
$$= a_0 + a_1 rkt$$

(2.9)

Solving this by (2.8)–(2.9)

$$e^{\lambda_1 t} - e^{\lambda_2 t} = -a\rho kt$$
$$a_1 = \frac{e^{\lambda_2 t} - e^{\lambda_1 t}}{\rho kt}.$$

(2.10)

Substitute (2.10) into (2.9)

$$e^{\lambda_2 t} = a_0 + \left[\frac{e^{\lambda_2 t} - e^{\lambda_1 t}}{\rho kt} \right] \rho kt$$
$$e^{\lambda_2 t} = a_0 + e^{\lambda_2 t} - e^{\lambda_1 t}$$
$$a_0 = e^{\lambda_1 t}.$$

Substitute $a_0 = e^{\lambda_1 t}$ and $a_1 = \frac{e^{\lambda_2 t} - e^{\lambda_1 t}}{\rho kt}$ into $\begin{bmatrix} A_{11} & A_{12} \\ A_{21} & A_{22} \end{bmatrix} = \begin{bmatrix} a_0 + a_1 rkt & 0 \\ a_1 r(1-k)t & a_0 \end{bmatrix}$

$$A_{11} = a_0 + a_1 \rho kt$$
$$= e^{\lambda_1 t} + \left[\frac{e^{\lambda_2 t} - e^{\lambda_1 t}}{\rho kt} \right] \rho kt$$
$$= e^{\lambda_1 t} + e^{\lambda_2 t} - e^{\lambda_1 t}$$
$$= e^{\lambda_2 t}$$
$$A_{12} = 0$$
$$A_{21} = \left[\frac{e^{\lambda_2 t} - e^{\lambda_1 t}}{\rho kt} \right] \rho(1-k)t$$
$$= \left[\frac{e^{\lambda_2 t} - e^{\lambda_1 t}}{k} \right] (1-k)$$
$$A_{22} = a_0$$
$$= e^{\lambda_1 t}$$

(2.11)

$$\begin{bmatrix} A_{11} & A_{12} \\ A_{21} & A_{22} \end{bmatrix} = \begin{bmatrix} e^{\lambda_2 t} & 0 \\ \left(\frac{e^{\lambda_2 t} - e^{\lambda_1 t}}{k} \right)(1-k) & e^{\lambda_1 t} \end{bmatrix}.$$

Then, substitute $\lambda_1 = 0$ and $\lambda_2 = rk$ to (2.11) and therefore,

$$e^{Nt} = \begin{bmatrix} e^{rkt} & 0 \\ \left(\frac{e^{rkt}-1}{k}\right)(1-k) & 1 \end{bmatrix}.$$

The solution of e^{Nt} seems similar to Aslina [1] and Aslina [1] solves the investment model using stochastic calculus. In this study, the working steps continue by using geometric Brownian motion that is the standard white noise. Hence, the quantity e^{Nt} is substituted in Eq. (2.7) in the form of matrices as follows:

$$\begin{bmatrix} E(t) \\ Q(t) \end{bmatrix} = \begin{bmatrix} e^{rkt} & 0 \\ \left(\frac{e^{rkt}-1}{k}\right)(1-k) & 1 \end{bmatrix} \begin{bmatrix} E(0) \\ Q(0) \end{bmatrix}.$$

The matrices can be converted to another form as Eqs. (2.12) and (2.13) below

$$E(t) = E(0)e^{rkt} \tag{2.12}$$

$$Q(t) = E(0)\left(\frac{e^{rkt} - 1}{k}\right)(1 - k) + Q(0) \tag{2.13}$$

This equation allows us to calculate in detail the value of the investment for both parties at any time and with initial value of the investment. The nature of the stock prices development is not fully known and it depends on the effect of random environment. The effect can be included in the rate of return or profit. Thus, the rate of return consists of deterministic profit rate $\alpha(t)$ and some white noise.

$$\begin{aligned} r &= \alpha(t) + \text{a function white noise} \\ &= \alpha(t) + f(p) \end{aligned} \tag{2.14}$$

where $f(p)$ is the white noise that involves systematic risk and unsystematic risk which the behavior is not clearly known but there exists the probability of the risk. According to Mohsen, Zabihallah and Heydari (2014), systematic risk is market risk which include interest rates, recession and war while unsystematic risk is company or industry specific risk includes financial risk and credit risk. Variable $\alpha(t)$ is a deterministic profit rate or profit rate that identified for risk free investment. Then substitute Eq. (2.14) into Eqs. (2.12) and (2.13) to obtain

$$E(t) = E(0)e^{(\alpha(t)+f(p))kt} \tag{2.15}$$

$$Q(t) = E(0)\left(\frac{e^{(\alpha(t)+f(p))kt} - 1}{k}\right)(1 - k) + Q(0) \tag{2.16}$$

According to Maheran [6], the standard white noise is more acceptable because it is recognized equal to the issue of Brownian motion. Thus variable $f(p)$ can be modeled as

$$f(p) = \lambda\phi \qquad (2.17)$$

where ϕ is the random value and λ is constant. Then Eq. (2.17) substitute into Eqs. (2.15) and (2.16) that will produce capital provider's *mudharabah* investment with risk model

$$E(t) = E(0)e^{(\alpha(t) + \lambda\phi)kt} \qquad (2.18)$$

and entrepreneur's *mudharabah* investment model is

$$Q(t) = E(0)\left(\frac{e^{(\alpha(t) + \lambda\phi)kt} - 1}{k}\right)(1 - k) + Q(0) \qquad (2.19)$$

where $\alpha(t)$ is the drift of stock prices, λ is the volatility of the stock prices and ϕ is a random value. The Eqs. (2.18) and (2.19) are used to forecast the *mudharabah* investment for two parties in risky investment.

The single party investment based on developed *mudharabah* model where it is the summation of the capital provider and entrepreneur investment model. It can be written in mathematics form as

$$S(t) = E(t) + Q(t)$$

where there is no profit sharing rate then the value of k is 1, the variable $Q(t)$ will become zero and variable $E(t)$ will equal to $S(t)$. Thus from Eqs. (2.18) and (2.19), the equation of $S(t)$ is as follows:

$$S(t) = E(0)e^{(\alpha(t) + \lambda\phi)t}.$$

3   Result

The accuracy of forecast investment was checked by using mean absolute percentage error (MAPE). In this study, the MAPE is used because it is suitable to compare forecast accuracy on several series with different scale [5]. The MAPE is calculated as follows:

$$MAPE = \frac{1}{M}\sum_{1}^{M}\frac{|X_A - X_f|}{X_A} \qquad (3.1)$$

where X_A is actual investment and X_f is the forecast investment. The forecast model is said to be accurate when the value of MAPE is lower. In order to judge the accuracy of forecast model, Lawrence, Kllimberg and Lawrence [5] had introduced a scale which is based on the MAPE measure as in Table 1 [6].

Then the average forecast profit difference, $A(n)$ between *mudharabah* investment and single party investment is obtained by using [3]

Table 1. A scale of judgement of forecast accuracy

MAPE	Judgement of forecast accuracy
<10 %	Highly accurate
11 % to 20 %	Good forecast
21 % to 50 %	Reasonable forecast
>51 %	Inaccurate forecast

$$A(n) = \frac{\sum_{t=1}^{n} \left| P_{S(t)} - \left(P_{E(t)} + P_{Q(t)} \right) \right|}{n} \tag{3.2}$$

where $P_{S(t)}$ is the forecast profit from single party investment at time t and $P_{E(t)} + P_{Q(t)}$ is total forecast profit for capital provider and entrepreneur at time t. The variable n represents the number of days to forecast. In this study, the number of days to forecast is 20 days or four weeks.

In order to verify the single party investment model is distributed accordingly between capital provider and entrepreneur using the developed *mudharabah* investment model, the percentage average forecast profit difference, $D(n)$ is calculated by using Eq. (3.3) [3].

$$D(n) = \frac{A(n)}{S(0)} \times 100\% \tag{3.3}$$

where the amount of capital investment $S(0)$ in this study is RM 12 000. The small value of $D(n)$ indicates that the profit of single party investment is distributed appropriately between capital provider and entrepreneur.

Table 2 shows that the value of MAPE of forecast *mudharabah* investment and single party investment are around 0.51 % until 3.68 %. Since the values of MAPE for these ten *syariah* counters are less than 10 %, thus it means that the proposed *mudharabah* investment with risk model is considered as accurately forecast. It also shows that the counter that produces the highest profit for both *mudharabah* and single investment is AJI counter then followed by DLADY, BKAWAN and PETGAS counters. The percentage forecast profit difference, $D(n)$ between *mudharabah* investment and single party investment for ten *syariah* counters is very small. Thus, it can be said that the profit of the single party investment model is divided appropriately to obtain the profit of capital provider and entrepreneur by using developed *mudharabah* model. This can be considering as the verification of the *mudharabah* model.

Table 2. Result summary of forecasted profit for ten *syariah* counters

COUNTERS	MAXIMUM FORECAST PROFIT (RM)				AVERAGE MAPE(%)		AVERAGE FORECAST PROFIT DIFFERENCE, $A(n)$ (RM)	PERCENTAGE FORECAST PROFIT DIFFERENCE,$D(n)$ (%)
	Mudharabah Investment			Single Investment $S(t)$	Mudharabah Investment	Single Investment		
	$E(t)$	$Q(t)$	$E(t)+Q(t)$					
AJI	379.49	162.64	542.12	545.78	3.11	3.68	1.26	0.001050
DLADY	353.05	151.31	504.36	507.22	1.27	1.81	1.02	0.003000
BKAWAN	298.03	127.73	428.01	425.76	2.10	2.28	0.83	0.006917
PETGAS	289.66	124.14	413.81	415.94	2.16	2.78	1.53	0.012750
TASEK	228.37	97.87	326.25	327.57	0.49	0.73	0.26	0.002167
TAKAFUL	226.40	97.03	323.43	324.73	3.64	3.34	0.39	0.003250
NESTLE	156.76	67.18	223.94	224.56	0.51	0.53	0.10	0.000833
SAPRESS	153.57	65.81	219.38	219.98	2.86	2.95	0.66	0.005500
PERSTIM	103.21	44.23	147.44	147.71	0.51	0.56	0.11	0.000917
LITRAK	102.18	43.79	145.97	146.24	1.62	1.64	0.08	0.000067
Average							0.62	0.003645

4 Conclusion

The study starts with the development of *mudharabah* with risk model for two parties which rate of return is considered as a total of a deterministic profit rate and a function of white noise. The white noise refers to systematic risk and unsystematic risk which is recognized as equal to the issue of Brownian motion.

The developed *mudharabah* investment model were used to forecast investment and profit for capital provider and entrepreneur. The total forecast profit for capital provider and entrepreneur was compared with forecast profit from single party investment. It was shown that the forecast profit for *mudharabah* investment was close to the forecast profit for single party investment. Then, the average forecast profit difference, $A(n)$ and percentage forecast profit difference, $D(n)$ were calculated for the ten best *syariah* counters. The smallest value of $D(n)$ indicated that profit of single party investment distributed accordingly by the developed *mudharabah* investment with risk model. Therefore it is verified that the developed *mudharabah* with risk model can be used in forecasting the investment and return for two parties.

In conclusion, the developed model can be used by investor to plan their risky investment wisely. Besides that, it is hope that the model can encourage more investors to make investment in *mudharabah* concept which follows *syariah* compliance. It is also hope that the development of model by using geometric Brownian motion is easier to understand its concept to investors. Furthermore, it is an honour if the result from this study can contribute and help future studies on *mudharabah* investment.

References

1. Aslina, O.: A Stochastic Mudharabah Model for Investment in Bursa Malaysia Syariah Counters. Master Thesis, Universiti Teknologi Mara (2013)
2. Azizah, M: Modeling Risky Investment By Considering Geometric Brownian Motion Using Musyarakah Concept. Master Thesis, Universiti Teknologi Mara (2014)
3. Faleel, J.: Islamic Finance for Dummies. John Wiley & Sons Inc., USA (2012)
4. Hyndman, R.J.: Measuring Forecast Accuracy, from http://robjhyndman.com
5. Lawrence, K.D., Kllimberg, R.K., Lawrence, S.M.: Fundamental of Forecasting Using Excel. Industrial Press Inc., USA (2009)
6. Maheran, M.J.: Model Pelaburan Musyarakah dan Mutanaqisah, Ph.D. thesis, Universiti Kebangsaan Malaysia (2006)
7. Maheran, M.J.: New musharakah model in managing Islamic investment. ISRA Int. J. Islamic Finan. 2(2), 25–36 (2010)
8. Mohamed, A.E.: Towards an Islamic stock market. Islamic Econ. Stud. 1(1), 1–20 (1993)
9. Muhammad, N.S.: Banking without interest. J. Res. Islamic Econ. 1(2), 85–90 (1983)
10. Wilmott, P.: Introduces Quantitative Finance. John Wiley and Son Ltd., England (2007)

Multi-script Text Detection and Classification from Natural Scenes

Zaidah Ibrahim[1]([⊠]), Zolidah Kasiran[1], Dino Isa[2], and Nurbaity Sabri[3]

[1] Faculty of Computer and Mathematical Sciences, Universiti Teknologi MARA,
40450 Shah Alam, Selangor, Malaysia
{Zaidah,zolidah}@tmsk.uitm.edu.my
[2] Faculty of Engineering, University of Nottingham Malaysia Campus, Jalan Broga,
43500 Semenyih, Selangor, Malaysia
dino.isa@nottingham.edu.my
[3] Faculty of Computer and Mathematical Sciences, Universiti Teknologi MARA Campus Jasin,
77300 Merlimau, Malacca, Malaysia
nurbaity_sabri@melaka.uitm.edu.my

Abstract. Most of the text detection and script classification approaches from natural scenes only cater for a single script whereas text in natural scenes may come in various scripts. This research proposes a gestalt-based approach for multi-script text detection and classification based on human perception. Human perceptual organization is where humans are able to organize visual input into meaningful information. This approach is based on the figure-ground articulation where we perceive the figure or text as standing in front of the background. Features extracted from wavelet coefficients and MSER is used as input to SVM for text detection and script classification. Experimental results indicate that this approach is competitive with the state of the art text detection and script classification approaches.

Keywords: Gestalt · MSER · Script classification · SVM · Text detection · Wavelet coefficients

1 Introduction

Text in images usually contains useful information. Currently, with the increase use of digital image capturing devices such as digital cameras and mobile phones, the automated procedures for the image understanding and interpretation are receiving great interest from researchers. The textual information present in these images can be used for various applications like license plate reading, road sign detection and translation and mobile text recognition. However, most of the existing text detection and script classification techniques only handle text in a single script.

Malaysia is a multi-racial country and three of the main races are Malay, Chinese and Indians. Even though the official language is Malay which is written in Latin characters, there are texts that are written in Chinese, Tamil and Jawi. The characters for these languages have different shapes and characteristics. Figure 1 illustrates some

© Springer Nature Singapore Pte Ltd. 2016
M.W. Berry et al. (Eds.): SCDS 2016, CCIS 652, pp. 200–210, 2016.
DOI: 10.1007/978-981-10-2777-2_18

sample images of these different scripts. According to [1], text script can be classified into Latin, Ideography and Arabic. Latin is written in languages like English or Malay. Ideography is written in languages like Chinese or Tamil while Jawi falls under the Arabic classification. These scripts can be differentiated based on height and density [1]. Latin and Ideography usually have similar height and pixel or stroke density while Arabic has lower height and density compared with the other two. But these rules may be applied for text in document images but not for scene text. Refer to Fig. 1 for an illustration of the multi-script text from natural scenes.

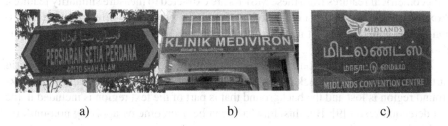

a)   b)   c)

Fig. 1. Sample images of text with multiple scripts from natural scenes.

Top-down and bottom-up are two popular approaches for text detection and usually these approaches only handle the detection of text from a single script. For top-down approach, the processing speed is relatively slow when sliding window is being used [2] since the size of the searched object is unknown in advance. Sliding windows of multiple sizes are applied to cater for multi-size font [3]. Bottom-up approach is sensitive to text alignment orientation whereas text in the natural scenes is not always in horizontal orientation. [4] applied gestalt-based approach that uses the similarity and proximity laws of human perceptual to detect text in various scripts. A gestalt is a region with significantly similar characteristics. The similarity laws utilize geometrical character-istics to classify text produce unsatisfactory results when testing with the scene images that consist of Chinese and Tamil scripts. This is due to the different geometrical char-acteristics between Chinese and Tamil scripts. More detail geometrical characteristics need to be further investigated for these multi-script text detection and classification.

By referring to Fig. 1, we can see that the multi-script characters do have some common features. They consist of similar and repetitive strokes and textures compared to the background. Even though we may not understand the meaning of the text but we can identify the existence of text in the image through these features. This research adopts the background/figure articulation from [5] since from personal observation, text does not always occupy the whole image and they stand out from the background. In this research, multi-script classification is based on a two-phase geometrical character-istics identification.

The paper is outlined as follows. In the next section, we discuss some of the related works. In Sect. 3, we present the proposed technique and the experimental results are discussed in Sect. 4. Section 5 concludes this paper and points to future work.

2 Related Work

Text detection usually involves two steps: feature extraction and text classification or detection. Various approaches have been applied for text detection and extraction and they can be categorized into gestalt-based or top-down and connected component-based or bottom-up approach.

[4] utilizes gestalt-based approach for proximity and similarity to detect text from natural scenes. First of all, Maximally Stable Extremal Regions (MSER) is constructed and geometrical features from these MSERs are extracted to measure similarity features. Proximity features are measured based on Helmholtz probability principle. One limitation of this approach is that non-character component within the defined proximity is also being detected as text.

Another alternative to construct text gestalt or region is to apply uniform image grid partition [6–8]. One drawback of this approach is that some text that is part of the background region is lost and the background that is part of the text region is included in the text detection result [9]. But this drawback can be overcome by applying non-uniform partition image grid.

Under the connected component approach, the character properties have been used to classify the connected components as text. According to [10], text can be distinguished from other elements in a scene image through its nearly constant stroke width. Using a logical geometric reasoning, components with similar stroke width can be grouped together into bigger components that are likely to be words. A connected component with a big stroke width variance is likely to be non-text. But this approach fails to detect text with strong highlights and excessive blur since there is not much uniformity of color distribution. This limitation is improved in [2] where the strokes are extracted in small image patches and five different corner strokes are identified in each patch. The stroke information in each patch provides better information about the existence of text. [11] extracts edges using Log-Gabor filter. The magnitude of the filter is extracted to produce an edge map. Then all the edge maps are integrated to form stroke maps. Since text component usually has a lot of strokes, the stroke map's coarseness is computed to detect the text lines. But, these approaches are not robust for images on complex background.

MSER extracts stable connected component regions from an image by looking at the change in the area with respect to the change in intensity of a connected component determined by thresholding the image at a given gray level. MSER is sensitive to image blur but [12] enhanced text detection performance in [13] by enhancing the blurred edges using Canny edge detector. [14] Improved the performance of their previous research in [13] by applying pruning exhaustive search to group the characters into words. A survey on text extraction from natural scene images can be found in [16]. But, all of these approaches only cater for single script.

Pixel density is computed within a bounding box for each character where Ideography script has higher pixel density compared to a Latin script while Arabic script has the lowest pixel density among the other two [17]. However problems occur in determining the bounding box since Arabic script does not have a consistent size of characters. It is also difficult to determine the threshold values for the densities of the different scripts. A review on script recognition for document images can be found in [15] where all the tested text is aligned horizontally whereas scene text images may not be aligned horizontally.

This paper proposes to utilize non-uniform partition image grid to construct gestalts for background and text for multi-script text. This is due to the fact that multi-script text has different geometrical features. Besides geometrical features, texture features are also extracted and act as input to Support Vector Machine (SVM) classifier for text detection and K-means clustering for script classification. Texture feature is applied because text usually has different texture compared to the background. The proposed approach will be discussed in the next section.

3   Multi-script Text Detection and Classification

3.1   Text Detection

A gestalt-based approach is presented for multi-script text detection where the extracted gestalts are detected based on geometrical, strokes and textures from a non-uniform partition image grid that represent either text or background. A two-phase approach is applied for multi-script classification. Figure 2 illustrates the block diagram of the proposed approach.

For an image with $m \times n$ partition grid, the partition parameter selection for the image grid needs to be identified in advance. If the value of m or n is small, each grid may not be able to cover the text region precisely whereas. If the value of m or n is large, the size of each grid may be too large and sensitive to noisy features. An experiment has been conducted to determine the size of the partition grid that is significant to be used for text/background gestalt classification. Table 2 illustrates the text/background gestalt classification result using SVM classifier based on the different sizes of each partition. A partition is classified as a text gestalt if there exists at least one partial character in that gestalt. SVM has been chosen as the classifier because of its popularity and performance in text detection [18–20].

Firstly, text image classification is being conducted to determine whether the image consist of text or not. This is significant for text frame classification in video images to eliminate the text detection process for frames that do not consist of text. For a given video image, it not necessary that all of the frames consists of text. The following features are extracted from discrete wavelet transform coefficients: mean, energy, and entropy. They are defined in Table 1 with 64×64 partition image grid size and they have been applied in [7] for text frame classification. In Table 1, $W(i,j)$ is the sub-partition image at pixel i,j. The proposed approach classifies better with the combination of features extracted from MSER, which are, area, aspect ratio, euler number and number of pixels on skeleton of image with 0.82 compared to 0.79 in [7]. A frame is classified as a text frame if it consists of at least one partition that has been classified as text. These features are extracted at three different levels of sub-bands to determine which level of sub-band leads to better text frame classification. The ICDAR2003 dataset [22] has been used for training and testing. The result is illustrated in Table 2.

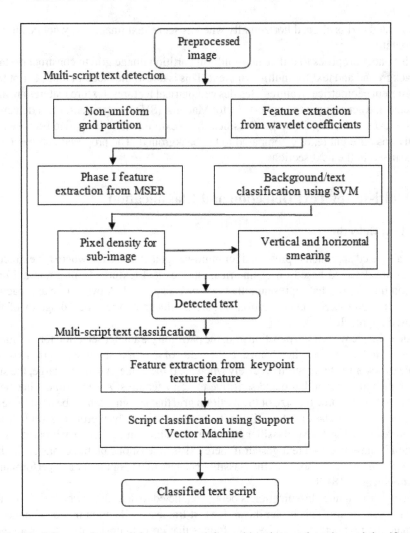

Fig. 2. Block diagram for gestalt-based approach for multi-script text detection and classification.

Table 1. Formula to compute mean, energy and entropy.

mean	$\dfrac{1}{NM}\sum\limits_{i=1}^{N}\sum\limits_{j=1}^{N} W(i,j)$
energy	$\sum\limits_{i,j} W^2(i,j)$
entropy	$\sum\limits_{i,j} W(i,j).\log W(i,j)$

Table 2. Result of text/background gestalt classification using SVM classifier with features from wavelet coefficients and MSER using ICDAR2003 dataset.

Wavelet level 1	Wavelet level 2	Wavelet level 3	MSER
0.6	0.65	0.7	0.58

Wavelet level 1 and MSER	Wavelet level 2 and MSER	Wavelet level 3 and MSER
0.65	0.71	0.82

By looking at the result in Table 2, we can see that the text/background frame classification result for the third level of wavelet coefficients sub-bands is better than the other two levels. Since text has distinct textual features compared to background, the use of wavelet coefficients lead to good text classification performance since they have good abilities to characterize texture features [20]. The use of MSER features do not perform as well as wavelet features because MSER is sensitive to illumination changes [21]. But the combination of the features extracted from wavelet coefficients and MSER performs better since wavelet coefficients are not as sensitive to image blur and noises compared to MSER.

Table 3. Result of text gestalt classification using SVM classifier based on the features extracted from MSER and wavelet coefficients.

SVM with features from wavelet coefficients and MSER	ICDAR 2003
32 × 32 image grid	0.79
16 × 16 image grid	0.75
64 × 16 image grid	0.78
64 × 32 image grid	0.84

A second experiment has been conducted by this research to determine a good partition image grid size. By looking at the result illustrated in Table 3, we can see that the text/background frame classification performance is better if the image is divided into non-symmetrical image grid of 64 × 32. This is because a 64 × 64 image grid consists of too many combinations of text and non-text objects and it does not cover the text region precisely. The occurrences of more combinations of the non-text objects similar to the text object seem more frequent. On the other hand, a smaller image grid fails to differentiate between the characters. As a result of the non-uniform partitioning, there are characters that have been partitioned into two separate partitions and this leads to misclassification. Figure 3 illustrates this example.

For every row of partition sub-images, if there exists a text partition sub-image and if its neighboring partition sub-image is classified as background sub-image, then it may be due to the break-up of the characters into two partition sub-images. Thus, to overcome this misclassification, the neighboring sub-image that has been classified as background

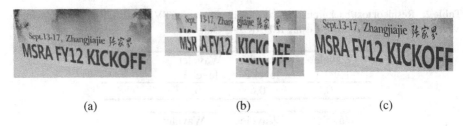

(a)   (b)   (c)

Fig. 3. (a) Original image from MSRA-TD500 (b) sub-image with partial characters image (c) detected text.

sub-image is further partitioned into two patterns as shown in Fig. 4 and the pixel density is computed for each sub-partitioned image. If the pixel density of one of the sub-partitioned image is more than a threshold value, then it is classified as text sub-image or text gestalt.

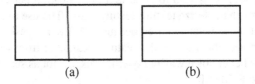

(a)   (b)

Fig. 4. Patterns to compute pixel density.

By looking at Fig. 3b, we can see that the fourth sub-image from the right has been misclassified as background gestalt even though it consists of a Chinese character. But since the third gestalt has been classified as a text gestalt, the fourth gestalt is further examined to reduce misclassification. The fourth gestalt is partitioned further for density computation and it has been classified as a text gestalt since the pixel density is more than a predefined threshold value.

To extract the text, vertical and horizontal smearing is applied for all the gestalts that have been classified as text. The result of the combination of the vertical and horizontal smearing is the detected text as shown in Fig. 3c. Figures 5 and 6 illustrate some sample results of the detected text.

Fig. 5. Sample results of the detected text from MSRA-TD500 dataset.

Fig. 6. Sample results of the detected text from personal collection.

3.2   Script Classification

The detected text is then further processed to classify the text script so that the result of this process can be recognized by the Optical Character Recognition (OCR) system. By referring to Fig. 6, we can see that the four scripts, that are, Latin for English and Malay, Chinese, Tamil and Jawi have distinct structural features that can be used to differentiate them. These structural features are listed in Table 4 which can be constructed as rules for the script classification.

Table 4. Structural features for script classification.

Script	Distinct structural features
Jawi	No consistency for the space between each character and the character itself, and no holes.
Tamil	No consistency for the space between each character and the character itself but it has a lot of holes, curves and horizontal strokes.
Chinese	Consistent size of character and inter-character space but each character does not consist of any holes
Malay and English	Consistent size of character and inter-character space, consist of a few holes, with horizontal, vertical and slanting strokes.

4   Experimental Results

Table 5 compares the result of the proposed approach with [4] using ICDAR2003 and personal data as the training data and MSRA-TD500 dataset as the testing data. The precision (P) and recall (R) rates are computed as follows: $P=|TP|/|E|$ and $R=|TP|/|T|$ where TP is true positive, E is estimated rectangle and T is ground truth rectangle. Figure 7 shows some sample images from MSRA-TD500 dataset [23] where the text detection fails. This is due to highlights, shadows and font size is too small.

Table 5. Text detection result.

Method	P	R	F-score
[4]	0.58	0.54	0.56
Proposed method	0.61	0.55	0.58

Fig. 7. Some sample images from MSRA-TD500 whose text cannot be detected by the proposed method.

Table 6 shows the result for script classification using personal dataset since the authors cannot find a suitable dataset that consists of the four different scripts. Classification rate is computed by dividing the correct script classification with the total number of word scripts. Figure 8a shows some scripts that have been correctly classified while Fig. 8b shows some scripts that have been misclassified by the proposed technique. Among the reason of misclassification is different script style, blurring and the different inter-character space of the words. More features need to be identified in order to cater these problems. Roman script achieves the highest classification rate because the number of different letters in this script is less compared to the other scripts. Besides that, Roman script does not consist of a lot of disconnected or small components as Jawi or Tamil where these scripts consist of a lot of periods that are small and disconnected from the major shape of the scripts.

Table 6. Script classification result.

Script	Correct classification [17]	Correct classification (proposed technique)
Jawi	0.5	0.6
Tamil	0.5	0.65
Chinese	0.6	0.7
Roman	0.75	0.85

(a)   (b)

Fig. 8. Sample scripts (a) correctly classified (b) misclassified

5 Conclusion

Detecting text from scene images is very challenging due to complex background, different orientation and multiple scripts. This research illustrates that a combination of features extracted from wavelet coefficients and MSER show promising result to detect text with the mentioned challenges. The use of structural features to classify the text scripts is also very promising. The text detection and script classification performance can be further improved by adding more features and more training data. This research plans to apply filtering techniques for background complexity, skew techniques for different text orientation and automatic adaptive partition image grid for better gestalt classification.

Acknowledgments. The authors thank the Ministry of Education and Universiti Teknologi MARA for sponsoring this research under the National Grant No 600-RMI/FRGS 5/3 (165/2013).

References

1. Suen, S.Y., Bergler, S., Nobile, N., Waked, B., Nadal, C.P., Bloch, A.: Categorizing document images into script and language classes. In: International Conference on Advances in Pattern Recognition, pp. 297–306 (1998)
2. Zhao, Y., Lu, T., Liao, W.: A Robust color-independent text detecting method from complex videos. In: ICDAR, pp. 374–378 (2011)
3. Pan, Y., Zhu, Y., Sun, J., Naoi, S.: Improving scene text detection by scale-adaptive segmentatin and weighted CRF varification. In: ICDAR, pp. 759–763 (2011)
4. Gomez, L., Karatzas, D.: Multi-script text extraction from natural scenes. In: 12th International Conference on Document Analysis and Recognition, pp. 467–471 (2013)
5. Gomez, L.: Perceptual organization for text extraction in natural scenes. Master Thesis, Computer Vision and Artificial Intelligence (2013)
6. Bouman, K.L., Abdollahian, G., Boutin, M., Delp, E.J.: A low complexity sign detection and text localization method for mobile applications. IEEE Trans. Multimedia **13**(5), 922–934 (2011)
7. Shivakumara, P., Dutta, A., Phan, T.Q., Tan, C.L., Pal, U.: A novel mutual nearest neighbor based symmetry for text frame classification in video. Pattern Recogn. **44**, 1671–1683 (2011)
8. Shivakumara, P., Dutta, A., Tan, C.L., Pal, U.: Multi-oriented scene text detection in video based on wavelet and angle projection boundary growing. Multimedia Tools Appl. **72**(1), 515–539 (2014)
9. Zhou, C., Yuan, J.: Arbitrary-shape object localization using adaptive image grids. In: Lee, K.M., Matsushita, Y., Rehg, J.M., Hu, Z. (eds.) ACCV 2012, Part I. LNCS, vol. 7724, pp. 71–84. Springer, Heidelberg (2013)
10. Epshtein, B., Ofek, E., Wexler, Y.: Detecting text in natural sceneswith stroke width transform. In: CVPR, pp. 2963–2970 (2010)
11. Huang, X., Ma, H.: Automatic detection and localization of natural scene text in video. In: ICPR, pp. 3216–3219 (2010)
12. Chen, H., Tsai, S.S., Schroth, G., Chen, D.M., Grzeszezuk, R., Girod, B.: Robust text detection in natural images with edge-enhanced maximally stable extremal regions. In: ICIP, pp. 2609–2612 (2011)

13. Neumann, L., Matas, J.: A method for text localization and recognition in real-world images. In: Kimmel, R., Klette, R., Sugimoto, A. (eds.) ACCV 2010, Part III. LNCS, vol. 6494, pp. 770–783. Springer, Heidelberg (2011)
14. Neumann, L., Matas, J.: Text localization in real-world images using efficiently pruned exhaustive search. In: ICDAR, pp. 687–691 (2011)
15. Ghosh, D., Dube, T., Shivaprasad, A.P.: Script recognition – a review. IEEE Trans. Pattern Anal. Mach. Intell. **32**(12), 2142–2161 (2009)
16. Zhang, H., Zhao, K., Song, Y., Guo, J.: Text extraction from natural scene image: a survey. Neurocomputing **122**, 310–323 (2013)
17. Hochberg, J., Kelly, P., Thomas, T., Kerns, L.: Automatic script identification from document images using cluster-based templates. IEEE Trans. Pattern Anal. Mach. Intell. **19**(2) (1997)
18. Anthimopoulos, M., Gatos, B., Pratikakis, I.: A two-stage scheme for text detection in video images. Image Vis. Comput. **28**, 1413–1426 (2010)
19. Steinwart, I., Christmann, A.: Support Vector Machine. Springer Science & Business Media, New York (2008)
20. Shivakumara, P., Phan, T.Q., Tan, C.L.: A robust wavelet transform based technique for video detection. In: ICDAR, pp. 1285–1289 (2009)
21. Zhen, W., Wei, Z.: A comparative study of feature selection for SVM in video text detection. In: Second International Symposium on Computational Intelligence and Design, vol. 2, pp. 552–556 (2009)
22. Lucas, S.M., Panaretos, A., Sosa, L., Tang, A., Wong, S., Young, R.: ICDAR 2003 robust reading competitions: entries, results, and future directions. IJDAR **7**(2–3), 105–122 (2005)
23. Yao, C., Bai, X., Liu, W., Ma, Y., Tu, Z.: Detecting texts of arbitrary orientations in natural images. In: Proceedings of the CVPR, pp. 1083–1090 (2012)

Fuzzy Logic

Fuzzy Logic

Algebraic and Graphical Interpretation of Complex Fuzzy Annulus (an Extension of Complex Fuzzy Sets)

Ganeshsree Selvachandran[1(✉)], Omar Awad Mashaan[2], and Abdul Ghafur Ahmad[2]

[1] Department of Actuarial Science and Applied Statistics, Faculty of Business and Information Science, UCSI University, Jalan Menara Gading, Cheras, 56000 Kuala Lumpur, Malaysia
ganeshsree86@yahoo.com
[2] School of Mathematical Sciences, Faculty of Science and Technology, Universiti Kebangsaan Malaysia, 43600 UKM Bangi, Selangor DE, Malaysia

Abstract. Complex fuzzy sets, which include complex-valued grades of memberships, are extensions of standard fuzzy sets that better represent time-periodic problem parameters. However, the membership functions of complex fuzzy sets are difficult to enumerate, as they are subject to personal preferences and bias. To overcome this problem, we generalize complex fuzzy sets to the *complex fuzzy annulus*, whose image is a sub-disk lying in the unit circle in the complex plane. The set theoretic operations of this concept are introduced and their algebraic properties are verified. The proposed model is then applied to a real-life problem, namely, the influencers of the Malaysian economy and the time lag between the occurrences of these influencers and their first manifestations in the economy.

Keywords: Complex fuzzy set · Fuzzy annulus · Interval-valued fuzzy set

1 Introduction

Complex fuzzy sets were conceptualized by Ramot et al. [1] for accurate representation of time-periodic problem parameters. Complex fuzzy sets are convenient for representing two-dimensional information; i.e. the description and periodicity of the problem parameters. In complex fuzzy set notation, this two-dimensional information is represented in a single set rather than as two or more ordinary fuzzy sets, reducing the execution time and the number of computations required in the problem solving. Among other notable research in this relatively unexplored area are the application of complex fuzzy sets to traditional fuzzy logic via the introduction of complex fuzzy logic by Ramot et al. [2] and Dick [3] who further expanded the theory of complex fuzzy logic by introducing several new operators pertaining to this theory.

However, assigning suitable values to the membership functions of the elements in complex fuzzy sets is subjective, and influenced by personal bias and/or preferences. To overcome this problem, many researchers have proposed hybrid structures of fuzzy

© Springer Nature Singapore Pte Ltd. 2016
M.W. Berry et al. (Eds.): SCDS 2016, CCIS 652, pp. 213–223, 2016.
DOI: 10.1007/978-981-10-2777-2_19

sets, such as interval-valued fuzzy sets [4], vague sets [5], intuitionistic fuzzy sets [6] and interval-valued intuitionistic fuzzy sets [7]. All of these set structures have improved the accuracy of assigning values to fuzzy set members.

Inspired by these researches and aiming to overcome similar problems, we extend the concept of complex fuzzy sets to the *complex fuzzy annulus*. Our complex fuzzy annulus more accurately represents the problem parameters in a complex setting, both algebraically and graphically. It functions similar to an ordinary annulus, which is characterized by two functions involving the radii of the inner and outer circles. Similarly, the complex fuzzy annulus is graphically described as an annulus within a unit circle in the complex plane. The lower and upper radii of the annulus represent the lower and upper bounds of the amplitude, respectively, whereas the lower and upper arguments of the complex number associated with the complex fuzzy annulus represent the lower and upper bounds of the phase term, respectively. By virtue of this feature, the proposed concept conveniently represents the parameters in real-world problems, especially when the membership function of an element cannot be assigned with absolute certainty, or is very difficult to assign. This difficulty arises because most of the attributes or parameters involved in real-world problems are subjective and humanistic in nature. Therefore, the assignment of membership functions is often dictated by individual perceptions, which are influenced by an individual's past experiences, geographic location, and family and educational background. By representing two-dimensional information as a complex fuzzy annulus, we can define a closed interval for the values of the membership functions, which cannot be assigned in complex fuzzy sets. This allows a more accurate representation of two-dimensional information.

In this paper, we introduce the concept of a complex fuzzy annulus and some special cases of this concept and graphically represent these sets in the complex plane. Furthermore, we define the basic operations (complement, union, and intersection) of the complex fuzzy annulus and verify the algebraic properties of these operations, namely, the commutative and associative laws and De Morgan's laws. The utility of the complex fuzzy analysis in representing periodic situations is demonstrated in an economic problem.

2 Preliminaries

In this section, we recapitulate the relevant concepts of fuzzy sets and aspects of their hybrid structure.

Definition 2.1 [8]. A *fuzzy set* A in a universe of discourse U is characterized by a membership function $\mu_A(x)$ that takes values in the interval $[0, 1]$, where $A \subset U$ and

$$\mu_A : x \rightarrow [0, 1], \forall x \in U. \tag{1}$$

Ramot et al. [1] introduced complex fuzzy sets as an extension of standard fuzzy sets. Complex fuzzy sets are fuzzy sets whose elements' membership functions are expressed as complex numbers with an amplitude and a phase term. The novelty of complex fuzzy sets lies in the range of values admitted by its membership function.

While the membership function of a classical fuzzy set is restricted to the interval [0, 1], the membership function range of complex fuzzy sets extends over a unit circle in the complex plane. The main definitions and the important set theoretic operations of complex fuzzy sets are presented below.

Definition 2.2 [1]. A *complex fuzzy set* A defined on a universe of discourse U is characterized by a membership function $\mu_A(x)$ that assigns a complex-valued grade of membership in A to any element $x \in U$. By definition, all values of $\mu_A(x)$ lie within the unit circle in the complex plane and are expressed by $\mu_A(x) = r_A(x)e^{i\omega_A(x)}$, where $i = \sqrt{-1}$, $r_A(x)$ and $\omega_A(x)$ are both real-valued, and $r_A(x) \in [0, 1]$. A complex fuzzy set A is the following set of ordered pairs:

$$A = \{(x, \mu_A(x)) : x \in U\} = \left\{\left(x, r_A(x)e^{i\omega_A(x)}\right) : x \in U\right\}. \tag{2}$$

Definition 2.3 [1]. Let A and B be two complex fuzzy sets on U with membership functions $\mu_A(x) = r_A(x)e^{i\omega_A(x)}$ and $\mu_B(x) = r_B(x)e^{i\omega_B(x)}$, respectively. The basic set theoretic operations on A and B are given below.

(i)   The *complement* of A, denoted by $\overline{A}$, is defined as

$$\overline{A} = \left\{(x, \mu_{\overline{A}}(x)) : x \in U\right\} = \left\{\left(x, r_{\overline{A}}(x)e^{i\omega_{\overline{A}}(x)}\right) : x \in U\right\}, \tag{3}$$

where $r_{\overline{A}}(x) = (1 - r_A(x))$ and $\omega_{\overline{A}}(x) = 2\pi - \omega_A(x)$.

(ii)   The *union* of A and B, denoted by $A \cup B$ is defined as

$$A \cup B = \{(x, \mu_{A \cup B}(x)) : x \in U\} = \left\{\left(x, r_{A \cup B}(x)e^{i\omega_{A \cup B}(x)}\right) : x \in U\right\}, \tag{4}$$

where $r_{A \cup B}(x) = \max\{r_A(x), r_B(x)\}$ and $\omega_{A \cup B}(x) = \max(\omega_A(x), \omega_B(x))$.

(iii)   The *intersection* of A and B, denoted by $A \cap B$ is defined as

$$A \cap B = \{(x, \mu_{A \cap B}(x)) : x \in U\} = \left\{\left(x, r_{A \cap B}(x)e^{i\omega_{A \cap B}(x)}\right) : x \in U\right\}, \tag{5}$$

where $r_{A \cap B}(x) = \min\{r_A(x), r_B(x)\}$ and $\omega_{A \cap B}(x) = \min(\omega_A(x), \omega_B(x))$.

3   Introduction to the Complex Fuzzy Annulus

An annulus is a region bounded by two concentric circles. In complex analysis, an annulus $ann(a; r, R)$ in the complex plane is an open region defined by $r < |z - a| < R$. This section develops our novel concept of the complex fuzzy annulus, which lies on a unit circle in the complex plane. To generalize the annulus in a complex plane, we replace the real-valued radii in the annulus with complex-valued fuzzy membership grades. This complex fuzzy annulus maps a universal set to a sub-disk of the unit circle in the complex plane. The complex fuzzy annulus is formally defined below.

Definition 3.1. Let A be a complex fuzzy set over a universe of discourse U. Then a *complex fuzzy annulus* of U, denoted by $A^\odot$, is characterized by two membership functions $\underline{\mu}_A^\odot$ and $\overline{\mu}_A^\odot$, representing the lower and upper bounds of the membership function of $A^\odot$, respectively. The lower bound of the membership function is the mapping $\underline{\mu}_A^\odot : U \to \{z_1 : z_1 \in \mathbb{C}\}$, defined as $\underline{\mu}_A^\odot = z_1 = \underline{r}_{A^\odot}(x)e^{i\underline{\omega}_{A^\odot}(x)}$. The upper bound is the mapping $\overline{\mu}_A^\odot : U \to \{z_2 : z_2 \in \mathbb{C}\}$, defined as $\overline{\mu}_A^\odot = z_2 - \overline{r}_{A^\odot}(x)e^{i\overline{\omega}_{A^\odot}(x)}$ with $0 \le |z_1| \le |z_2| \le 1$ and $i = \sqrt{-1}$. Both the amplitude terms $\underline{r}$ and $\overline{r}$ and the phase terms $\underline{\omega}$ and $\overline{\omega}$ are real-valued with $\underline{r} < \overline{r}$ and $\underline{\omega} < \overline{\omega}$. If $\underline{r} = \overline{r}$ and $\underline{\omega} = \overline{\omega}$, the complex fuzzy annulus reduces to a complex fuzzy set as defined in Eq. (2). If $\underline{w} = \overline{w} = 0\pi$, then the complex fuzzy annulus degenerates to an ordinary fuzzy set as defined in Eq. (1). Thus the ordinary fuzzy set and complex fuzzy set are special cases of the complex fuzzy annulus.

By definition, the values of $\underline{\mu}_A^\odot$ and $\overline{\mu}_A^\odot$ lie on a disk inside the unit circle centered at the origin in the complex plane. Thus, the complex fuzzy annulus $A^\odot$ may be represented as the set of ordered pairs

$$A^\odot(x) = \left\{ \left(x, ann\left(0, \underline{\mu}_A^\odot, \overline{\mu}_A^\odot\right)\right) : x \in U \right\}.$$

The amplitude terms of $\underline{\mu}_A^\odot$ and $\overline{\mu}_A^\odot$ represent the radii of the inner and outer bounded circles, respectively. They also signify the lower and upper bounds of the degree of belongingness of the elements with respect to certain criteria. Conversely, the phase terms $\underline{\mu}_A^\odot$ and $\overline{\mu}_A^\odot$ represent the lower and upper bounds of the *phase*, or the periodic changes in the elements over time (called the *periodicity* of the elements). For example, the term $\left(x, ann\left(0, 0.2e^{0.3\pi i}, 0.5e^{0.5\pi i}\right)\right)$ tells us that the membership function and the phase of element x is between 0.2 and 0.5 and between 0.3π and 0.5π, respectively.

The complex fuzzy annulus is graphically conceptualized in Fig. 1.

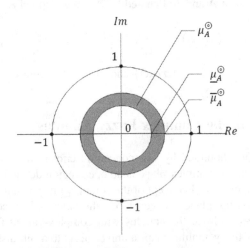

Fig. 1. A complex fuzzy annulus.

There are two other special cases for the complex fuzzy annulus:

(i)   A *fully fuzzy punctured disk* occurs when $\underline{\mu}_A^{\odot}(x) = 0$ and $\overline{\mu}_A^{\odot}(x) = 1$ for all $x \in U$.
(ii) A *null fuzzy punctured disk* occurs when $\underline{\mu}_A^{\odot}(x) = \overline{\mu}_A^{\odot}(x) = 0$ for all $x \in U$.

These two cases are graphically represented in Figs. 2 and 3, respectively.

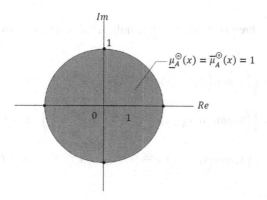

Fig. 2. A fully fuzzy punctured disk

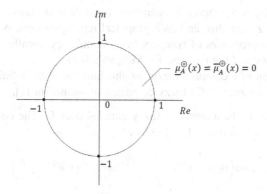

Fig. 3. A null fuzzy punctured disk

Definition 3.2. Let $A^{\odot}$ and $B^{\odot}$ be complex fuzzy annuli over a common universe of discourse U. Then, $A^{\odot}$ is a *subset* of $B^{\odot}$, denoted by $A^{\odot} \subseteq B^{\odot}$, if and only if A is a complex fuzzy subset of B; that is, the following conditions are satisfied for all $x \in U$:

(a) $\underline{r}_{A^{\odot}}(x) \leq \underline{r}_{B^{\odot}}(x)$ and $\overline{r}_{A^{\odot}}(x) \leq \overline{r}_{B^{\odot}}(x)$ for the amplitude terms
(b) $\underline{\omega}_{A^{\odot}}(x) \leq \underline{\omega}_{B^{\odot}}(x)$ and $\overline{\omega}_{A^{\odot}}(x) \leq \overline{\omega}_{B^{\odot}}(x)$ for the phase terms.

Definition 3.3. Let $A^\odot$ and $B^\odot$ be complex fuzzy annuli over a common universe of discourse U. Then, $A^\odot$ is *equal* to $B^\odot$ denoted by $A^\odot \equiv B^\odot$, if and only if for all $x \in U$, A is a complex fuzzy subset of B and B is a complex fuzzy subset of A, that is the following conditions must be satisfied for all $x \in U$:

(a) $\underline{r}_{A\odot}(x) = \underline{r}_{B\odot}(x)$ and $\overline{r}_{A\odot}(x) = \overline{r}_{B\odot}(x)$ for the amplitude terms
(b) $\underline{\omega}_{A\odot}(x) = \underline{\omega}_{B\odot}(x)$ and $\overline{\omega}_{A\odot}(x) = \overline{\omega}_{B\odot}(x)$ for the phase terms.

We now define three complex fuzzy annuli, $A^\odot, B^\odot$, and $C^\odot$, over a universe of discourse U:

$$A^\odot = \left\{ \left(x, ann\left(0, \underline{r}_{A\odot}(x)e^{i\underline{\omega}_{A\odot}(x)}, \overline{r}_{A\odot}(x)e^{i\overline{\omega}_{A\odot}(x)} \right) \right) : x \in U \right\},$$

$$B^\odot = \left\{ \left(x, ann\left(0, \underline{r}_{B\odot}(x)e^{i\underline{\omega}_{B\odot}(x)}, \overline{r}_{B\odot}(x)e^{i\overline{\omega}_{B\odot}(x)} \right) \right) : x \in U \right\},$$

$$C^\odot = \left\{ \left(x, ann\left(0, \underline{r}_{C\odot}(x)e^{i\underline{\omega}_{C\odot}(x)}, \overline{r}_{C\odot}(x)e^{i\overline{\omega}_{C\odot}(x)} \right) \right) : x \in U \right\}.$$

4   Set Theoretic Operations on the Complex Fuzzy Annulus

This section introduces the theoretic operations (complement, union, and intersection) of the complex fuzzy annulus, and their graphical representations. We then study and verify the algebraic properties of complex fuzzy annuli, specifically, the commutative and associative laws and the relevant De Morgan's laws.

The complement of a complex fuzzy annulus introduced in Definition 4.1 is based on the directional complex (DC) fuzzy complement defined in [1].

Definition 4.1. Let $A^\odot$ be a complex fuzzy annulus over U. The *complement* of $A^\odot$, denoted by $(A^\odot)^c$ is then defined as follows:

$$(A^\odot)^c = \left\{ \left(x, ann\left(0, (\underline{r}_{A\odot})^c(x)e^{i(\underline{\omega}_{A\odot})^c(x)}, (\overline{r}_{A\odot})^c(x)e^{i(\overline{\omega}_{A\odot})(x)} \right) \right) : \forall x \in U \right\}, \quad (6)$$

where the complements of the amplitude terms are $(\underline{r}_{A\odot})^c(x) = 1 - \overline{r}_{A\odot}(x)$ and $(\overline{r}_{A\odot})^c(x) = 1 - r_{A\odot}(x)$ and the complements of the phase terms are $(\underline{\omega}_{A\odot})^c(x) = 2\pi - \overline{\omega}_{A\odot}(x)$ and $(\overline{\omega}_{A\odot})^c(x) = 2\pi - \underline{\omega}_{A\odot}(x)$.

Proposition 4.2 *Let $A^\odot$ be a complex fuzzy annulus over a universe U. Then* $((A^\odot)^c)^c = A^\odot$.

Proof. Let $(A^\odot)^c = B^\odot$. Then, for all $x \in U$, it follows that

$$\left(B^{\odot}\right)^{c}=\left\{\left(x,ann\left(\left(1-\overline{r}_{B^{\odot}}(x)\right)e^{i\left(2\pi-\overline{\omega}_{B^{\odot}}(x)\right)},\left(1-\underline{r}_{B^{\odot}}(x)\right)e^{i\left(2\pi-\underline{\omega}_{B^{\odot}}(x)\right)}\right)\right)\right\}$$

$$=\left\{\left(x,ann\left(\begin{array}{c}\left(1-\left(1-\underline{r}_{A^{\odot}}(x)\right)\right)e^{i\left(2\pi-\left(2\pi-\underline{\omega}_{A^{\odot}}(x)\right)\right)},\\\left(1-\left(1-\overline{r}_{A^{\odot}}(x)\right)\right)e^{i\left(2\pi-\left(2\pi-\overline{\omega}_{A^{\odot}}(x)\right)\right)}\end{array}\right)\right)\right\}$$

$$=\left\{\left(x,ann\left(0,\underline{r}_{A^{\odot}}(x)e^{i\underline{\omega}_{A^{\odot}}(x)},\overline{r}_{A^{\odot}}(x)e^{i\overline{\omega}_{A^{\odot}}(x)}\right)\right):\forall x\in U\right\}$$

$$=A^{\odot}.$$

This completes the proof.   ∎

Definition 4.3. Let $A^{\odot}$ and $B^{\odot}$ be two complex fuzzy annuli over U. The basic set theoretic operations on $A^{\odot}$ and $B^{\odot}$ are as follows:

(i) The *union* of $A^{\odot}$ and $B^{\odot}$ is a complex fuzzy annulus $C^{\odot}$, defined as $A^{\odot}\cup B^{\odot}=C^{\odot}$. The membership function of $C^{\odot}$ is

$$C^{\odot}=\left\{\left(x,ann(0,\underline{r}_{A\cup B^{\odot}}(x)e^{i\underline{\omega}_{A\cup B^{\odot}}(x)},\overline{r}_{A\cup B^{\odot}}(x)e^{i\overline{\omega}_{A\cup B^{\odot}}(x)}\right):x\in U\right\}$$

$$=\left\{\left(x,ann\left(0,\max(\underline{r}_{A^{\odot}}(x),\underline{r}_{B^{\odot}}(x))e^{i\left(\max\left(\underline{\omega}_{A^{\odot}}(x),\underline{\omega}_{B^{\odot}}(x)\right)\right)},\right.\right. \tag{7}$$

$$\left.\left.\max(\overline{r}_{A^{\odot}}(x),\overline{r}_{B^{\odot}}(x))e^{i\left(\max\left(\overline{\omega}_{A^{\odot}}(x),\overline{\omega}_{B^{\odot}}(x)\right)\right)}\right)\right):x\in U\right\}.$$

(ii) The *intersection* of $A^{\odot}$ and $B^{\odot}$ is a complex fuzzy annulus $D^{\odot}$, defined as $A^{\odot}\cap B^{\odot}=D^{\odot}$. The membership function of $D^{\odot}$ is

$$D^{\odot}=\left\{\left(x,ann(0,\underline{r}_{A\cap B^{\odot}}(x)e^{i\underline{\omega}_{A\cap B^{\odot}}(x)},\overline{r}_{A\cap B^{\odot}}(x)e^{i\overline{\omega}_{A\cap B^{\odot}}(x)}\right):x\in U\right\}$$

$$=\left\{\left(x,ann\left(0,\min(\underline{r}_{A^{\odot}}(x),\underline{r}_{B^{\odot}}(x))e^{i\left(\min\left(\underline{\omega}_{A^{\odot}}(x),\underline{\omega}_{B^{\odot}}(x)\right)\right)},\right.\right. \tag{8}$$

$$\left.\left.\min(\overline{r}_{A^{\odot}}(x),\overline{r}_{B^{\odot}}(x))e^{i\left(\min\left(\overline{\omega}_{A^{\odot}}(x),\overline{\omega}_{B^{\odot}}(x)\right)\right)}\right)\right):x\in U\right\}.$$

These operator definitions are analogous to the original definitions in [1]. The max and min operators act on the amplitude terms of the union and intersection of the complex fuzzy annulus, respectively. The most commonly used operators for the phase term are also the max and min operators. Meanwhile, there are several other possibilities for the operators on the phase terms $\underline{\omega}_{A\cup B^{\odot}}$ and $\overline{\omega}_{A\cup B^{\odot}}$. Choice of these operators largely depends on the application or the context in which the union or intersection of the complex fuzzy annulus is used. Some commonly used operators on the phase terms are given below. In deriving these operators, we generalized the original definitions of Ramot et al. [1] to the concept of complex fuzzy annuli.

(i) Sum: $\underline{\omega}_{A \cup B^\circ} = \underline{\omega}_{A^\circ} + \underline{\omega}_{B^\circ}$ and $\bar{\omega}_{A \cup B^\circ} = \bar{\omega}_{A^\circ} + \bar{\omega}_{B^\circ}$

(ii)   "Winner takes all" : $\underline{\omega}_{A \cup B^\circ} = \begin{cases} \underline{\omega}_{A^\circ} & \text{if } \underline{r}_{A^\circ} > \underline{r}_{B^\circ} \\ \underline{\omega}_{B^\circ} & \text{if } \underline{r}_{B^\circ} > \underline{r}_{A^\circ} \end{cases}$

            and $\bar{\omega}_{A \cap B^\circ} = \begin{cases} \bar{\omega}_{A^\circ} & \text{if } \bar{r}_{A^\circ} > \bar{r}_{B^\circ} \\ \bar{\omega}_{B^\circ} & \text{if } \bar{r}_{B^\circ} > \bar{r}_{A^\circ} \end{cases}$

Thus, the choice of operators for the phase terms almost always depends on the context of the application. Further information is provided in [1].

All of the regular commutative and associative laws pertaining to these operations hold, and follow directly from the commutative and associative properties of the maximum and minimum operators used in the union and intersection operations. Therefore, we omit their proofs, and prove only the relevant De Morgan's laws for the union and intersection of two complex fuzzy annuli.

Proposition 4.4. Let A° and B° be two complex fuzzy annuli over U. Then, the following De Morgan's laws hold:

(i) $(A^\circ \cup B^\circ)^c = (A^\circ)^c \cap (B^\circ)^c$

(ii) $(A^\circ \cap B^\circ)^c = (A^\circ)^c \cup (B^\circ)^c$

(iii) $((A^\circ)^c \cup (B^\circ)^c)^c = A^\circ \cap B^\circ$

(iv) $((A^\circ)^c \cap (B^\circ)^c)^c = A^\circ \cup B^\circ.$

Proof. To prove these De Morgan's laws, we use the standard maximum and minimum operators. These laws also hold for the sum, winner-takes-all, and all other commonly used operators.

(i) $(A^\circ \cup B^\circ)^c$

$= \left\{ \left(x, \left(ann\left(0, (1 - \max(\underline{r}_{A^\circ}(x), \underline{r}_{B^\circ}(x))) \right) e^{i\left(1 - \max\left(\underline{\omega}_{A^\circ}(x), \underline{\omega}_{B^\circ}(x)\right)\right)}, \right. \right. \right.$
$\left. \left. \left. (1 - \max(\bar{r}_{A^\circ}(x), \bar{r}_{B^\circ}(x))) e^{i\left(1 - \max\left(\bar{\omega}_{A^\circ}(x), \bar{\omega}_{B^\circ}(x)\right)\right)} \right) \right) \right\}$

$= \left\{ \left(x, \left(ann\left(0, (\min(1 - \bar{r}_{A^\circ}(x), 1 - \bar{r}_{B^\circ}(x))) \right) e^{i\left(\min\left(2\pi - \bar{\omega}_{A^\circ}(x), 2\pi - \bar{\omega}_{B^\circ}(x)\right)\right)}, \right. \right. \right.$
$\left. \left. \left. \min(1 - \underline{r}_{A^\circ}(x), 1 - \underline{r}_{B^\circ}(x)) e^{i\left(\min\left(2\pi - \underline{\omega}_{A^\circ}(x), 2\pi - \underline{\omega}_{B^\circ}(x)\right)\right)} \right) \right) \right\}$

$= \left\{ \left(x, \left(ann\left(0, (\min((\underline{r}_{A^\circ}(x))^c, (\underline{r}_{B^\circ}(x))^c)) \right) e^{i\left(\min\left((\underline{\omega}_{A^\circ}(x))^c, (\underline{\omega}_{B^\circ}(x))^c\right)\right)}, \right. \right. \right.$
$\left. \left. \left. (\min((\bar{r}_{A^\circ}(x))^c, (\bar{r}_{B^\circ}(x))^c)) e^{i\left(\min\left((\bar{\omega}_{A^\circ}(x))^c, (\bar{\omega}_{B^\circ}(x))^c\right)\right)} \right) \right) \right\}$

$= (A^\circ)^c \cap (B^\circ)^c$

This completes the proof.

(ii), (iii), and (iv): The proofs are omitted, as they are similar to that of part (i). ∎

5 Application of Complex Fuzzy Annulus in an Economic Problem

In this section, we present a brief case study of an example of a complex fuzzy annulus and the possible interpretations of the membership functions of its elements based on real-life occurrences in the Malaysian economy.

Case Study

Consider a universal set $U = \{x_1 =$ unemployment rate, $x_2 =$ inflation rate, $x_3 =$ exchange rate, $x_4 =$ GDP, $x_5 =$ global oil prices$\}$, whose elements are financial indicators or indices describing the Malaysian economy. Let $A = \{a_1 =$ no *influence*, $a_2 =$ minimal influence, $a_3 =$ average influence, $a_4 =$ great *influence*$\}$ be a set of parameters describing the degree of influence of these financial indicators on the Malaysian economy. The interaction between this set of attributes and the Malaysian economy is measured over a limited time span (one year). In this problem, the complex fuzzy annulus $A^{\odot}$ takes the form of Eq. (6) as given below:

$$A^{\odot} = \left\{ \left(x_i, ann(0, \underline{r}_{A\odot}(x_i)e^{i\underline{\omega}_{A\odot}(x_i)}, \overline{r}_{A\odot}(x_i)e^{i\overline{\omega}_{A\odot}(x_i)}) \right) : x_i \in U \right\}.$$

In the context of this example, the lower and upper amplitude terms measure the degree of influence of a financial indicator on the Malaysian economy, and the phase term represents the time lag between the occurrence of a financial indicator and the first effects of that indicator in the Malaysian economy. These values can be determined or calculated through data obtained from various agencies, such as the Central Bank of Malaysia, the Kuala Lumpur Stock Exchange (KLSE), and the International Monetary Fund (IMF), or by experts such as economists and financial analysts.

We now describe scenarios that might occur in this context. For example, the statement "Global oil prices have a very large influence on the Malaysian economy, and this influence becomes evident in the Malaysian economy within three to five months" can be expressed as the following complex fuzzy annulus:

$$A(a_4) = \left(x_5, ann(0, 0.8e^{\frac{3}{12}(2\pi)i}, 0.9e^{\frac{5}{12}(2\pi)i}) \right),$$

which simplifies to $A(a_4) = \left(x_5, ann(0, 0.8e^{\frac{\pi}{2}i}, 0.9e^{\frac{5\pi}{6}i}) \right)$.

The amplitude interval $[0.8, 0.9]$ signifies the very large influence of global oil prices on the Malaysian economy, whereas the phase interval $\left[\frac{\pi}{2}, \frac{5\pi}{6}\right]$ corresponds to the 3- to 5-month time lag (in this notation, the measurement interval of one year is represented by 2π) between the occurrence and effect of global oil prices on the Malaysian economy. These values are reasonable, as the Malaysian government sources 30 %–40 % of its revenue from oil and gas. Consequently, any change in global oil prices would greatly affect (positively or adversely) the Malaysian economy. Moreover, the time lag of 3–5 months between a change in global oil prices and its effect on the Malaysian economy reflects the recent actual situation in Malaysia. Similarly, in complex fuzzy annulus notation, the statement "The exchange rate has a very large influence on the Malaysian economy, which becomes evident within two to three months" becomes $A(a_1) = \left(x_3, ann(0, 0e^{\frac{2}{12}(2\pi)i}, 0.05e^{\frac{3}{12}(2\pi)i}\right)$, which then simplifies to $A(a_1) = \left(x_3, ann(0, 0e^{\frac{\pi i}{3}}, 0.05e^{\frac{\pi i}{2}}\right)$. Again, this statement is reasonable as the degree of membership of the exchange rate is zero or close to zero, corresponding to the attribute "no influence" in set A.

Comparison between existing models in the literature
The case study presented above involves the degree of influence of a set of financial indicators on the Malaysian economy and the time it takes for the effect of these indicators to become evident on the Malaysian economy. These information can also be modelled using interval-valued fuzzy sets or even ordinary fuzzy sets. However, in this case, both aspects of these information would have to be represented separately by defining two fuzzy or interval-valued fuzzy sets, with the first set describing the degree of influence of the set of financial indicators on the Malaysian economy and the second set describing the time it takes for the effects of these indicators to manifest themselves on the Malaysian economy. Although this can be easily executed, both aspects of these information (the degree of influence and the time taken) would lose its significance when represented separately. Combining the information from both of these sets by linking them, would necessitate finding the relations between the fuzzy or interval-valued fuzzy sets. This would in turn, require the user to perform tedious set theoretic or composition operations between the two sets, which would only result in additional computations which does not in any way increase the accuracy of the representation of the two-dimensional information. As such, the complex fuzzy annulus proposed in this paper, is superior to ordinary fuzzy sets and interval-valued fuzzy sets in terms of computational efficiency, ease and accuracy of representation of time-periodic information and also in terms of the graphical representation of the information in the form of an annulus in a complex plane, all of which are not possible with ordinary fuzzy sets and interval-valued fuzzy sets.

6 Conclusion

This paper proposes a novel concept called the complex fuzzy annulus, which extends the concept of classical complex fuzzy sets to better describe time-periodic phenomena (both algebraically and graphically). The proposed complex fuzzy annulus maps a universal set to a sub-disk of the unit circle in the complex plane. The radii of the circles forming the annulus represent the lower and upper bounds of the amplitude and phase terms of the complex fuzzy annulus, thus providing a closed interval for the membership grades of each element. The basic set theoretic operations (complement, union, and intersection) pertaining to the complex fuzzy annulus were introduced and the fundamental algebraic properties of these operations were verified. Lastly, the application of the complex fuzzy annulus was illustrated by applying it in a time-periodic problem relating to the major factors affecting the Malaysian economy.

Acknowledgments. The author would like to gratefully acknowledge the financial assistance received from the Ministry of Education, Malaysia and UCSI University, Malaysia under Grant no. FRGS/1/2014/ST06/UCSI/03/1.

References

1. Ramot, D., Milo, R., Friedman, M., Kandel, A.: Complex fuzzy sets. IEEE Trans. Fuzzy Syst. **10**(2), 171–186 (2002)
2. Ramot, D., Friedman, M., Langholz, G., Kandel, A.: Complex fuzzy logic. IEEE Trans. Fuzzy Syst. **11**(4), 450–461 (2003)
3. Dick, S.: Toward complex fuzzy logic. IEEE Trans. Fuzzy Syst. **13**(3), 405–414 (2005)
4. Turksen, I.B.: Interval valued fuzzy sets based on normal forms. Fuzzy Sets Syst. **20**(2), 191–210 (1986)
5. Gau, W.L., Buehrer, D.J.: Vague Sets. IEEE Trans. Syst. Man Cybern. **23**(2), 610–614 (1993)
6. Atanassov, K.T.: Intuitionistic fuzzy sets. Fuzzy Sets Syst. **20**(1), 87–96 (1986)
7. Atanassov, K.T., Gargov, G.: Interval valued intuitionistic fuzzy sets. Fuzzy Sets Syst. **31**(3), 343–349 (1989)
8. Zadeh, L.A.: Fuzzy sets. Inf. Control **8**, 338–353 (1965)

A Hierarchical Fuzzy Logic Control System for Malaysian Motor Tariff with Risk Factors

Daud Mohamad$^{(\boxtimes)}$ and Lina Diyana Mohd Jamal

Mathematics Department, Faculty of Computer and Mathematical Sciences,
Universiti Teknologi MARA, Shah Alam 40450, Selangor, Malaysia
daud@tmsk.uitm.edu.my, mj_diana@yahoo.com

Abstract. In many countries, including Malaysia, it is made compulsory to have a motor insurance policy and the premium is determined based on the Motor Tariff which ensures that a standard premium is imposed to the policyholders. At present, the premium in Malaysia includes only two factors which are the sum insured and the cubic capacity of the engine. Many existing methods used to calculate the tariff depend solely on the data and does not enable the experts to provide their input into the system. In contrast, the rule based system which is used in the Fuzzy Logic Control System could cater for the experts' input. This research aims to develop a system that can determine the motor tariff using the Hierarchical Fuzzy Logic Control System. Besides the sum insured and the cubic capacity of the engine, the system will also incorporate the risk level of policyholders into the Motor Tariff. As a prototype, two selected risk factors are used, namely the age of drivers and the age of cars. The risk premium subsystem is developed before combining it with the main tariff premium system that constitute the Hierarchical Fuzzy Logic Control System. The result confirmed that the premium is loaded when the risk level is high and discounted when the risk level is low. The finding is in tandem with Bank Negara Malaysia (BNM) impending detariffication exercise for determining the motor insurance policy.

Keywords: Fuzzy logic · Hierarchical fuzzy logic control system · Motor insurance policy · Risk factor

1 Introduction

In Malaysia, the motor premium paid to the general insurer is based on a tariff known as the Motor Tariff (MT), formulated by the General Insurance Association of Malaysia [1]. The tariff takes into account only two rating factors which may not suffice to arrive at a suitable premium. There are other rating factors which can contribute to the calculation of the premium that could be considered as well. The vehicles are classified into three categories which are private car, motorcycle and commercial vehicle. The MT takes into account only two rating factors which are Cubic Capacity (CC) and Sum Insured (market value) of the vehicle in determining the premium. The main weakness of this mode of pricing is that it only takes into account the factors associated to the vehicle whilst ignoring the factors associated to the driver. A study by [2] stated

© Springer Nature Singapore Pte Ltd. 2016
M.W. Berry et al. (Eds.): SCDS 2016, CCIS 652, pp. 224–236, 2016.
DOI: 10.1007/978-981-10-2777-2_20

that when pricing the motor premium, among the factors which should be considered and in line with the practice of many countries are the factors relating to the vehicle and driver. For example, risk related factors connected to the driver could be age, gender, marital status, occupation and credit rating while factors related to the vehicle may be age of vehicle, make, model and type. This initiative is endorsed by the central bank of Malaysia Bank Negara Malaysia (BNM) by the middle of 2016 [3, 4].

The Association of British Insurers [5] noted that the higher the risk level, the higher the premium and vice versa. The fragmentation of grouping risk with similar characteristic is called risk classification [1]. Premium rating differs between countries in whereby some have a market where each risk is priced individually according to the rating factors and others have a market where everyone pays the same premium. The most widely used models in the insurance practice to estimate the claim frequency and severity are the Poisson Regression Model and Gamma Regression Model correspondingly which were used by [6] to compute the risk premium. A system has been proposed by [1] to model the Malaysian motor insurance claim using Artificial Neural Network (ANN) and Adaptive Neuro Fuzzy Inference System (ANFIS) that basically broken down into the frequency model and severity model.

The current practice in Malaysia in calculating the premium is based on the Motor Tariff. The tariff takes into account only two rating factors which may not suffice to arrive at a suitable premium. There are other rating factors which can contribute to the calculation of the premium that could be considered as well. The common practice in other countries such as the UK, USA, Canada, Sweden and Singapore is that a risk is priced individually using several rating factors. This research is aimed to produce an output which is the premium amount by using a rule based hierarchical fuzzy logic control system (HFLCS). This method is chosen due its ability to cater for human (expert) input in the validation process rather than depending solely on the trend of the data. The advantages of a rule based system are its interpretability, comprehensibility, robustness and its ability to provide for qualitative information. This study involves developing a hierarchical fuzzy logic system which includes other rating factors on top of the existing rating factors used in the tariff.

The scope of study is limited to only general insurers in Malaysia as different countries may have different demographic and environmental factors. Moreover, the data used will be obtained from one of the renowned insurer in Malaysia hence the claim trend may not be reflective of other countries. Furthermore, the additional factors evaluated are not only based on the literature review conducted on the common practice in other countries, but also based on the available factors in the data at hand. The chosen analyzed factors depends on both the availability of rating factors in the data as well as the common practice in other countries. In addition, the available data used for the analysis are limited to the only private car and the data are comprised of only Comprehensive cover. The main assumption in this study is that the claim trend of the data obtained from one insurer is reflective of the players in the industry. Another assumption is that the current claim trend is the same as the claim trend in the historical data used. Finally, the premium derived from the system developed does not include the commissions, expenses, profit and contingencies.

2 Basic Concepts

In this section, some basic concepts that will be used in the paper are explained.

Definition 2.1 [7]. A fuzzy set A is defined by a set of ordered pairs, a binary relation,

$$A = \{(x, \mu_A(x) \,|\, x \in A, \mu_A(x) \in [0,1]\}$$

where $\mu_A(x)$ is known as the membership function of A.

Definition 2.2 [8]. A fuzzy set is said to be normal when $\exists\ x \in R$, $\mu_A(x) = 1$.

Definition 2.3 [7]. Consider the universe U to be the set of real number R. A subset S of R is said to be convex if and only if, for all x_1, x_2 in S and for every real number λ satisfying $0 \leq \lambda \leq 1$, we have

$$\lambda x_1 + (1 - \lambda)x_2 \in S$$

Definition 2.4 [9]. A fuzzy number is a fuzzy set which is convex, normalized, defined in the real number and has a membership function which is piecewise continuous.

Definition 2.5 [9]. A trapezoidal fuzzy number can be defined as $A = (a_1, a_2, a_3, a_4)$ that has a membership function in the form

$$\mu_A(x) = \begin{cases} \frac{x-a_1}{a_2-a_1} & a_1 \leq x \leq a_2 \\ 1 & a_2 \leq x \leq a_3 \\ \frac{a_4-x}{a_4-a_3} & a_3 \leq x \leq a_4 \\ 0 & elsewhere \end{cases}$$

If $a_2 = a_3$, then A is called a triangular fuzzy number, denoted by a 3-tuple ($a_1, a_2 = a_3, a_4$).

Definition 2.6 [8]. A linguistic variable is characterized by (X, T, U, M), where:

- X is the name of a linguistic variable, for example speed of a car.
- T is the set of linguistic terms that X can take for example $T = \{$slow, medium, fast$\}$ which is known as a linguistic term.
- U is the actual physical domain in which the linguistic variable X takes it quantitative (crisp) values.
- M is a semantic rule that relates each linguistic term in T with a fuzzy set in U.

A fuzzy linguistic term consists of two main parts which are the fuzzy predicate and the fuzzy modifier. Some examples of the former are *expensive*, *old* and *good* and that of the latter are *very*, *extremely* and *likely*.

2.1 Fuzzy Logic Control System

Fuzzy logic is a logic which is not exact, but approximate in nature, hence very useful since most modes of human reasoning and common sense are ambiguous rather than precise [10]. The concept of fuzzy logic becomes very practical in the context where human perception is involved. In [11], the fuzzy logic control system (FLCS) is used to develop a power controller for a parallel hybrid vehicle that will optimize fuel economy and in [12] the FLCS is applied to design a fuzzy car which successfully followed a crank shaped track and parked itself in a garage [12].

The advantage of the FLCS is its interpretability that ensures comprehensibility as the fuzzy rule can be interpreted linguistically [13]. The number of rules in the FLCS is decided by experts who are familiar with the system to be modelled [14], hence the utilization of the if-then rules can model the qualitative aspect of human knowledge without the need for accurate qualitative analysis [15]. However, there is no standard method available to convert human knowledge and experience into fuzzy rules of FLCS and no effective method of deriving the membership function which will produce minimum output error [14].

2.2 Hierarchical Fuzzy Logic Control System (HFLCS)

The problem with FLCS is that the rules grow exponentially with the number of variables hence affecting the real time performance of the system. However, this issue could be decomposed by hierarchically breaking down the inputs of the system according to its type of employment before integrating them via another higher layer [16]. Hence the HFLCS has a main component and at least one subsystem. Some theoretical development and applications of HFLCS are discussed in [17, 18]. According to [19], the total number of fuzzy rules in the FLCS is determined by the following formula $k = m^n$ where k is the number of rules, m is the number of fuzzy sets and n is the number of inputs into the system. As the number of fuzzy sets increase, the rules increase exponentially. As for the HFLCS, the number of rules is reduced to only $k = \sum_{i=1}^{L} m^{n_i}$ where L is the number of hierarchical levels with n_i is the number of variables contained in the ith level.

3 The Proposed Model

The proposed prototype model of the motor insurance tariff using HFLCS starts with the data collection, obtained from one of the general insurers in Malaysia and the following parameters are considered shown in Table 1.

The data is cleansed and the risk premium analysis is done to choose the additional rating factors besides the vehicle *CC* and *Sum covered*. As in Table 1, the risk factors can be many, however, only two additional rating factors were considered in the development of the prototype; *Age of Driver* and *Age of Vehicle* as they have the highest risk premium difference between each category. Figures 1 and 2 illustrates the risk

Table 1. Parameters consideration in model development

Parameter	Remarks
New IC No.	To calculate the age and gender
Vehicle CC	To calculate Tariff Premium
Sum covered	To calculate Tariff Premium
Vehicle age	Potential additional rating factor
Vehicle make	Potential additional rating factor
State	Potential additional rating factor
No Claim Discount (NCD)	Potential additional rating factor
Gender	Potential additional rating factor
Age	Potential additional rating factor
Vehicle use	To filter for only private car own use
Gross premium	To calculate the policy count
Claim paid amount	To calculate the claim cost
Claim estimated amount	To calculate the claim cost
Claim closed flag	To calculate the claim cost

premium of *Age of Driver* and *Age of Vehicle*. The rating factor *CC (Cubic Capacity)* and *Sum Insured* is chosen automatically as they are used in the Motor Tariff in Malaysia which is used to calculate the Tariff Premium based on these rating factors. From the data, the distribution of the drivers' age and vehicle's age as compared to the risk premium are depicted in Figs. 1 and 2.

The data filtration is then done to ensure that the extreme values are not considered as their exposure base is low. This may distort the frequency value which in turn will affect the risk premium value. Hence, the data for each rating factor is filtered to 95th percentile. For *CC* and *Sum Insured*, both the left (2.5 %) and right (2.5 %) tails are excluded as the distributions are assumed to be not skewed. However, for the *Age of*

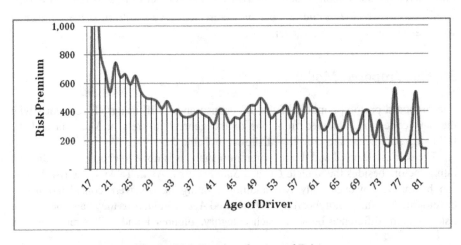

Fig. 1. Risk Premium for Age of Driver

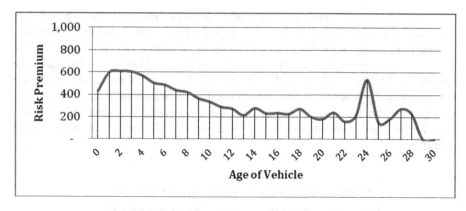

Fig. 2. Risk Premium for Age of Vehicle

Driver and *Age of Vehicle*, only the right tail (5 %) is deleted as the distributions are assumed to be positively skewed.

The band of each rating factor except for the *CC* is determined by computing the multiple of 20th percentile as five membership functions will be defined for each linguistic variable *Sum Insured, Age of Driver* and *Age of Vehicle*. As for the linguistic variable for the *CC*, the percentiles need not be computed as the banding is determined in accordance to the bands available in the Motor Tariff. The percentile range is chosen to be the same, so that the exposure becomes roughly the same size. In other words, each band will consist roughly twenty percent of the data. The final banding selected for the rating factors are as in Table 2.

Table 2. Range of each band of the rating factors

Band	CC	Sum Insured	Age of Driver	Age of Vehicle
1	<=1400	<=15,000	<=25	<3
2	1401–1650	15,001–25,000	26–35	3–4
3	1651–2200	25,001–35,000	36–40	5–6
4	>2200	35,001–45,000	41–45	7–9
5	n/a	>45,000	>45	>9

3.1   Development of Risk Premium FLCS (Subsystem)

For the Risk Premium FLCS, the inputs are Age of Driver and Age of Vehicle while the output is Risk Premium. The membership function for *Age of Driver* and *Age of Vehicle* is based on the midpoint of the middle bands of the rating factor and the minimum point of the lowest band as well as the maximum point of the highest band. For example, the midpoints of the categories 26–35, 36–40 and 41–45 are 30, 38 and 43 correspondingly while the minimum point of the category <=25 is 17 and the maximum point of the category >45 is 56.

The multiples of 20th percentile are used to determine the number of bands *Age of Driver* and *Age of Vehicle* as well as the range for *Risk Premium* correspondingly as five linguistic terms have been set up for each linguistic variable. However, for the output membership function in Risk Premium FLCS which is *Risk Premium*, the multiples of 25th percentile, including 0th percentile are calculated to determine the peak (membership function of 1) and the lowest point (membership function of 0) of the linguistic variable. Table 3 summarizes the membership functions and linguistic terms for the Risk Premium FLCS. The notation [x y z] denotes a triangular fuzzy number.

Table 3. Membership functions for Risk Premium FLCS

Linguistic variable	Linguistic term	Membership function
Age of Driver	Very Young	[17 17 30]
	Young	[17 30 38]
	Marginally Old	[30 38 43]
	Old	[38 43 56]
	Elderly	[43 56 56]
Age of Vehicle	New	[0 0 3.5]
	Relatively New	[0 3.5 5.5]
	Middle	[3.5 5.5 8]
	Relatively Old	[5.5 8 14]
	Old	[8 14 14]
Risk Premium	Very Low (VL)	[272 272 362]
	Low (L)	[272 362 424]
	Standard (S)	[362 424 536]
	High (H)	[424 536 689]
	Very High (VH)	[536 689 689]

In order to define the fuzzy rules, the *Risk Premium* for each band of the rating factors need to be computed and Table 4 illustrates the output:-

Table 4. Risk Premium by Age of Driver band and Age of Vehicle band

Risk Premium	Age of Vehicle				
Age of Driver	<3	3–4	5–6	7–9	>=10
<=25	689	601	638	568	542
26–35	558	520	468	396	360
36–40	394	462	403	318	272
41–45	362	424	384	355	277
>45	479	536	454	379	281

Next, the multiple of 20th percentile is calculated to determine the five output membership functions and this is depicted in Table 5. The Fuzzy Associate Memory (FAM) will then be constructed based on information in Tables 4 and 5 to make the process of defining the rules easier.

Table 5. Percentile values for linguistic terms of Risk Premium

Percentile	Percentile value	Range	Linguistic term
20th	359	<359	Very Low (VL)
40th	395	359–394	Low (L)
60th	464	395–463	Standard (S)
80th	545	464–544	High (H)
100th	689	>=545	Very High (H)

The rules are validated and approved by experts in the insurance company. The actual FAM is slightly modified (suggested by experts) to make the fuzzy rules in tandem with the trend for Age Band of Driver and Age Band of Vehicle and is illustrated in Table 6.

Table 6. Modified FAM for Risk Premium FLCS

Age of Driver	Age of Vehicle				
	New	Relatively new	Middle	Relatively old	Old
Very Young	VH	VH	VH	VH	H
Young	VH	H	H	S	L
Marginally Old	S	S	S	VL	VL
Old	S	S	L	VL	VL
Elderly	H	H	S	L	VL

An example of the if-then rule from Table 6 can be stated as: IF <the age driver is Very Young> and <the age of the vehicle is New> THEN <the risk premium is Very High(VH)>.

3.2 Development of the Main System HFLCS

There are four linguistic variables involved in the Main System (HFLCS) which are *Cubic Capacity, Sum Insured, Risk Premium* and *Premium with Risk Factors*. Figure 3 shows the configuration of the HFLCS.

Before embedding the sub-system to the main system, the membership functions for *Cubic Capacity, Sum Insured* and *Tariff Premium* (Premium without Risk Factors) need to be initially defined. The membership functions for each of them are obtained by simulating the data using Adaptive Neuro Fuzzy Inference System (ANFIS) method [14].

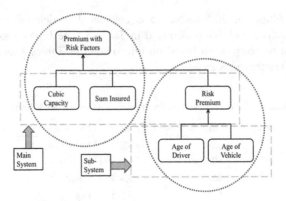

Fig. 3. Input and Output for main system

Table 7. Linguistic terms and its abbreviation for HFLCS output

Linguistic term for premium with risk factors	Symbol	Linguistic term for premium with risk factors	Symbol
Double loaded very low	VL++	Double loaded high	H++
Loaded very low	VL+	Loaded high	H+
Very low	VL	High	H
Discounted very low	VL−	Discounted high	H−
Double discounted very low	VL−−	Double discounted high	H−−
Double loaded low	L++	Double loaded very high	VH++
Loaded low	L+	Loaded very high	VH+
Low	L	Very high	VH
Discounted low	L−	Discounted very high	VH−
Double discounted low	L−−	Double discounted very high	VH−−
Double loaded standard	S++		
Loaded standard	S+		
Standard	S		
Discounted standard	S-		
Double discounted standard	S−		

There are twenty five linguistic terms for *Premium with Risk Factors* as there are five linguistic terms (VL, L, S, H and VH) for *Risk Premium* and each linguistic term is further broken down into five categories by applying either loading, double loading, discount, double discount or no loading/discount. Table 7 lists the linguistic terms involved in the *Premium with Risk Factors*.

Table 8 shows the loading and discount percentages charged on the *Tariff Premium* (Premium without Risk Factors) to obtain the *Premium with Risk Factors*. The percentages 5 % and 10 % used are chosen arbitrarily.

Table 8. Loading and discount percentages for fuzzy quantifier

Fuzzy quantifier	Symbol	Percentage
Loaded	+	5 %
Double loaded	++	10 %
Discounted	--	-5 %
Double discounted	--	-0 %

After the loading/discount percentages are incorporated to the *Tariff Premium* membership function, the rest of the membership functions are obtained. The screenshot of the membership functions is demonstrated in Fig. 4.

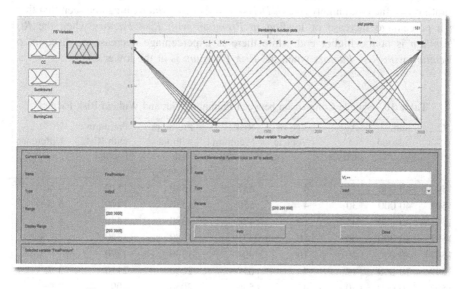

Fig. 4. Membership functions of Premium with Risk Factors

The Fuzzy Associate Memory (FAM) is constructed as in Table 9 to make the process of defining the rules easier. Altogether there are one hundred rules for the Main System HFLCS. The same twenty five rules apply for each linguistic term in *CC* which are *Small, Fairly Small, Fairly Large* and *Large*.

The output of the *Premium with Risk Factors* is compared to the expected Premium with Risk Factors calculated using Microsoft Excel. The accuracy is measured using Average Percentage Error (APE). The membership function is modified until an acceptable APE is obtained. The details of the output are presented in the next section.

Table 9. Fuzzy Associative Memory (FAM) for HFLCS

Risk Premium			Sum Insured		
	Low	Reasonably low	Average	Reasonably high	High
VL	VL--	L--	S--	H-	VH--
L	VL-	L-	S-	H-	VH-
S	VL	L	S	H	VH
H	VL+	L+	S+	H+	VH+
VH	VL++	L++	S++	H++	VH++

4 Results and Discussions

Table 10 shows all the input involved as well as the output with and without incorporating risk factors, namely *Age of Driver* and *Age of Vehicle*. It can be seen that there is a percentage increase or loading in *Premium with Risk Factors* when the *Risk Premium* is at the higher end while there is a percentage decrease or discount on *Premium with Risk Factors* when the *Risk Premium* is at the lower end.

Table 10. Sample Comparison between Premium With and Without Risk Factors

CC	Sum Insured	Age of Driver	Age of Vehicle	Risk Premium	Premium without Risk Factors	Premium with Risk Factors	% Difference
1597	40 000	17	4	637	1380	1510	9 %
1597	40 000	30	4	550	1380	1470	6 %
1597	40 000	45	1	478	1380	1440	4 %
1597	40 000	36	10	371	1380	1340	−3 %
1597	40 000	40	10	304	1380	1290	−7 %
1597	70 000	17	4	637	2050	2110	3 %
1597	70 000	30	4	550	2050	2090	2 %
1597	70 000	45	1	478	2050	2050	0 %
1597	70 000	36	10	371	2050	1970	−4 %
1597	70 000	40	10	304	2050	1900	−8 %
1597	90 000	17	4	637	2420	2460	2 %
1597	90 000	30	4	550	2420	2450	1 %
1597	90 000	45	1	478	2420	2410	0 %
1597	90 000	36	10	371	2420	2380	−2 %
1597	90 000	40	10	304	2420	2340	−3 %

5 Conclusion and Recommendation

In this paper a prototype motor insurance tariff system has been developed using the HFLCS which is more efficient than the standard FLCS as the number of rules is reduced significantly. The system is developed using additional risk factors, namely the *Age of Driver* and *Age of Vehicle* chosen. The inclusion allows an insurer to charge a fairer rate to the policyholders as the level of risk is computed and tagged to the premium amount in which the premium increases as the risk increases. This study is in tandem with BNM's objective of partial detariffication and premium bond requirement on motor insurance premium pricing because the existing rating factors act as the backbone of the motor premium pricing which controls the deviation of the premium change. The additional risk factors incorporated enable the insurers to charge a fairer rate to the policyholders as a risk is loaded or discounted according to their risk level. The configuration of the Main System HFLCS ensures that the *Premium with Risk Factors* does not deviate too much from the *Tariff Premium*. For future development, other significant risk factors such as the gender, the geographical location and the residential address where the vehicle is most frequently driven maybe included for improvement purposes and comprehensiveness of the tariff system.

References

1. Mohd Yunos, Z., Shamsuddin, S., Ismail, N., Sallehuddin, R.: Modeling the Malaysian Motor Insurance Claim Using Artificial Neural Network and Adaptive Neuro Fuzzy Inference System. In: 20th National Symposium on Mathematical Sciences, pp. 1431–1436. AIP Publishing, Kuala Lumpur (2013)
2. Baker, V., Kumar, S.: Motor premium rating. In: 5th Global Conference of Actuaries, pp. 52–58 (2003)
3. Dhesi, D.: De-tariffication of Motor and Fire Insurance Premiums Expected. Business News, Petaling Jaya (2015)
4. Sabhlok, R., Malattia, R.: Detariffication in the Malaysian general insurance sector. Towers Watson, Malaysia (2014)
5. Association of British Insurers (ABI): Insurance in the UK: The Benefits of Pricing Risk (2008)
6. Cheong, P., Jemain, A., Ismail, N.: Practice and pricing in non-life insurance: the malaysian experience. J. Qual. Meas. Anal., 11–24 (2008)
7. Bojadziev, G., Bojadziev, M.: Fuzzy Logic for Business, Finance and Management. World Scientific Publishing Co., Pte. Ltd., Singapore (2007)
8. Wang, L.: A Course in Fuzzy Systems and Control. Prentice Hall Internationl Inc. (1996)
9. Kwang, H.: First Course of Fuzzy Theory and Applicaitions. Springer, Germany (2005)
10. Berhan, E., Abraham, A.: Hierarchical Fuzzy Logic System for Manuscript Evaluation. Middle-East J. Sci. Res. **19**(9), 1235–1245 (2004)
11. Schouten, N., Salman, M., Kheir, N.: Fuzzy logic control system in hybrid vehicles. IEEE Trans. Contr. Syst. Technol. (2002)
12. Chuen, C.: Fuzzy logic in control systems: fuzzy logic controller, Part II. IEEE Trans. Syst., 419–433 (1990)

13. Fakhrahmad, S.M., Zare, A., Jahromi, M.Z.: Constructing accurate fuzzy rule-based classification systems using apriori principles and rule-weighting. In: Yin, H., Tino, P., Corchado, E., Byrne, W., Yao, X. (eds.) IDEAL 2007. LNCS, vol. 4881, pp. 547–556. Springer, Heidelberg (2007)
14. Jang, J.: ANFIS: adaptive-network-based fuzzy inference system. IEEE Trans. Syst., Man Cybern. 23(3), 665–685 (1993)
15. Chai, Y., Jia, L., Zhang, Z.: Mamdani model based adaptive neural fuzzy inference system and its application. Int. J. Comp. Intell. 5(1), 22–29 (2009)
16. Hagras, H.: A hierarchical type-2 fuzzy logic control architecture for autonomous mobile robots. IEEE Trans. Fuzz. Syst. (2004)
17. Raju, G.V.S., Zhou, J., Kisner, R.A.: Hierarchical Fuzzy Control. Int. J. Control 54, 1201–1216 (1991)
18. Renkas, K., Niewiadomski, A.: Hierarchical fuzzy logic systems: current research and perspectives. In: Rutkowski, L., Korytkowski, M., Scherer, R., Tadeusiewicz, R., Zadeh, L.A., Zurada, J.M. (eds.) ICAISC 2014, Part I. LNCS, vol. 8467, pp. 295–306. Springer, Heidelberg (2014)
19. Mohammadian, M.: Designing customized hierarchical fuzzy logic system for modelling and prediction. In: 4th Asian Pacific Conference on Simulated Evolution and Learning, Singapore (2002)

Modeling Steam Generator System of Pressurized Water Reactor Using Fuzzy Arithmetic

Wan Munirah Wan Mohamad[1], Tahir Ahmad[1,2(⊠)], and Azmirul Ashaari[1]

[1] Department of Mathematical Science, Faculty of Science, Universiti Teknologi Malaysia, 81310 UTM Skudai, Johor, Malaysia
[2] Centre for Sustainable Nanomaterials, Ibnu Sina Institute for Scientific and Industrial Research, Universiti Teknologi Malaysia, 81310 UTM Skudai, Johor, Malaysia
tahir@ibnusina.utm.my

Abstract. Steam generator system is known as the bridge between the primary and secondary systems for phase changes from water into steam. The aim of this paper is to identify the best input that influence the steam generator system in the process of changing from water to steam, to ensure the process is efficient. The method consists of the transformation method of fuzzy arithmetic which is to compute the measure of influence for each parameter in the model system. The result is then verified against simulation and analysis.

Keywords: Fuzzy arithmetic · Transformation method · Measure of influence and steam generator

1 Introduction

A steam generator is used in pressurized water reactor (PWR) as heat exchangers in order to convert water into steam between primary and secondary systems [1]. In a nuclear power the pressurized steam feed into a steam turbine to activate [2]. The secondary coolant is cooled down and condensed in a condenser. The condenser transforms the steam to a liquid to be pumped back into the steam generator. Therefore, a vacuum is maintained at the turbine outlet to ensure the pressure drops across the turbine. Hence the energy extracted from the steam is maximized. However, before entered into the steam generator, the condensed steam is preheated to minimize the thermal shock.

In this paper, a general transformation method is introduced to model a steam generator in a pressurized water reactor. The analytical solution is obtained in order to reduce the number of parameters of the system. The transformation method is described in Sect. 2. Section 3 exhibits the analytical solution of the state equation. Section 4 is the modeling of a Steam Generator System of Pressurized Water Reactor. The conclusion for the modeling is given in the last section of this paper.

© Springer Nature Singapore Pte Ltd. 2016
M.W. Berry et al. (Eds.): SCDS 2016, CCIS 652, pp. 237–246, 2016.
DOI: 10.1007/978-981-10-2777-2_21

2 The Transformation Method

This method can be used to evaluate fuzzy rational expressions and also can simulate static or dynamic systems with fuzzy valued parameters. The coefficients for general transformation method were proposed in [3–8].

2.1 Simulation of a System with Uncertain Parameters: General Transformation Method

The uncertain parameters can be represented by fuzzy numbers $\tilde{p}_i$ with $i = 1, 2, \ldots, n$.

$$P_i = \left\{ X_i^{(0)}, X_i^{(1)}, \ldots, X_i^{(m)} \right\} \tag{1}$$

$$X_i^{(j)} = \left[a_i^{(j)}, b_i^{(j)} \right], a_i^{(j)} \le b_i^{(j)}, \tag{2}$$

$i = 1, 2, \ldots, n, j = 0, 1, \ldots, m.$

A fuzzy parameterized model is expected to show non-monotonic behavior with respect to n, for $n > 1$ and fuzzy value parameter $\tilde{p}_i$ with $i = 1, 2, \ldots, n$. The intervals $X_i^{(j)}$, $i = 1, 2, \ldots, n, j = 0, 1, \ldots, m-2$, are considered for the transformation scheme. The intervals are now transformed into arrays $\hat{X}_i^{(j)}$ in the form

$$X_i^{(j)} = \underbrace{\left(\left(\gamma_{1,i}^{(j)}, \gamma_{2,i}^{(j)}, \ldots, \gamma_{(m+1-j),i}^{(j)} \right), \ldots, \left(\gamma_{1,i}^{(j)} \gamma_{2,i}^{(j)}, \ldots, \gamma_{(m+1-j),i}^{(j)} \right) \right)}_{(m+1-j)^{i-j}(m+1-j)^{-tuples}} \tag{3}$$

$$\gamma_{l,i}^{(j)} = \underbrace{\left(c_{l,i}^{(j)}, \ldots, c_{l,i}^{(j)} \right)}_{(m+1-j)^{n-i} elements} \tag{4}$$

$$c_{l,i}^{(j)} = \begin{cases} a_i^{(j)} \text{ for } l = 1 \text{ and } j = 0, 1, \ldots, m, \\ \frac{1}{2} \left(c_{l-1,i}^{(j+1)} + c_{l,i}^{(j+1)} \right) \text{ for } l = 2, 3, \ldots, m-j \text{ and } j = 0, 1, \ldots, m-2, \\ b_i^{(j)} \text{ for } l = m-j+1 \text{ and } j = 0, 1, \ldots, m, \end{cases} \tag{5}$$

The arithmetical expression of F is given in the form of

$$\tilde{q} = F(\tilde{p}_1, \tilde{p}_2, \ldots, \tilde{p}_n) \tag{6}$$

The evaluation is then carried out by evaluating the expression separately at each of 2^n positions of the combination using the conventional arithmetic for crisp numbers. The result of the problem can be expressed in its decomposed and transformed form by the combination $\hat{Z}^{(j)}$, $j = 0, 1, \ldots, m$, the kth element $^k \hat{Z}^{(j)}$ of the array $\hat{Z}^{(j)}$ given by

$$^k\hat{Z}^{(j)} = F(k\hat{x}_1^{(j)}, {}^k\hat{x}_2^{(j)}, \ldots, {}^k\hat{x}_n^{(j)}) \tag{7}$$

$k = 1, 2, \ldots, 2^n$,

Finally, the fuzzy-valued result $\tilde{q}$ of the expression can be achieved in its decomposed form

$$Z^{(j)} = \left[a^{(j)}, b^{(j)} \right], \quad j = 0, 1, \ldots, m \tag{8}$$

By retransforming $\hat{Z}^{(j)}$ with a correction procedure namely by the recursive formulas

$$a^{(j)} = \min_k \left(a^{(j+1)}, {}^k\hat{z}^{(j)} \right), \tag{9}$$

$$b^{(j)} = \min_k \left(b^{(j+1)}, {}^k\hat{z}^{(j)} \right), \quad j = 0, 1, \ldots, m-1, \tag{10}$$

$$a^{(m)} = \min_k({}^k\hat{z}^{(m)}) = \max_k({}^k\hat{z}^{(m)}) = b^{(m)} \tag{11}$$

2.2 Analysis of a System with Uncertain Parameters: General Transformation Method

The coefficients eta, $\eta_i^{(j)}$, $i = 1, 2, \ldots, n$, $j = 0, 1, \ldots, m-1$, are determined

$$\eta_i^{(j)} = \frac{1}{(m-j+1)^{n-1}(b_i^{(j)} - a_i^{(j)})} \sum_{k=1}^{(m-j+1)^{n-i}} \sum_{l=1}^{(m-j+1)^{i-1}} ({}^{s_2}\hat{z}^{(j)} - {}^{s_1}\hat{z}^{(j)}) \tag{12}$$

$$s_1(k, l) = k + (l-1)(m-j+1)^{n-i+1} \tag{13}$$

$$s_2(k, l) = k + [(m-j+1)l - 1](m-j+1)^{n-i} \tag{14}$$

The values $a_i^{(j)}$ and $b_i^{(j)}$ denote the lower and upper bound of the interval $X_i^{(j)}$, and $^k\hat{Z}^{(j)}$ is the kth elements of the array $\hat{Z}^{(j)}$. The coefficients $\eta_i^{(j)}$ are gain factors of the effect of the uncertainty of the ith parameter on the uncertainty of the output z of the membership level μ_j. The mean gain factors $\eta_i^{(j)}$ is given as

$$\bar{\eta}_i = \frac{\sum_{j=1}^{m-1} \mu_j \eta_i^{(j)}}{\sum_{j=1}^{m-1} \mu_j} \tag{15}$$

Finally, the degree of influence ρ_i is determined for $i = 1, 2, \ldots, n$, using

$$\rho_i = \frac{\sum_{j=1}^{m-1} \mu_j \left| \eta_i^{(j)} \left(a_i^{(j)} + b_i^{(j)} \right) \right|}{\sum_{q=1}^{n} \sum_{j=1}^{m-1} \mu_j \left| \eta_q^{(j)} \left(a_q^{(j)} + b_q^{(j)} \right) \right|} \tag{16}$$

$$\sum_{i=1}^{n} \rho_i = 1 \tag{17}$$

2.3   Famous

In this paper, a software package called FAMOUS is used to simulate and analyze the model of steam generator system. FAMOUS is stands for *Fuzzy Arithmetical Modeling of Uncertain Systems*.

The software FAMOUS is an analysis tool in MATLAB that implements fuzzy arithmetic for simulation and analysis of uncertain systems [9]. The toolbox provides a framework to handle parametric uncertainties in the analysis systems.

3   Analytical Solution of State Equation

Recall the state space equation

$$\dot{x} = Ax(t) + Bu(t) \tag{18}$$

$$y(t) = C\,x(t) + D\,u(t) \tag{19}$$

Equation (18) is deduced from initial condition x $(t_0) = x_0$ and input (t) for t $\geq$ 0 [10].

$$\dot{x} - Ax(t) = Bu(t) \tag{20}$$

Multiplying Eq. (20) by the factor e^{-At} leads to

$$e^{-At}\dot{x}(t) - e^{-At}Ax(t) = e^{-At}B\,u(t) \tag{21}$$

Therefore,

$$\frac{d}{dt}\left[e^{-At}x(t)\right] = e^{-At}B\,u(t) \tag{22}$$

Integrating both sides of this equation over λ from t_0 to t gives

$$e^{-At}x(t) - e^{-At_0}x(t_0) = \int_{t_0}^{t} e^{-A\lambda}Bu(\lambda)d\lambda \qquad (23)$$

Or

$$x(t) = e^{At}e^{-At_0}x(t_0) + e^{At}\int_{t_0}^{t} e^{-A\lambda}Bu(\lambda)d\lambda \qquad (24)$$

Hence,

$$x(t) = e^{A(t-t_0)}x(t_0) + \int_{t_0}^{t} e^{A(t-\lambda)}Bu(\lambda)d\lambda \qquad (25)$$

Substituting Eq. (25) in the output equation of (19),

$$y(t) = Ce^{A(t-t_0)}x(t_0) + C\int_{t_0}^{t} e^{A(t-\lambda)}Bu(\lambda)d\lambda + Du(t) \text{ for } t \geq 0 \qquad (26)$$

Where e^{At} is the state-transition matrix of the system

$$e^{-At} = I + At + \frac{A^2t^2}{2!} + \frac{A^3t^3}{3!} + \ldots = \sum_{k=0}^{\infty} \frac{A^k t^k}{k!} \qquad (27)$$

In general, each entry in the matrix is an infinite series that will converge. The term $Ce^{A(t-t_0)}x(t_0)$ from Eq. (26) is the zero-input or initial condition response since it is the complete output response when the input u(t) is zero. The term $C\int_{t_0}^{t} e^{A(t-\lambda)}$ $Bu(\lambda)d\lambda + Du(t) + Du(t)$ is the zero state or force response, since it is the output response when the initial state $x(0)$ is zero. It is also response to input u(t) with no initial energy in the system at time $t = 0$ [10].

4 Modeling a Steam Generator System of Pressurized of Water Reactor

The assumptions are developed in order to have a low order dynamic model of a steam generator system [1]:

i. The dynamics of the steam generators is assumed in a quasi-steady state and the conservation balances are not being constructed. Since, the primary system is speedy than secondary system in steam generator.
ii. The dynamics of the secondary system is assumed to be equilibrium between the water and the steam phases. In order, the steam phase is moving quickly than in a liquid phase for steam generator.
iii. The physical properties of the secondary system in steam generators are assumed to be constant.

iv. The controllers on the secondary water level and steam pressure are assumed to be ideal.

The state space equation is deduced by [1];

$$\begin{pmatrix} dM_{SG}/dt \\ dT_{SG}/dt \end{pmatrix} = \begin{pmatrix} \frac{m_{SG,in}}{M_{SG}} & \frac{-m_{SG,out}}{T_{SG}} \\ \frac{T_{SG,sw}}{M_{SG}} - \frac{(C^v_{p,SG}m_{SG},T_{SG}+E_{evap,SG})}{C^L_{p,SG}M_{SG}} & \frac{-K_{T,Sg}}{C^L_{p,SG}M_{SG}} \end{pmatrix} \begin{pmatrix} M_{SG} \\ T_{SG} \end{pmatrix}$$
$$+ \begin{pmatrix} 0 & 0 \\ \frac{-K_{T,Sg}}{C^L_{p,SG}M_{SG}} & \frac{-1}{C^L_{p,SG}M_{SG}} \end{pmatrix} \begin{pmatrix} T_{PC} \\ W_{loss,PC} \end{pmatrix} \tag{28}$$

The Output is:

$$\begin{pmatrix} U_{SG} \\ p_{SG} \end{pmatrix} = \begin{pmatrix} C^L_{p,SG}T_{SG} & 0 \\ 0 & p^T_* \end{pmatrix} \begin{pmatrix} M_{SG} \\ T_{SG} \end{pmatrix} \tag{29}$$

Equation (28) is the state of steam generator system, and Eq. (29) is the output of steam generator system. The state of the steam generator model M_{SG} is the secondary steam mass flow rate and T_{SG} is the temperature. The two output parameters are U_{SG} and p_{SG}.

4.1   Model Reduction of a Steam Generator System of Pressurized of Water Reactor

In order to reduce computing time, a model reduction approach is applied on the model equations. The output is selected from Eq. (29).

$$p_{SG} = p^T_* T_{SG} \tag{30}$$

From the Eq. (30) we need to identify T_{SG}.

$$dT_{SG}/dt = \frac{T_{SG,sw}}{M_{SG}} - \frac{(C^v_{p,SG}m_{SG},T_{SG}+E_{evap,SG})}{C^L_{p,SG}M_{SG}}M_{SG} - \frac{K_{T,Sg}}{C^L_{p,SG}M_{SG}}T_{SG} - \frac{K_{T,Sg}}{C^L_{p,SG}M_{SG}}T_{PC}$$
$$- \frac{1}{C^L_{p,SG}M_{SG}}W_{loss,PC}$$
$$\tag{31}$$

$$\mu = e^{\int \frac{K_{T,Sg}}{C^L_{p,SG}M_{SG}}dt} \tag{32}$$

$$\mu = e^{\frac{K_{T,Sg}}{C^L_{p,SG}M_{SG}}t} \tag{33}$$

$$\dot{T}_{SG} + \frac{K_{T,Sg}}{C_{p,SG}^L M_{SG}} T_{SG} = \frac{T_{SG,sw}}{M_{SG}} - \frac{(C_{p,SG}^v m_{SG}, T_{SG} + E_{evap,SG})}{C_{p,SG}^L M_{SG}} M_{SG} - \frac{K_{T,Sg}}{C_{p,SG}^L M_{SG}} T_{PC}$$
$$- \frac{1}{C_{p,SG}^L M_{SG}} W_{loss,PC}$$

(34)

$$\int \dot{T}_{SG} \left(T_{SG} e^{\frac{K_{T,Sg}}{C_{p,SG}^L M_{SG}}t} \right) = \int \frac{T_{SG,sw}}{M_{SG}} - \frac{(C_{p,SG}^v m_{SG}, T_{SG} + E_{evap,SG})}{C_{p,SG}^L M_{SG}} M_{SG} - \frac{K_{T,Sg}}{C_{p,SG}^L M_{SG}} T_{PC}$$
$$- \frac{1}{C_{p,SG}^L M_{SG}} W_{loss,PC} dt$$

(35)

$$T_{SG} e^{\frac{K_{T,Sg}}{C_{p,SG}^L M_{SG}}t} = \frac{\left(e^{\frac{K_{T,Sg}}{C_{p,SG}^L M_{SG}}t} \right) \left(\frac{T_{SG,sw}}{M_{SG}} - \frac{(C_{p,SG}^v m_{SG}, T_{SG} + E_{evap,SG})}{C_{p,SG}^L M_{SG}} M_{SG} - \frac{K_{T,Sg}}{C_{p,SG}^L M_{SG}} T_{PC} - \frac{1}{C_{p,SG}^L M_{SG}} W_{loss,PC} \right) + C}{\frac{K_{T,Sg}}{C_{p,SG}^L M_{SG}}}$$

(36)

$$T_{SG} = \frac{\left(\frac{T_{SG,sw}}{M_{SG}} - \frac{(C_{p,SG}^v m_{SG}, T_{SG} + E_{evap,SG})}{C_{p,SG}^L M_{SG}} M_{SG} - \frac{K_{T,Sg}}{C_{p,SG}^L M_{SG}} T_{PC} - \frac{1}{C_{p,SG}^L M_{SG}} W_{loss,PC} \right)}{\frac{K_{T,Sg}}{C_{p,SG}^L M_{SG}}} + C e^{-e^{\frac{K_{T,Sg}}{C_{p,SG}^L M_{SG}}t}}$$

(37)

$$T_{SG} = \frac{\left(\frac{T_{SG,sw}}{M_{SG}} - \frac{(C_{p,SG}^v m_{SG}, T_{SG} + E_{evap,SG})}{C_{p,SG}^L M_{SG}} M_{SG} - \frac{K_{T,Sg}}{C_{p,SG}^L M_{SG}} T_{PC} - \frac{1}{C_{p,SG}^L M_{SG}} W_{loss,PC} \right)}{\frac{K_{T,Sg}}{C_{p,SG}^L M_{SG}}} + C e^{-e^{\frac{K_{T,Sg}}{C_{p,SG}^L M_{SG}}t}}$$

(38)

Substitute Eq. (38) into (30)

$$p_{SG} = p_*^T \left(\frac{\left(\frac{T_{SG,sw}}{M_{SG}} - \frac{(C_{p,SG}^v m_{SG}, T_{SG} + E_{evap,SG})}{C_{p,SG}^L M_{SG}} M_{SG} - \frac{K_{T,Sg}}{C_{p,SG}^L M_{SG}} T_{PC} - \frac{1}{C_{p,SG}^L M_{SG}} W_{loss,PC} \right)}{\frac{K_{T,Sg}}{C_{p,SG}^L M_{SG}}} + C e^{-e^{\frac{K_{T,Sg}}{C_{p,SG}^L M_{SG}}t}} \right)$$

(39)

4.2   Simulation Results of Steam Generator System of Pressurized of Water Reactor

The parameter of input steam generator system will be expressed by fuzzy numbers ρ_1 and ρ_2 with membership functions are derived from experimental data or expert knowledge as illustrated in Figs. 1 and 2.

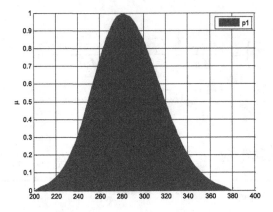

Fig. 1. Uncertain model parameter of $T_{PC} = \rho_1$ of Steam Generator System.

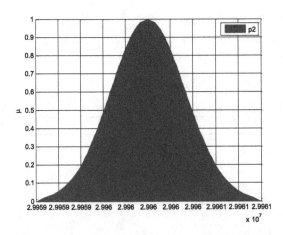

Fig. 2. Uncertain model parameter of $W_{loss,PC} = \rho_2$ of Steam Generator System.

4.3   Analysis Results of Steam Generator System of Pressurized of Water Reactor

Outputs $\tilde{q}(t) = p_{SG}(t)$ are obtained for steam generator system. Figure 3 is the fuzzy valued results of $p_{SG}(t)$. The graph do not show any variation in shape compared to the

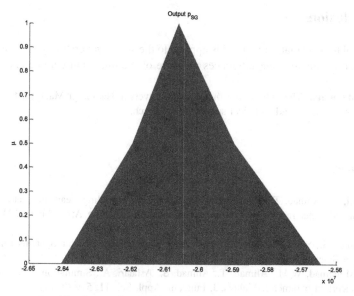

Fig. 3. Output $p_{SG}(t)$ of steam generator system.

original symmetric quasi-Gaussian shape of the uncertain model parameters ρ_1 and ρ_2. The nonlinearities of the model show a moderate effect under the given operating conditions.

The analysis of the transformation method of the steam generator model is given in Fig. 4. The model output is induced by the uncertainty ρ_1 (T_{PC}) with the percentage of the degree of influence is 99 %. The measure of influence of uncertainties for $\rho_2(W_{loss,PC})$ is 1 % have low impact and can be considered as negligible.

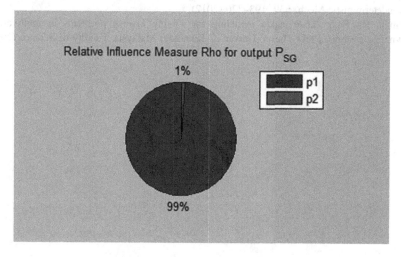

Fig. 4. Measure of influence for output $p_{SG}(t)$ of steam generator system.

5 Conclusions

The general transformation method is applied to the steam generator systems in order to analyze and to quantify the influences measure of the uncertain parameters.

Acknowledgments. The authors are thankful to Universiti Teknologi Malaysia for providing necessary environment and technical support for research.

References

1. Ashaari, A., Ahmad, T., Shamsuddin, M., Omar, N.: Modeling steam generator system of pressurized water reactor using fuzzy state space. Int. J. Pure Appl. Math. **103**, 106–115 (2015)
2. Glasstone, S., Sesonske, A.: Nuclear Reactor Engineering: Reactor Systems Engineering. Springer Science & Business Media (2012)
3. Wan Mohamad, W.M., Ahmad, T., Ahmad, S., Ashaari, A.: Simulation of furnace system with uncertain parameter. Malays. J. Fundam. Appl. Sci. **11**, 5–9 (2015)
4. Wan Mohamad, W.M., Ahmad, T., Ashaari, A., Abdullah, A.: Modeling fuzzy state space of reheater system for simulation and analysis. AIP Conf. Proc. **1605**, 488–493 (2014)
5. Hanss, M.: The transformation method for the simulation and analysis of systems with uncertain parameters. Fuzzy Sets Syst. **130**, 277–289 (2002)
6. Hanss, M.: Applied Fuzzy Arithmetic. Springer, Heidelberg (2005)
7. Hanss, M., Oliver, N.: Simulation of the human glucose metabolism using fuzzy arithmetic. In: 19th International Conference of the North American Fuzzy Information Processing Society, NAFIPS, pp. 201–205. IEEE (2000)
8. Hanss, M., Oliver, N.: Enhanced parameter identification for complex biomedical models on the basis of fuzzy arithmetic. In IFSA World Congress and 20th NAFIPS International Conference, Joint 9th, pp. 1631–1636. IEEE (2001)
9. Haag, T., Hanss, M.: Comprehensive modeling of uncertain systems using fuzzy set theory. Nondeterministic Mech. **539**, 193–226 (2012)
10. Ismail, R.: Fuzzy state space modeling for solving inverse problems in multivariable dynamic systems. Ph.D Thesis, Universiti Teknologi Malaysia, Faculty of Science (2005)

Edge Detection of Flat Electroencephalography Image via Classical and Fuzzy Approach

Suzelawati Zenian[1,3], Tahir Ahmad[1,2(✉)], and Amidora Idris[1]

[1] Department of Mathematical Sciences, Faculty of Science, Universiti Teknologi Malaysia,
81310 UTM Johor Bahru, Malaysia
`amidora@utm.my`, `suzela@ums.edu.my`

[2] Centre for Sustainable Nanomaterials, Ibnu Sina Institute for Scientific and Industrial Research,
Universiti Teknologi Malaysia, 81310 UTM Johor Bahru, Malaysia
`tahir@ibnusina.utm.my`

[3] Department of Mathematics with Computer Graphics, Faculty of Science and Natural
Resources, Universiti Malaysia Sabah, Jalan UMS, 88400 Kota Kinabalu, Sabah, Malaysia

Abstract. Edge detection is a crucial step in image processing in order to mark the point where the light intensity changed significantly. It is widely used to detect gray-scale and colour images in various fields such as medical image processing, machine vision system and remote sensing. The classical edge detectors such as Prewitt, Robert, and Sobel are quite sensitive towards noise and sometimes inaccurate. In this paper, the boundary of the epileptic foci of Flat EEG (fEEG) is determined by implementing some of the methods ranging from classical to fuzzy approach. There are two methods being applied for the fuzzy edge detector technique which are Minimum Constructor and Maximum Constructor methods; and Fuzzy Mathematical Morphology approach.

Keywords: Flat EEG · Fuzzy set · Fuzzy image · Grayscale morphology · Edge detection · Mathematical morphology

1 Introduction

An image plays an important role in human perception. It is a result from the combination between reflections of lighting sources with elements in the scene of imaging. Image processing is a process that takes image as an input and produced the desired image as an output [1].

Edge in an image refers to the boundaries between two regions that separate the object and its background clearly. It is not a physical entity and displays just like a shadow which shows the outline of an object. Edge detection is the study of changes in pixel values which is discontinuous in gray level, colour, and texture. It is a type of image segmentation which aims to recognize structure and properties of objects [2, 3].

There are three main algorithms in edge detection, namely, gradient measurement, smoothing, and localization. The gradient magnitude in an image is calculated to determine edges whereby a grayscale intensity is to measure of an image changes. Smoothing in edge detection is to remove noise. The localization is a process to eliminate number of false edges [4, 5].

© Springer Nature Singapore Pte Ltd. 2016
M.W. Berry et al. (Eds.): SCDS 2016, CCIS 652, pp. 247–255, 2016.
DOI: 10.1007/978-981-10-2777-2_22

Some of the applications of edge detection are in fingerprint recognition, human facial appearance, segmentation of a lung region, computer-aided diagnosis of medical images, and remote sensing image. A medical image poses challenges to distinguish the exact edge from the noise since it contains object boundaries, object shadows, and noise [3]. There are different methods on edge detection such as classical and fuzzy approaches. Fuzzy approach is an alternative way to the classical approach.

Fuzzy sets was introduced by Zadeh in 1965. It is a generalization of the classical sets theory. A fuzzy set is designed to accommodate uncertainties and to provide a formal way of describing real-world phenomena [6].

Mathematical morphology provides the way to analyze and process geometrical structures based on set theory, lattice theory, topology, and random functions. It extracts different image components and has been widely used in image detection and segmentation. Morphological operators transform the original image into another image of certain shape and size which is known as structuring element. It is used to find the morphological gradient or to denoise the image. The erosion and dilation operations aim to find the image edge by taking the difference between the dilated and eroded images. Binary mathematical morphology depends only on two basic operations without taking into account the gray value or colour of the image pixel [7–9].

In this paper, the clusters of epileptic foci for the fEEG images are detected using classical edge detector such as Sobel, Prewitt, and Robert operator. Two fuzzy methods are implemented, namely, the constructor (i.e. Minimum and Maximum) and morphological methods.

2   Basic Concepts

Fuzzy Topographic Topological Mapping (FTTM) is a non-invasive technique for solving neuromagnetic inverse problem. It is based on mathematical concepts namely topology and fuzzy. It aims to accommodate static and simulated, experimental magnetoencephalography (MEG), and recorded electroencephalography (EEG) signals. FTTM consists of three algorithms that link four components of the model as shown in Fig. 1. The four components namely the Magnetic Contour Plane (MC), Base Magnetic Plane (BM), Fuzzy Magnetic Field (FM), and Topographic Magnetic Field (TM) [10].

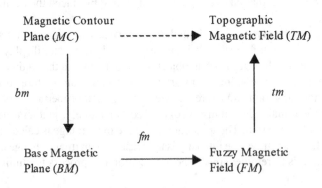

Fig. 1. FTTM [10].

EEG is a system that widely applied for measuring and recording the electrical activity of the brain in graphic form [11]. It reads voltage differences on the head relative to a given point. Figure 2 shows a sample of EEG signal during seizure.

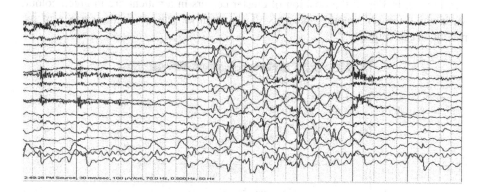

Fig. 2. EEG signal [12].

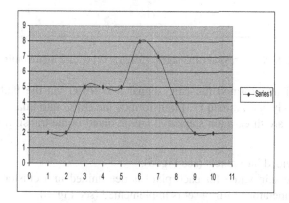

Fig. 3. Analysed EEG signal [12].

In order to map the high dimensional signal, namely EEG, into low dimensional space, a method which is known as fEEG was developed by the Fuzzy Research Group of UTM in 1999 [12]. This method has been used purely for visualization and able to preserve information recorded during seizure. The details of fEEG is given as follows:

Fauziah's EEG coordinate system is defined as

$$C_{EEG} = \left\{ \left((x, y, z), e_p \right) : x, y, z, e_p \in \Re \text{ and } x^2 + y^2 + z^2 = r^2 \right\} \qquad (1)$$

whereby r is the radius of a patient head. The mapping of C_{EEG} to a plane is defined as follows. $S_t : C_{EEG} \rightarrow MC$ such that

$$S_t \left((x, y, z), e_p \right) = \left(\frac{rx + iry}{r + z}, e_p \right) = \left(\frac{rx}{r + z}, \frac{ry}{r + z} \right)_{e_p(x, y, z)} \qquad (2)$$

where $MC = \{((x, y)_0, e_p) : x, y, e_p \in R\}$ is the first component of FTTM. Both C_{EEG} and MC were designed and proven as 2-manifolds.

The EEG signal during seizure can be compressed and analyzed second by second (see Fig. 4). In Fig. 4, the position of cluster centers in a patient are in green colour. Meanwhile, the red colour represents the location of sensors on the surface of the patient's head (Fig. 3).

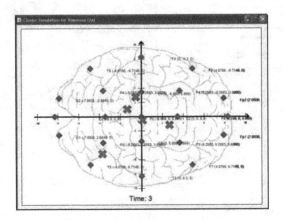

Fig. 4. Compressed EEG signal (fEEG) [12] (Color figure online).

Fauziah [12] transformed the EEG signal into fEEG via the flattening method. Furthermore, Abdy and Ahmad [13] transformed the fEEG into image by using fuzzy approach. There are three main steps that are involved in the transformation of fEEG into image [13].

(a) fEEG is divided into pixels (see Fig. 5).
(b) The membership value for each pixel is determined in a cluster centre and the maximum operator of fuzzy set is implemented (see Fig. 6).

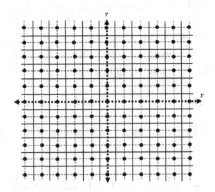

Fig. 5. fEEG pixels [13].

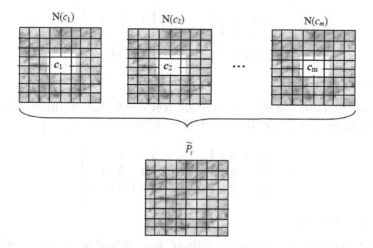

Fig. 6. Fuzzy neighborhood of each cluster centre c_j of a fEEG [13].

(c) The membership value of pixel is transformed into image data (see Fig. 7).

Fig. 7.   fEEG image (input image) [13].

3   Methodology

In this section, an explanation on the classical and fuzzy approaches that used in this work are discussed briefly.

A. Traditional Edge Detector.
(i) Sobel Operator.

Sobel operator is less sensitive to noise and reduce it more efficiently. The kernel G_x and G_y are as follows [3].

$$G_x = \begin{bmatrix} -1 & 0 & 1 \\ -2 & 0 & 2 \\ -1 & 0 & 1 \end{bmatrix} \quad G_y = \begin{bmatrix} 1 & 2 & 1 \\ 0 & 0 & 0 \\ -1 & -2 & -1 \end{bmatrix}$$

(ii) Prewitt Operator.

This operator also use the 3×3 kernels. It is usually used to detect vertical and horizontal edges in an image. The kernels are as follows [3].

$$G_x = \begin{bmatrix} -1 & 0 & 1 \\ -1 & 0 & 1 \\ -1 & 0 & 1 \end{bmatrix} \quad G_y = \begin{bmatrix} 1 & 1 & 1 \\ 1 & 0 & 0 \\ -1 & -1 & -1 \end{bmatrix}$$

(iii) Robert Operator.

Robert operator separates the horizontal and vertical edges individually and combines them again using the approximation equation for the detection results. The kernel is as follows [3].

$$G_x = \begin{bmatrix} 0 & 1 \\ -1 & 0 \end{bmatrix} \quad G_y = \begin{bmatrix} 1 & 0 \\ 0 & -1 \end{bmatrix}$$

B. Fuzzy Edge Detector by Minimum and Maximum Constructors.

Detecting edges in most images is quite challenging since some images may have unclear, damaged, or blurred edges. Fuzzy edge detection method is an approach to detect edges based on the combination of classical and fuzzy methods in image processing. In this work, the minimum and maximum construction methods are used to detect the boundary of fEEG image. Both constructors are based on the concept of triangular norms (t-norm). Basic t-norm that is used for minimum and maximum constructors are given in Eqs. (3) and (4), respectively [14, 15].

$$\textit{Minimum } T_M(x, y) = \min(x, y) \tag{3}$$

$$\textit{Maximum } T_M(x, y) = \max(x, y) \tag{4}$$

C. Fuzzy Mathematical Morphology Approach.

Fuzzy mathematical morphology is developed as an extension of the binary morphology to grayscale morphology. It soften the ordinary morphology by using fuzzy sets, so that the operators is less sensitive to image imprecision. In fuzzy morphological operations, the union operation is replaced by a maximum operation. On the other hand the intersection operation is replaced by a minimum operation [7, 8].

In this work, the gradient image is obtained by using dilation-erosion whereby Hamacher t-norm and t-conorm are used in the computation as in Chaira [7]. The structuring element B may take any size or shape depends on the content of the image and the aim of the morphological operation.

The dilation and erosion operations are opposite in manner. Dilation is defined as the maximum value of the image in the window outlined by the structuring element *B*. It is expected to produce brighter image than the original image. Moreover, any small size and dark details of the image have been reduced or eliminated. Furthermore, erosion is defined as the minimum value of the image in the window outlined by the structuring element *B*. It produces darker image and any small size and bright features are reduced [7–9].

The general steps of the method are as follows [7]

a. The image is initially fuzzified by using any membership function.
b. The erosion and dilation operations are performed by using the structuring element.
c. The gradient image is obtained by computing the difference between the eroded and dilated image.

4 Results

In this section, the output of fEEG images during epileptic seizure are compared based on classical and fuzzy methods for edge detection. Figure 7 is the original fEEG image. The resultant images are shown in Figs. 8, 9 and 10.

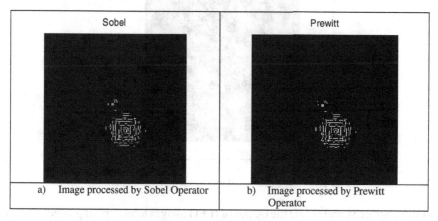

a)   Image processed by Sobel Operator   b)   Image processed by Prewitt Operator

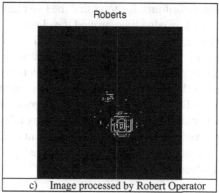

c)   Image processed by Robert Operator

Fig. 8. Classical operator

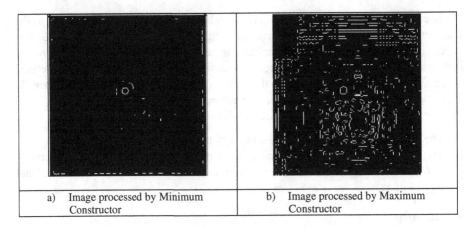

a) Image processed by Minimum Constructor	b) Image processed by Maximum Constructor

Fig. 9. Minimum and Maximum constructor

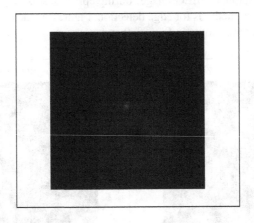

Fig. 10. Image processed by Dilation-erosion operator

Figure 8(a), (b), and (c) are the output of fEEG images after applying Sobel, Prewitt, and Robert edge detection algorithms. It shows that the two clusters are detected by the bright edged but with many tiny edge lines around the clusters.

Moreover, Fig. 9(a) and (b) are the results of fEEG image by using Minimum constructor and Maximum constructor, respectively. Minimum constructor is able to detect only one cluster clearly. Meanwhile, one cluster is detected clearly in Maximum constructor but with many other unwanted edge lines in the image.

The output image for fEEG by using morphological gradient operation is given in Fig. 10. It shows that both clusters are detected without any other unwanted edge lines around the clusters. However, one of the clusters appears in brighter intensity than the other one. It shows that the morphological operator managed to identify the clusters but in different levels of brightness.

5   Conclusions

In this work, the boundary of cluster centre in the original fEEG image can be considered as vague since the sharp boundary was not clearly determined. The vague boundary is represented by different intensity brightness around the clusters. Therefore, the boundary of the cluster centre of fEEG image is detected by using different edge detection algorithms ranging from the classical and fuzzy methods. Each methods gave different edge detection results. Among all the outputs, morphological operator gave better result in detecting the cluster centre with no other false edges around the clusters.

Acknowledgments. The authors would like to thank their family members for their support and encouragement, the members of the Fuzzy Research Group (FRG), Department of Mathematics and Ibnu Sina Institute for Scientific and Industrial Research, UTM, and Universiti Malaysia Sabah (UMS) for their assistance and cooperation.

References

1. Jahne, B.: Digital Image Processing, 6th edn. Springer, Berlin (2005)
2. Zhai, L., Dong, S.P., Ma, H.L.: Recent methods and application on image edge detection. In: International Workshop on Education Technology and Training & International Workshop on Geoscience and Remote Sensing, vol. 1, pp. 332–335 (2008)
3. Rafael, C.G., Richard, E.W.: Digital Image Processing. Pearson Prentice Hall, New Jersey (2007)
4. Ziou, D., Tabbone, S.: Edge Detection Techniques-An Overview, Canada (2000)
5. Jayaraman, S., Esakkirajan, S., Veerakumar, T.: Digital Image Processing. McGraw Hill Education, India (2009)
6. Zadeh, L.A.: Fuzzy Set. Inf. Control **8**, 338–353 (1965)
7. Chaira, T.: Medical Image Processing-Advanced Fuzzy Set Theoretic Techniques. CRC Press (2015)
8. Yee, Y.H., Khaing, K.A.: Fuzzy Mathematical Morphology Approach in Image Processing. Int. Sci. Index, Comput. Inf. Eng. **6**(2) (2008)
9. Mehena, J.: Medical images edge detection based on mathematical morphology. Int. J. Comput. Commun. Technol. (IJCCT) **2** (2011)
10. Ahmad, T., Ahmad, R.S., Zakaria, F., Yun, L.L.: Development of Detection Model for Neuromagnetic Fields. In: Proceeding of BIOMED, Kuala Lumpur, pp. 119–121 (2000)
11. Rudman, J.: EEG Technician. National Learning Corporation (2012)
12. Zakaria, F.: Dynamic Profiling of EEG Data during Seizure using Fuzzy Information. Ph.D thesis, Universiti Teknologi Malaysia (2008)
13. Abdy, M., Ahmad, T.: Transformation of EEG signals into image form during epileptic seizure. Int. J. Basic Appl. Sci. **11**, 18–23 (2011)
14. Jena, K.K.: Result Analysis of Different Image Edges by Applying Existing and New Technique. Int. J. Comput. Sci. Inf. Technol. Secur. (IJCSITS) **5**(1), 183–189 (2015)
15. Mishra, S., Jena, K.K., Mishra, S.: Result analysis of edge detection of brighter satellite images: a comparative study of new and existing techniques. Int. J. Adv. Res. Comput. Eng. Technol. (IJARCET) **4**(2), 491–495 (2015)

Information and Sentiment Analytics

Feel-Phy: An Intelligent Web-Based Physics QA System

Kwong Seng Fong[1(✉)], Chih How Bong[1],
Zahrah Binti Ahmad[2], and Norisma Binti Idris[3]

[1] Faculty of Computer Science and Information Technology, Universiti Malaysia Sarawak,
94300 Kota Samarahan, Sarawak, Malaysia
ksfong@siswa.unimas.my, chbong@fit.unimas.my
[2] Centre for Foundation Studies in Science, University of Malaya,
50603 Kuala Lumpur, Malaysia
zahrah@um.edu.my
[3] Faculty of Computer Science and Information Technology, University of Malaya,
50603 Kuala Lumpur, Malaysia
norisma@um.edu.my

Abstract. Feel-Phy is a computerized and unmanned question answering system which is able to solve open-ended Physics problems, providing adaptive guidance and retrieve relevant resources to user inputs. Latent Semantic Indexing (LSI) is employed to process the user inputs and retrieve relevant references. The proposed architecture for Feel-Phy constitutes of four basic modules: data extraction, question classification, solution identification and answer formulation. The data extraction module is used to construct a Physics knowledge base. The question classification module is used to identify question type and understand the question. The solution identification module computes the answer to the question and also retrieve the top n most relevant resource references to the users. Finally, the last module, answer formulation is to present the results to the users. Our preliminary experiments have shown that this proposed method is able to solve well-structure Physics question and retrieve relevant references to the users.

Keywords: LSI · QA system · Calculation · Information retrieval · Physics problem · Open text question

1 Introduction

Question Answering (QA) system is information retrieval system that aims to respond to the natural language question by returning concise and precise answers rather than informative documents. The QA system is different if compared with search engine where the later is used to find documents or paragraphs in the document that are likely to contain the keywords to the search text or query. QA system is meant to return answers in natural language to the question posed by humans.

Currently, there have various types of QA systems existed online or offline which focusing on different tasks and domains. For example, "Ask.com", "Cha Cha" and "Yahoo! Answers", all focus on multiple topics in English, including Physics. Most of the QA systems are suitable to search answers for simple questions but unsuitable for

© Springer Nature Singapore Pte Ltd. 2016
M.W. Berry et al. (Eds.): SCDS 2016, CCIS 652, pp. 259–270, 2016.
DOI: 10.1007/978-981-10-2777-2_23

problems that need computation likes START natural language QA system. It just can answer most of the simple questions about factual information such as "Who first discovered radiocarbon dating?", but it cannot perform the calculation. For "BlikBook", "Brilliant.org", "Chegg" and "Transtutors", they focus on academic subjects including Physics as well. However, none of the reviewed systems are able to directly solve the open-ended Physics problems automatically. For example, "Yahoo! Answers" is an online system which allows users to post any question but relies on the community to response answers. The drawback with such system is the question can take the indefinite length of time for the answers to arrive yet the answers are not always warranted.

In this paper, we proposed an intelligent Physics QA system known as Feel-Phy. It could automatically provide guided learning environment and to help the users, which are Malaysia Form 4 and Form 5 secondary students, to solve open-ended Physics problems and retrieve relevant references according to the queries were given. All queries will be accompanied by solution and references, which can be served as learning materials.

Section 2 will present some literatures on the existing QA systems. Section 3 will detail out the proposed QA system architecture. Section 4 will present the evaluation results follows by a discussion. Section 5 will conclude our work.

2   Related Physics QA System

According to Hirschman and Gaizauskas, the current search engines can only return ranked lists of searched documents, but they could not deliver an exact answer to the question quickly and sufficiently [7]. On the other hand, question answering systems could overcome this problem. Due to the user demands, the research community has started to discuss the performance, requirements, users and challenges of the QA systems, especially Text Retrieval Conference (TREC) [7]. Since 1999, TREC has sponsored a question answering track to evaluate QA system which only focused on factual questions [7]. Das et al. had mentioned that factual question was based on given fact to get direct answer [4].

BASEBALL [6] and LUNAR [14] were two early QA systems. Baseball was a computer program that able to answer what were the month, day, place, teams and scores for each game in the American baseball league up to a year [6]. LUNAR, on the other hand, able to provide answer to the questions on rocks that were analyzed during the Apollo Lunar missions [15].

"Yahoo Answers", which is a large and diverse community base question answer system, acting as a medium for knowledge sharing and also serves as a place to seek advice, gather options and satisfy one's curiosity about things which may not have a single best answer [1]. It also included various Physics questions. The format of interaction of the system was entirely achieved through questions and answers: a user posted a question and others replied directly the question with their answers. The questions and their answers were stored according to the categories. However, the user may not get the answer if no one replied the answer.

Currently there are not have any existed intelligent QA systems which can automatically solve any types of Physics questions without human helping especially the calculation type of question (Physics problem). Even the community based QA system, it still needs others users help to reply the answers. Therefore, the user can use other types of system to find the Physics question such as online tutoring system and online e-learning system.

Tutorvista.com is one of the online tutoring systems which involves the subjects of Chemistry, Math, Statistic and Physics. The system has provided one-on-one tutoring session between the professional tutor and the user and also users the virtual whiteboard workspace to draw and write in order to explain the problems and solutions [13]. However, it is paid service and the user requires to register and book for tutoring sessions first before the user asked the questions.

Furthermore, one of the online e-learning system is called as Score A. It is used for primary school to secondary school students. The students can select the programs depends on their levels to practice on the set of predefined questions and try out the trial or past year examination questions [9]. For the Physics subject, the system currently just only provides the multiple choice questions accompanied with the explanation for the students to do the exercises. In other words, the system just only provides a set of limited exercises for the users. Besides, these all limited exercises are predefined and the user cannot ask their own Physics questions.

There are various different methods used to construct QA systems. Kamdi and Agrawal proposed a closed domain QA system for the legal documents of Indian Penal Code (IPC) Sections and Indian Laws by adopting machine learning approach and information retrieval [8]. The algorithm facilitates syntactic information and keywords from the question to match the term in Index Term Dictionary to identify the candidate results from a knowledge-base. Jaccard Coefficient mechanism is used to calculate the relative similarity of the results to the question.

AskMSR, a web-based QA system is created based on the search engine and ranking model in order to find the best matching answers [12]. The system passed the question to a search engine and then extract candidate answers from the top retrieved snippets. The information from Wikipedia and WordNet were used to train the ranking model.

Besides that, Cui and Wang have developed an intelligent QA system based on the Latent Dirichlet Allocation (LDA) (also known as topic model) for teaching of database principles [3]. It is based on topics-documentation-knowledge points. Normally, the knowledge tree is formed according to the structure of the chapter, section and knowledge point. Therefore, the system will depend on the topic involved in the question to returns the answer. Their QA system includes five fundamental modules, which were question understanding to analyse the question, FAQ pretreatment was applied to obtain the feature words for each of questions, determining a candidate set of questions according to question type and features words, question similarity calculation between user question and the problems in the candidate problems that retrieve from FAQ library to return the most similar answers and FAQ library expansion is used to add new question as candidate question into FAQ library.

In this paper, we based on the advantages and limitation of the discussed related works to propose QA system by using Python as programming language. We apply Latent

Semantic Indexing (LSI) to provide most related resource to the user, including text, images and videos regardless sometime no answer found by the system. At the same time, we apply semantic and syntactic information to the system, which can automatically compute and return the answer by guideline for the Physics problem without any human helping. At the time of writing, there is no literature to support the Physics learning units which able to solve Physics problems using corresponding equations through human-computer interaction system.

3   Proposed System Architecture

Feel-Phy in general consists of four modules which are: data extraction, question classification, solution identification and answer formulation as depicted in Fig. 1. The QA system able to show guided step-by-step solution leading to the final answer, accompanying the answer with relevant references.

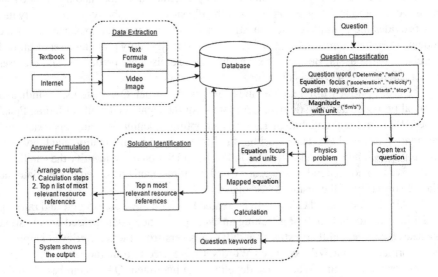

Fig. 1.   Architecture of an intelligent web-based Physics QA system

3.1   Data Extraction

The purpose of the data extraction module is to construct a semantic Physics knowledge base. The knowledge base consists of resources extracted from selected four high school to collegial level Physics textbooks, which are: Physics textbooks: Newtonian Physics written by Benjamin Crowell, The Free High School Texts: A Textbook for High School Students Physics, College Physics written by Dr. Paul Peter Urone, Dr. Roger Hinrichs, Dr. Kim Dirks and Dr. Manjula Sharma and Physics: Principles with Applications written by Douglas C Giancoli. For the sake of brevity, we restricted our study on five topics as shown in Fig. 2 below.

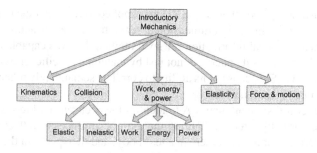

Fig. 2. Physics topics

First, we extracted paragraphs, and images from the textbooks and stored them in a relational database. For each topic, we also extracted a list of equations corresponding to its LHS (Left-hand Side) variables (equation focus) and RHS (Right-hand side) variables. The total number of collected equations are 247. Table 1 is shown as some of the examples of equations in equation table. Besides, we also facilitated a table with 78 total number of rows to associate the LHS variable, the RHS variable and the Standard Imperial (SI) unit with the learning concepts. We have shown three examples in Table 2 below. With these, the system is able to use LHS variable, RHS variables and the units identified from the user question to determine the most suitable equation to solve the problem.

Table 1. Equation table in relational database.

ID	Equation	RHS variable	LHS variable (equation focus)
1	$a = (v - u)/t$	v, u, t	a
2	$t = (2s)/(u + v)$	s, u, v	t
3	$v = s/t$	s, t	v

Table 2. Quantity table in relational database

ID	Learning concept	LHS/RHS variable	LHS variable (equation focus)
1	acceleration	a	m/s^2
2	initial velocity	u	m/s
3	final velocity	v	m/s

If a particular paragraph can be represented by using online videos or images, then the Uniform Resource Locator (URL) of the videos (mainly from Youtube) and images (mainly from Google Image) were stored corresponding to the paragraph with its topic in the relational database. The questions were stored to solve Physics problems.

Once all the data were collected, we employed Latent Semantic Indexing (LSI) on all paragraphs to create a reduced dimensional semantic space. LSI is an alias of Latent Semantic Analysis (LSA). According to Deshmukh and Hegde, LSI was an information retrieval technique that uses a mathematical technique called Singular Value Decomposition (SVD), which identified the pattern in an unstructured collection of text and found relationships between them [5]. SVD was applied to represent the terms from the paragraphs as vectors in a high-dimensional semantic space. In addition, Bradford defined LSI as a statistical information retrieval method that was capable of retrieving text based on the concepts it contained, not just by matching specific terms or keywords [2]. It was called as LSI because of its ability to correlate semantically related terms that were "latent" in a collection of text.

LSI uses a term-document matrix A to identify the occurrence of terms within a set of documents, normally a term weighting the term occurrences to reflect the fact that some terms were essential than others in a text body, and then perform the SVD on the matrix to determine patterns in the relationship between the terms and concepts used in the documents [2].

$$A \approx A' = U_k \sum_k V_k^T \tag{1}$$

In Eq. (1), A is term-document matrix of terms normalized with frequency-inverse document frequency (TFIDF). We then employed SVD to convert A into three components: type vectors, U (term coordinate in vector space), document vectors V (document coordinate in vector space) and weight represented by a singular value, $\sum$, where the diagonal entries of $\sum$ is k = 300 [10]. LSI used a mathematical transform technique to decrease the dimension number in the term space of the matrix to make it more useable and efficient [2].

$$TFIDF_{ij} = \left(N_{ij}/N_{*j}\right) * log\left(D/D_i\right) \tag{2}$$

where N_{ij} is the number of times word i appears in document j (the original cell count), N_{*j} is the number of total words in document j (just add the counts in column j), D is the number of documents (the number of columns) and D_i is the number of documents in which word i appears (the number of non-zero columns in row i).

Given a question, q, in order to find similar documents in the semantic space, the terms of the question is used to create a new column in A, normalized with TFIDF, thus the vector representation of the question q is

$$q' = q^T U_k \sum_k^{-1} \tag{3}$$

where q^T is the vector of words in the question. As a result, queries against a set of documents that have undergone LSI will return results that were conceptually similar in meaning to the query, even if they do not have shared a specific term in the query. The similarity is computed using cosine similarity, $cos(\theta)$.

$$similarity = cos(\theta) = (A.B)/(\|A\| \|B\|) \tag{4}$$

Given the similarity score is x, where $-1.0 \leq x \leq 1.0$. If the similarity score is 1.0, it indicates exact match and -1.0 means totally not match.

3.2 Question Classification

The question classification module is used to analyze the user input and identify question word, equation focus, question keywords, magnitudes with units and question type. The QA system able to provide answers to two type of queries: **open text questions** and **Physics problems**.

Open text questions are the question-like text that just have the natural language which needs to have the explanation type of solution, likes "What is force?". The system will show the most relevant resources, including text, images and videos to the user. For this type of question, we treated it as a search engine ranking problem where the query will be projected into the semantic space to return the top n most similar results.

On the other hand, Physics problem is a solvable question which consists of the magnitudes and units where requires calculation. For example, "What is the acceleration of the car when it starts from 10 m/s to 50 m/s within 2 min?".

In order to solve the problem, firstly, the system performs word segmentation, stop-words removal and keywords extraction for the user question to get the question keywords. Then, it will look for keywords such as "determine", "what", "how", "find" and "calculate" and if the keywords is existed in the list of question keywords. Besides, we also have a set of rules to determine the equation focus from the keywords which is used to determine the equation focus. Currently, we just focused the 16 types equation focus and Table 3 below have shown four examples of them. Based on the Physics problem given above, "acceleration" word is found after the keyword "what", from the list of question keywords, so "acceleration" is considered as equation focus.

Table 3. List of equation focus identification

Equation focus	Possible keywords in question
distance	high, length, far, distance, displacement, position
final velocity	final velocity, new speed, velocity
initial velocity	takeoff speed, initial velocity
time	long, time

Next, the system identifies magnitudes (float values) and the corresponding units to differentiate the question type. If any of the mentioned indicators exist, the question will be treated as a Physics problem, otherwise, an open text question. The system will also check other keywords that are exist in the questions in order to find any "latent" variables in the questions. For instance, if "from stationary" exists in question, it means the "latent" variable is "0 m/s" which indicates an initial velocity. If "to a stop" exists in question, it means the hidden term is "0 m/s" and it is final velocity.

Based on the given question above, the module analyzes question type and word, equation focus, keywords, magnitudes with units and possible "latent" variables as depicted in Table 4 below.

Table 4. Question analyzation result

Question	What is the acceleration of the car when it starts from 10 m/s to 50 m/s within 2 min?
Question word	what
Question focus	acceleration
Question keywords	car, starts, from, 10 m/s, to, 50 m/s, within, 2 min
Magnitude with units	10 m/s, 50 m/s, 2 min
Question type	Physics problem
Latent variable	None

3.3 Solution Identification

The main purpose of the solution identification is used to compute solution as well as return the top n (n = 100) most relevant resources from the knowledge base. For Physics problem, the system uses equation focus and units to identify mapped equation and then perform the calculation to get the answer as shown in Fig. 3.

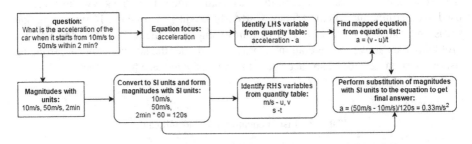

Fig. 3. Overall process for Physics problem in calculation part

First, the system examines if every unit from the question is Standard Imperial (SI) unit or derived unit. For derived units, the system converts them to the SI units. For example, the SI unit of time is second, so if the unit of time is minutes, the system need to convert it to second. Next, the system finds the mapped equation from the equation list to obtain the corresponding units. For the problem mentioned above, since the equation focus is "acceleration" and the magnitudes with units detected by the system at the question classification phase are "10 m/s", "50 m/s" and "2 min". With these, the system examines if there is a need to convert them into SI units. Since the first two magnitudes with units are SI units, so no conversion needed. However, for "2 min", it is derived unit, and it needed to be converted it to SI unit, which is "120 s".

Next, the system identifies all the possible RHS variables given the SI units in the question. The question has an initial velocity, u (its SI unit is "m/s"), a final velocity, v (its SI unit is "m/s") and a time, t (its SI unit is "s"). At the same time, the system determines the LHS variable (equation focus) from quantity table, which is a. Using these variables, the system finds the suitable equation according to equation focus. Hence, the most suitable equation found is:

$$a = (v - u)/t \tag{5}$$

All the magnitudes with SI units are substituted into the found equation and to compute the final answer. For example, "from 10 m/s" is treated as initial velocity, u and "to 50 m/s" are treated as final velocity, v. Therefore, the final answer is:

$$a = (v - u)/t = (50m/s - 10m/s)/120s = 0.33m/s^2 \tag{6}$$

For open text questions, the system skips the calculation part and perform references extraction. The system projects the open text question into the semantic space and retrieves the most similar related resources, including references text, images, videos and equations.

3.4 Answer Formulation

The main role of answer formulation is to organize the solutions or results to users. For Physics problems, the system able to shows all the steps which start from getting the magnitudes with units until getting the final answer to the users. Its purpose is to provide an adaptive guidance for the students and not just to show the final answer as shown in Fig. 4 below.

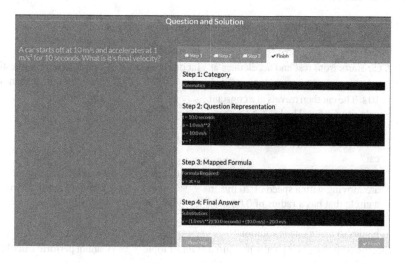

Fig. 4. Calculation solution given a Physics problem

Polya had mentioned user needed to solve the problem step by step and the first step is to understand the problem which by determining the relevant and unknown data [11]. It was meaningless if the user get the final answer without knowing the steps of solution.

Firstly, the system indicates the topic of the question it belongs to. Next, it returns all the question representations, which are the LSH, RHS variables and their magnitudes with units. Meanwhile, if the system has performed the conversion for these derived units to SI units, then the system shows the steps of the conversion. Next, it shows mapped equation and then performs the substitution for the equation with the magnitudes and units to return the final answer. In addition, the top n most relevant resource references, images and videos also displayed to the user.

On the other hand, if it is open text question, the system just returns the top n most relevant resource references, images and videos.

4   Evaluation and Testing

We have identified 70 questions with answers covering five topics to be used a benchmarking data to evaluate our system efficacy. Table 5 showed the overall results.

Table 5. Overall result of the system

	Result of the system			
	Correct	Incorrect	Out of topic	Unable to solve
Total number of answers	45	3	8	14

According to the table above, it is obvious that the system able to answer most of the question correctly. However, there are 25 number of questions cannot be solved due to for a number of reasons. We have listed out and discussed five examples of unsolvable questions as shown in Table 6.

Table 6. Some unsolvable questions analyzation

No	Question	Unsolvable type	Reason
1	A car starts from rest and accelerates at a constant acceleration of 3 m/s^2 for 10 s. The car then travels at a constant velocity for 5 s. The brakes are then applied and the car stops in 5 s. What is the total distance traveled by the car?	Incorrect answer	Complexity of the question, involves more than one equation
2	The tips of the blades in a food blender are moving with a speed of 20 m/s in a circle that has a radius of 0.06 m. How much time does it take for the blades to make one revolution?	Out of topic	It was circular motion topic and currently no involve this topic
3	Electric resistance is typically measured in what units?	Out of topic	Cannot perform calculation
4	A rock is thrown straight upward with an initial velocity of 9.6 m/s in a location where the acceleration due to gravity has a magnitude of 9.8 m/s^2. To what height does it rise?	Unable to solve	Cannot detect the hidden value, so no mapped equation has found
5	A car traveled 88 m/s, 90 km in the next 2 min, and then 76 m/s before reaching its destination. What was the car's average speed?	Unable to solve	Too many magnitudes with units and cannot find mapped equation

For the first unsolvable question, it was due to the complexity of the question which causes the system cannot detect the mapped equations and calculate correctly. This question involved four equations and they were interrelated. It consisted of three parts that needed to separately calculate the distances by using the Eq. (7):

$$s = ut + at^2 \tag{7}$$

$$\sum s = s_1 + s_2 + s_3 \tag{8}$$

Then, the Eq. (8) was applied to calculate the total distance traveled by the car. However, currently the system cannot perform very complicated calculation. Therefore, the system was calculated and returned the wrong answer.

For the second unsolvable question above, since it was the question for the circular motion topic and currently we did not involved this topic, we considered it as out of topic.

For the third unsolvable question, it was asked about the unit of electric resistance. It was not a type of Physics problem. It was not needed to perform any calculation, so we also considered it as out of topic.

For the fourth unsolvable question, it consisted of "latent" value, which is the final velocity. Based on this question, once the rock reached the highest point, the final velocity of the rock will automatically become 0 m/s, but the system cannot detect it. Therefore, the system cannot find the mapped equation to perform the calculation.

$$\sum t = t_1 + t_2 + t_3 \tag{9}$$

$$v = s/t \tag{10}$$

For the fifth unsolvable question, the system not able to detect the magnitude and calculate it correctly. This is especially true for the question which have the latent variables and equations. This question needed to add all the distances by using the Eq. (8) and also needed to add all the time spent which was used the Eq. (9). Finally, all the magnitudes needed to be averaged by applying the Eq. (10). Due to these latent equations, the system could not produce any solution.

5   Conclusion and Future Work

The QA system presented in the paper can be used to solve the Physics problems. It consisted of four modules which were data extraction, question classification, solution identification and answer formulation. After the data have been extracted, the system started to identify the question whether it was an open question or a problem. It then found the solution by identify the parameters and units. For the top n most relevant resource references, it relied on LSI for extracting the data from the knowledge base.

The overall results showed that most of the questions can be solved. We also discovered some of the problems cannot be solved correctly and this was due to the complexity of the question. Besides, some of the questions were out of the scope that was not covered in the knowledge base.

For future work, we plan to improve the system by including freebase database or DBPedia database by using the Quepy framework. With the combination of these databases, the system should have more information to improve the accuracy of the answers.

Acknowledgment. This research is supported by research grant RACE/b(6)/1098/2013(06), KPM. Besides, I would like to express my deepest gratitude to my supervisor, Dr. Bong Chih How for his extraordinary efforts in providing guidance and motivation. Besides, I also want to thank Associate Prof. Dr. Zahrah Binti Ahmad and Dr. Norisma Binti Idris for suggesting a lot of ideas and comments about the system in order to improve my system. I also wish to take this opportunity to acknowledge my heartiest thanks to all my friends who involved directly or indirectly towards the accomplishment of my system.

References

1. Adamic, L.A., Zhang, J., Bakshy, E., Ackerman, M.S.: Knowledge sharing and yahoo answers: everyone knows something. In: Proceedings of the 17th International Conference on World Wide Web, pp. 665–674. ACM (2008)
2. Bradford, R.: Why lsi? Latent semantic indexing and information retrieval (2009)
3. Cui, L., Wang, C.: An intelligent q&a system based on the lda topic model for the teaching of database principles. World Trans. Eng. Technol. Educ. **12**(1), 26–30 (2014)
4. Das, S., Catterjee, R., Mandal, J.K.: An approach for creating framework for automated question generation from instructional objective. In: Satapathy, S.C., et al. (eds.) Proceedings of the Second International Conference on Computer and Communication Technologies. AISC, vol. 379, pp. 527–535. Springer, India (2016)
5. Deshmukh, A., Hegde, G., Lathi, R., Govikarn, S.: A literature survey on latent semantic indexing. In: International Conference on Computing, p. 100 (2012)
6. Green Jr., B.F., Wolf, A.K., Chomsky, C., Laughery, K.: Baseball: an automatic question-answerer. Papers Presented at the 9–11 May 1961, Western Joint IREAIEE-ACM Computer Conference, pp. 219–224. ACM (1961)
7. Hirschman, L., Gaizauskas, R.: Natural language question answering: the view from here. Nat. Lang. Eng. **7**(4), 275–300 (2001)
8. Kamdi, R.P., Agrawal, A.J.: Keywords based closed domain question answering system for indian penal code sections and indian amendment laws. Int. J. Intell. Syst. Appl. (IJISA) **7**(12), 57 (2015)
9. Lobo, R.: Score a programme product features, December 2009. http://www.slideshare.net/scoreasifu/score-a-programme-product-features
10. Martin, D.I., Berry, M.W.: Mathematical foundation behind latent semantic analysis. In: Handbook of Latent Semantic Analysis, pp. 35–56 (2007)
11. Polya, G.: How to Solve It: A New Aspect of Mathematical Method. Princeton University Press, Princeton (2014)
12. Tsai, C., Yih, W.T., Burges, C.: Web-based question answering: Revisiting askmsr. Technical report MSR-TR-2015-20, Microsoft Research (2015)
13. TutorVista.com: How it works (2016). http://www.tutorvista.com/howitworks.php
14. Woods, W.A.: Progress in natural language understanding: an application to lunar geology. In: Proceeding of the 4–8 June 1973, National Computer Conference and Exposition, pp. 441–450. ACM (1973)
15. Woods, W.A.: Semantics and quantification in natural language question answering. Adv. Comput. **17**(3), 1–87 (1978)

Usability Evaluation of Secondary School Websites in Malaysia: Case of Federal Territories of Kuala Lumpur, Putrajaya and Labuan

Wan Abdul Rahim Wan Mohd Isa[✉], Zawaliah Ishak, and Siti Zulaiha Shabirin

Faculty of Computer and Mathematical Sciences, Universiti Teknologi MARA,
40450 Shah Alam, Selangor, Malaysia
wrahim2@tmsk.uitm.edu.my, cikliah@gmail.com,
hijrahonebeauty@yahoo.com

Abstract. The main objective of this study is to investigate the usability of secondary schools websites in Federal Territories Kuala Lumpur, Putrajaya and Labuan, Malaysia. The evaluation was done by using three automated tools; (i) Web Page Analyzer (websiteoptimization.com), (ii) DeadLink Checker and (iii) Broken Link Checker. The samples include 53 secondary schools websites in Malaysia. The data was analyzed based from Nielson usability guidelines for (i) ideal size of web pages, (ii) number of broken links and the (iii) webpage size. The result of this study shows that the secondary schools websites in Federal Territories Kuala Lumpur, Putrajaya and Labuan, Malaysia had few usability issues. This study provides recommendations for usability improvement for secondary schools website in Malaysia. Future work may involve accessibility evaluation of the secondary school websites in Malaysia.

Keywords: Usability evaluation · Secondary school websites

1 Introduction

Researches show millions of dollars loss from the industry due to lack of usability of the website [1]. Usability of the website need to be assess from time to time. The assessment is crucial to ensure the usability is consistent from time to time. It will further guide for future research and enhancement of the website. This includes the maintenance of the website on regular basis. The assessment need to involve with the end users so that the satisfaction of them can be measured accurately from time to time [2]. There are growing researches of usability and accessibility evaluations with different genre of websites in Malaysia [3–10]. There are also growing demands that the usability and accessibility evaluations are being conducted on regular basis. Tables 1 and 2 show the examples of usability and accessibility evaluations on different website genres in Malaysia.

© Springer Nature Singapore Pte Ltd. 2016
M.W. Berry et al. (Eds.): SCDS 2016, CCIS 652, pp. 271–279, 2016.
DOI: 10.1007/978-981-10-2777-2_24

Table 1. Example of usability evaluations on different website genre in Malaysia

Website genre	Evaluation	Reference
Homestay	Usability	[3]
Malaysia Higher Education	Usability	[4]
E-Government	Usability	[5]
E-Commerce	Usability	[6, 7]
Portal	Usability	[8]

Table 2. Example of accessibility evaluations on different website genre in Malaysia

Website genre	Evaluation	Reference
Homestay	Accessibility	[9]
Small & Medium Enterprise (SME)	Accessibility	[10]
Malaysia Higher Education	Accessibility	[4]
E-Government	Accessibility	[5]

2   Method

There are about 122 secondary schools in Federal Territories of Kuala Lumpur, Putrajaya and Labuan from eight types of school. The type of schools are; (i) Kolej, (ii) Sekolah Sukan, (iii) SM Agama, (iv) SM Berasrama Penuh, (v) SM Kebangsaan, (vi) SM Khas, (vii) SM Teknik and (viii) SMK Agama. However, there are only 82 secondary schools within the area with official websites. From 82 secondary schools websites, only 53 websites suitable to be sampled because the rest of the websites return no data when the usability test using automated tools run over it. Thus, there are only 53 secondary schools websites that located in Kuala Lumpur, Putrajaya and Labuan which were selected as sample in this study. The list of secondary schools involved in this study was selected from Education Management Information System (EMIS Online), a web portal developed and maintained by Ministry of Education from https://emis-portal1.moe.gov.my/emis/emis2/emisportal2/.

These are three tools are used by this study to investigate web usability aspect of 53 secondary school websites:-

(a) Web Page Analyzer
     http://www.websiteoptimization.com/services/analyze/
(b) DeadLink
     http://www.deadlinkchecker.com
(c) BrokenLinkChecker
     http://www.brokenlinkcheck.com/

Applying Nielson's usability guideline [11], Table 3 shows the Usability Traits Filters which determine the usability issues of the sample websites of Malaysia Secondary Schools in Federal Territories of Kuala Lumpur, Putrajaya and Labuan.

Table 3. Usability traits filters (usability issues)

Usability traits filters
Speed > 10 s
Broken Link >= 1
Page Size > 37 KB

3   Analysis and Findings

3.1   Usability Analysis

The usability analysis results of Malaysia Federal Territories Kuala Lumpur, Putrajaya and Labuan secondary schools websites are grouped into eight type of secondary schools which consist of:-

1. Kolej
2. Sekolah Sukan
3. SM Agama
4. SM Berasrama Penuh
5. SM Kebangsaan
6. SM Khas
7. SM Teknik
8. SMK Agama

The analysis results of usability consist of Broken Links, Page Size and modem Speed. Table 3 shows the usability issues being predetermined as usability traits filters. The predetermined usability issues are (i) the broken links, (ii) Page Size >37 kb and (iii) the download time using six different modem speeds that are 14.4 k, 28.8 k, 33.6 k, 56 k, 128 k and 1.44 Mbps which are more the 10 s. The data of the websites download time then are filtered to the categories that need more than 10 s to be downloaded for each type of modem speed.

Table 4 shows the usability analysis on broken link, page size and speed based on different types of secondary schools in federal territories of Kuala Lumpur, Putrajaya and Labuan. The results in Table 4 shows the secondary schools from SM Kebangsaan school type websites contribute to the highest number of websites that have broken links with the total of 29 schools websites. The second highest of websites with broken links are from SM Berasrama Penuh and Kolej schools type which contributes three and two websites with broken links, accordingly. The other types of schools which are Sekolah Sukan, SM Agama, SM Khas, SM Teknik and SMK Agama contribute one websites for each school type.

For the 'page size' analysis, the results in Table 4 shows the secondary schools from SM Kebangsaan school type websites contribute to the highest websites that exceed the size of 37 kb with the total of 42 websites. The second highest websites that exceed the size of 37 kb are Kolej and Sekolah Berasrama Penuh school types with two websites, respectively. This is followed with the following school types; (i) Sekolah Sukan, (ii) SM Agama, (iii) SM Khas, (iv) SM Teknik and (v) SMK Agama with one website, each.

Table 4. Usability analysis on broken link, page size and download speed.

Type of secondary school	Broken link >= 1	Page size > 37 kb	Download times >10 s					
			14.4 k	28.8 k	33.6 k	56 k	128 k	1.44 MB
Kolej	2	2	2	2	2	2	2	2
Sekolah Sukan	1	1	1	1	1	1	1	1
SM Agama	1	1	1	1	1	1	1	1
SM Berasrama Penuh	3	2	2	2	2	2	2	2
SM Kebangsaan	29	42	42	42	42	42	42	31
SM Khas	1	1	1	1	1	1	1	1
SM Teknik	1	1	1	1	1	1	1	1
SMK Agama	1	1	1	1	1	1	1	1

For the 'download time >10 s' usability issue analysis in the Table 4, the results of usability analysis for the Malaysia Federal Territories Kuala Lumpur, Putrajaya and Labuan Secondary Schools Websites shows download time more than 10 s based from the six types of modem speed (14.4 kbps, 28.8 kbps, 33.6 kbps, 56 kbps, 128 kbps, 1.44 Mbps). For the modem speed 14.4 kbps analysis, Table 4 shows that the school type SMK Kebangsaan had the highest numbers of websites that requires more than 10 s to be downloaded with 42 websites. It is followed with Kolej and Sekolah Berasrama Penuh school type with 2 websites, for each type. This is followed with the rest of the school types which are (i) Sekolah Sukan, (ii) SM Agama, (iii) SM Khas, (iv) SM Teknik and (v) SMK Agama with one website, each.

Furthermore, the result of usability analysis of download time for the secondary schools websites more than 10 s using the modem speed 28.8 kbps still shows the SMK Kebangsaan school type still shows the highest websites with the usability issue. The second highest of usability issue are Kolej and Sekolah Berasrama Penuh school type with two websites, respectively that, the download time is more than 10 s. This is followed with school types of (i) Sekolah Sukan, (ii) SM Agama, (iii) SM Khas, (iv) SM Teknik and (v) SMK Agama with one website with more than 10 s download time using this modem speed.

Table 4 also shows the usability analysis result of modem with speed 33.6 kbps. The results show that Malaysia Federal Territories Kuala Lumpur, Putrajaya and Labuan Secondary School websites from SMK Kebangsaan had the highest numbers of websites with this usability issue as 42 numbers of the websites requires more than 10 s of download time. It is followed by websites of secondary school type of Kolej and Sekolah Berasrama Penuh, with two websites, each and school types of (i) Sekolah Sukan, (ii) SM Agama, (iii) SM Khas, (iv) SM Teknik and (v) SMK Agama with one website with more than 10 s download time using this modem speed.

In addition, for the 56 kbps speed modem, the results in Table 4 also shows SMK Kebangsaan with the highest websites that had downloaded time more than 10 s, with a total number of 42 websites. This is followed with Kolej and Sekolah Berasrama Penuh school type with two websites, each. This is followed with school types of (i) Sekolah Sukan, (ii) SM Agama, (iii) SM Khas, (iv) SM Teknik and (v) SMK Agama with one website, each. Table 4 also shows, for the 128kbps modem, SMK Kebangsaan school type had the highest number of websites with more than 10 s download time with 42 websites. This is followed with Kolej and Sekolah Berasrama Penuh school types with two websites, each and school types of (i) Sekolah Sukan, (ii) SM Agama, (iii) SM Khas, (iv) SM Teknik and (v) SMK Agama with one website, type.

For the speed size 1.44 Mbps modem, the result shows that website from school type SMK Kebangsaan had the highest number of websites with 31 websites download with more than 10 s. This is followed with (i) Kolej and (ii) Sekolah Berasrama Penuh school type with two websites with download time more than 10 s. This is followed with (i) Sekolah Sukan, (ii) SM Agama, (iii) SM Khas, (iv) SM Teknik and (v) SMK Agama with one number of website each. Based from Fig. 1, there are 79 % (42/53) of secondary schools in Federal Territories of Kuala Lumpur, Putrajaya and Labuan that have websites with broken links and only 21 % (11/53) of the websites with no broken link. About 96 % (51/53) of Malaysia Federal Territories Kuala Lumpur, Putrajaya and Labuan Secondary Schools Websites have websites larger than 37 Kb and only 4 % (2/53) of the secondary school websites with page size of below than 37 Kb. This show that most of Malaysia Federal Territories Kuala Lumpur, Putrajaya and Labuan Secondary Schools Websites which is 96 % (51/53) may encounter problem of slow download page because of the large page size.

Figure 1 shows only 4 % (2/53) of the Malaysia Federal Territories Kuala Lumpur, Putrajaya and Labuan Secondary Schools Websites had downloaded time of less than 10 s by using 14.4 kbps modem speed. While 96 % (51/53) of the total website require more than 10 s download time. This situation caused users who accessed the websites by using this modem speed may need much longer time. This situation is similar to 28.8 kbps modem speed, which only 4 % (2/53) of the websites can be accessed less than 10 s while the other 96 % (51/53) requires more than 10 s download time to enable user to view the page. This situation also applicable to users with the modem speed 33.6 kbps, 56 kbps and 128 kbps. This means that all users are using the modem speed 33.6 kbps, 56 kbps and 128 kbps will also experience longer download time when open the web page.

However, there are different in percentages with regards to gaining access the websites with 1.44 Mbps modem speed. By using this kind of modem, 76 % (40/53) of the websites had downloaded time of more than 10 s. The remaining 25 % (13/53) of the websites can be downloaded less than 10 s. Somehow, this figure still indicate users that trying to connect to most of the Malaysia Federal Territories Kuala Lumpur, Putra-jaya and Labuan Secondary Schools websites still have to face the slower download time than appropriate download time (less than 10 s). As a summary, the Malaysia Federal Territories Kuala Lumpur, Putrajaya and Labuan Secondary schools websites may face slow download time to be gained access or connected with various type of modem. This is because the users may need to wait much longer time to access secondary schools websites with modem speed of 14.4 k kbps, 28.8 kbps, 33.6 kbps, 56 kbps, 128 kbps and 1.44 Mbps. This is also indicate that Malaysia Federal Territories Kuala Lumpur, Putrajaya and Labuan Secondary Schools websites have usability issue of download speed.

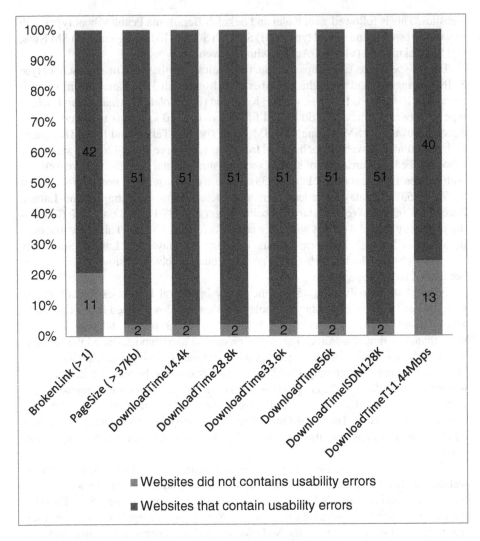

Fig. 1. Usability analysis of Malaysia federal territories Kuala Lumpur, Putrajaya and Labuan secondary school websites (Page size, Broken Link and Download Speed).

3.2 Discussions on Findings

This study had investigated the usability of 53 samples of Malaysia Federal Territories Kuala Lumpur, Putrajaya and Labuan Secondary Schools websites. From the analysis conducted on the sample, it was found that the Malaysia Federal Territories Kuala Lumpur, Putrajaya and Labuan Secondary Schools websites contain several usability issues. The results of the study, is presented in this report together with the clarification and explanation. There are three usability problems identified in this study. The identified three usability problems are described as below:-

1. Page Size >37 kb

   The recommended webpage size is 37 kb and below. The size of webpage will determine the webpage download time. The bigger size of the webpage, the time took to download it will be much longer. The unreasonable download time will affect users' mood to browse the page that they will leave and decide not to use or explore the website or use the services provided by the website.

2. Broken Link >=1

   Broken link will prevent users to reach the required information. This is due to the unmaintained website or not updated. This issue will give negative perception or bad reputation to the website. Users will get frustrated if they cannot access the information needed. Broken link also will affect the website traffic and more broken link means it will slow down the website traffic. Users will leave the website if this problem keep occurs while they want to use it.

3. Speed >10 s

   The general recommended time to download a page is 10 s or less. If the time took to download a web page longer than 10 s, users will not be patient to keep waiting. Users come with different speed of modem, so the webmaster should consider all type of modem speed that is used to access to the webpage. Users who are using slower modem will get frustrated if download time more than 10 s. This may cause them not to visit the website again.

The scope of this research had focused on only 53 samples of Malaysia Federal Territories Kuala Lumpur, Putrajaya and Labuan Secondary Schools websites from eight type of secondary school; (i) Kolej, (ii) Sekolah Sukan, (iii) SM Agama, (iv) SM Berasrama Penuh, (v) SM Kebangsaan, (vi) SM Khas, (vii) SM Teknik and (viii) SMK Agama. There are about 101 secondary schools in Federal Territory of Kuala Lumpur, 11 secondary schools located in Federal Territory of Putrajaya and another 10 in Federal Territory of Labuan, making a total of 122 secondary schools available. Unfortunately 40 of the secondary schools do not own any website. This resulted in only the total of 82 websites, possibly to be sampled. However, out of 82 websites available, only 53 of the websites turn to be useful by using the automatic usability evaluation in providing data needed in this survey. All these 53 websites samples had been evaluated by using free automated tools; (i) Dead Links and (ii) Broken Link Checker (Broken link checker tools) and (iii) Web optimization (web usability checker tool).

The limitation also comes from the web usability checking tools that have limitation used per day. This will take such a long time to finish the query for the long list of the samples. Another difficulty is the tools sometimes cannot read the URL of the sample website. Then the query has to be rerun for a few times to confirm that the URL really cannot be read by the tools in order to get better result. The usability surveys rely on automatic test single method cannot deal with all the usability concerns of website. Some usability errors may obstructs and not be able to be trace by automatic checking tools. From this study, it is found that websites that owned by school category from SM Berasrama Penuh are well maintained by the webmasters as compared to other school categories. Some of the schools website from other categories still use blogspot or other unofficial site to host their homepage rather than maintained their own web server to store their web pages.

4   Conclusion

The main objective of this study is to investigate the usability of secondary schools websites in Federal Territories Kuala Lumpur, Putrajaya and Labuan, Malaysia. The evaluation was done by using three automated tools; (i) Web Page Analyzer (websiteoptimization.com), (ii) DeadLink Checker and (iii) Broken Link Checker. The samples include 53 secondary schools websites in Malaysia. The data was analyzed based from Nielson usability guidelines for (i) ideal size of web pages, (ii) number of broken links and the (iii) webpage size. The result of this study shows that the secondary schools websites in Federal Territories Kuala Lumpur, Putrajaya and Labuan, Malaysia had few usability issues. This study provides recommendations for usability improvement for secondary schools website in Malaysia.

As a summary, most of the Malaysia Federal Territories of Kuala Lumpur, Putrajaya and Labuan Secondary Schools websites are facing slow download time to be gained access or connected with various type of modem. Users may need to wait for much longer (>10 s) to access Secondary Schools website despite using different the kind of modems of modem speed 14.4 k kbps, 28.8 kbps, 33.6 kbps, 56 kbps, 128 kbps and 1.44 Mbps. This show that the Malaysia Federal Territories Kuala Lumpur, Putrajaya and Labuan Secondary Schools websites still face usability issue with regards to download time (Speed >10 s). The Ministry of Education in Malaysia may need to focus this issue seriously since the usability of the Secondary Schools website is an important criterion for supporting the learning and information dissemination. Certain percentage of fund need to be allocated for maintenance and upgrading the website to increase the usability levels. To ensure the usability levels are being sustained, different initiatives such as IT audit program, surveys with the users, at any time should be regularly conducted.

The usability policy should be developed at relevant agencies to safeguard the practice. This is to ensure that all schools have same similar standard of website usability and the web designer may need to design according to establish relevant policies. The usability policies should be understood at all levels, including students, teachers and parents. Furthermore, frequent surveys need to be imposed on the website's user. This survey will help to assist policy makers and practitioners to improve the usability of secondary schools websites in Kuala Lumpur, Putrajaya and Labuan.

For better result of evaluation, further research of this topic is strongly recommended to use multiple methods of usability investigations such as automatic tools and heuristic evaluation. The heuristic evaluation might be the best resolution to tackle the short-comings of automatic test. However it is not means to restrict with only this recommended of methods. Other procedures that deemed as a good method are welcome to apply. This study strongly recommended using more than one evaluation method because one method able to tackle the drawbacks of another method and as the result, it is hope that the investigation result becomes meaningful. Future studies may also include usability and accessibility evaluations of websites for secondary schools from all states in Malaysia. Thus, this may provide better result to present the real situation about website usability issues of secondary schools in Malaysia.

To overcome the issue on most of secondary schools other than SM Berasrama Penuh school type websites that are not well maintain, it is recommended that the Ministry of Education to allocate a special budget to all secondary schools to have their own web server to host their official school homepages. The budgets also need to cover the cost of maintenance and webmaster salary or wages to maintain the websites. Thus, it is hope that the quality of the secondary school website will be well maintained and the usability issues of the secondary schools website in Malaysia will be handled properly. There are also needs to have more usability and accessibility evaluations on pre-school website, primary school website, higher education institution website and other different genre website in Malaysia.

References

1. Lais, S.: How to stop web shopper flight. Computerworld **36**, 44–45 (2002)
2. Hinchliffe, A., Mummery, W.K.: Applying usability testing techniques to improve health promotion websites. Health Promot. J. Austral. **19**(1), 29–35 (2008)
3. Wan Mohd Isa, W.A.R., Md Yusoff, M., Ag Nordin, D.A.: Evaluating the usability of homestay websites in Malaysia using automated tools. In: Berry, M.W., et al. (eds.) SCDS 2015. CCIS, vol. 545, pp. 221–230. Springer, Heidelberg (2015). doi:10.1007/978-981-287-936-3_21
4. Abdul Aziz, M., Wan Mohd Isa, W.A.R., Nordin, N.: Assessing the accessibility and usability of Malaysia higher education website. In: 2010 International Conference on User Science Engineering (i-USEr), pp. 203–208. IEEE (2010)
5. Wan Abdul Rahim, W.M.I., Muhammad Rashideen, S., Noor Ilyani, S., Siti Suhada, S.: Assessing the usability and accessibility of Malaysia e-government website. Am. J. Econ. Bus. Adm. **3**(1), 40–46 (2011). Science Publications
6. Goh, K.N., Chen, Y.Y., Lai, F.W., Daud, S.C., Sivaji, A., Soo, S.T.: A comparison of usability testing methods for an e-commerce website: a case study on a Malaysia online gift shop. In: 2013 Tenth International Conference on Information Technology: New Generations (ITNG), pp. 143–150 (2013)
7. Zainudin, N.M., Ahmad, W.F.W., Nee, G.K.: Evaluating C2C e-commerce website usability in Malaysia from users' perspective: a case study. In: 2010 International Symposium on Information Technology, Kuala Lumpur, pp. 151–156 (2010)
8. Rozali, N.B.N., Said, M.Y.B.: Usability testing on government agencies web portal: a study on ministry of education Malaysia (MOE) web portal. In: 2015 9th Malaysian Software Engineering Conference (MySEC), Kuala Lumpur, pp. 37–42 (2015)
9. Sabaruddin, S.A., Abdullah, N.H., Jamal, S.A., Tarmudi, S.: Exploring the accessibility and content quality of the Go2homestay website. In: Radzi, et al. (eds.) Theory and Practice in Hospitality and Tourism Research, pp. 327–332. Taylor & Francis Group, London (2015)
10. Wan Mohd Isa, W.A.R., Abdul Aziz, M., Abdul Razak, M.R.: Evaluating the accessibility of small and medium enterprise (SME) websites in Malaysia. In: 2011 International Conference on User Science and Engineering (i-USEr), pp. 135–140. IEEE (2011)
11. Nielson, J.: Usability Engineering. Morgan Kaufmann, San Francisco (1994)

Judgment of Slang Based on Character Feature and Feature Expression Based on Slang's Context Feature

Kazuyuki Matsumoto[1($\boxtimes$)], Seiji Tsuchiya[2], Minoru Yoshida[1], and Kenji Kita[1]

[1] Faculty of Science and Engineering, Tokushima University,
Minamijosanjima-cho, 2-1, Tokushima, 7708506, Japan
matumoto@is.tokushima-u.ac.jp
[2] Faculty of Science and Engineering, Doshisha University, Kyo-Tanabe,
Kyoto 610-0394, Japan

Abstract. Our research aim was to develop the means to automatically identify a particular character string as slang and then connect the detected slang word to words with similar meaning in order to successfully process the sentence in which the word appears. By recognizing a slang word in this way, one can apply different processing to the word and avoid the distinctive problems associated with processing slang words. This paper proposes a method to distinguish standard words from slang words using information from the characters comprising the character string. An experiment testing the effectiveness of our method showed a 30 % or more improvement in classification accuracy compared to the baseline method. We also use a contextual feature related to emotion to expand the unregistered slang word in the training data into other expressions and propose an emotion estimation method based on the expanded expressions. In our experiment, successful emotion estimation was obtained in nearly 54 % of the cases, a notably higher rate than with the baseline method. Our proposed method was shown to have validity.

Keywords: Slang · Character feature · Context feature · Unknown expression

1 Introduction

Recently the availability of mobile devices such as smart phones and tablet PCs has rapidly expanded, creating more and more users of Social Networking Services (SNS) such as Twitter and Facebook. Most users of these services are younger people - from teenagers to those in their thirties. Such younger users tend to use distinctive words in communicating with their friends. These distinctive words are often labeled 'youth slang.' Youth slang is considered a specific type of slang, although there is no clear definition to distinguish youth slang from other types of slang. For example, "*gekioko*" is youth slang meaning "being very angry." Because most of these slang expressions are naturally generated and eventually fall into disuse, it is highly inefficient to register them manually into dictionaries each time a new expression appears. There are numerous studies on the automatic expansion of lexicons, focusing mainly on the extraction of

© Springer Nature Singapore Pte Ltd. 2016
M.W. Berry et al. (Eds.): SCDS 2016, CCIS 652, pp. 280–288, 2016.
DOI: 10.1007/978-981-10-2777-2_25

new words from the sentences in which they appear. If key attributes of the new words could be estimated when they are detected, it would be relatively easy to register them into a dictionary. It would also be helpful in estimating emotion from text.

In this paper, we propose a method to determine whether an unknown word is youth slang by calculating the similarity between the unknown word and existing youth slang terms based on various features. We also propose a method to expand youth slang words by using expressions of emotion around the youth slang word as clues.

2 Previous Works

2.1 Youth Slang Extraction/Classification Method

Previous research has proposed judging the emotions expressed in youth slang words by referring to co-occurrence words. Matsumoto and Ren [1], for example, considered several features: the kind of characters used (i.e., katakana, kanji or hiragana), the stroke counts of the characters, the difficulty level of the characters and the content words obtained from the definitions of the youth slang word. They then estimated the emotions expressed in the youth slang words using the k-nearest neighbor method.

Matsumoto et al. [2] considered the emotional vectors of co-occurring content words and used the Earth Mover's Distance to identify similar youth slang. They also proposed a method to extract slang words from sentences based on CRFs by using the kind of characters and the stroke counts of the characters as key features [3].

Slang words, including youth slang, tend to be used to describe how one feels or to change one's impression of something. Such uses suggest that slang words might have distinctive features not found in other newly created words. In this research, we examined the possibility that Japanese youth slang has certain distinguishing characteristics in the character string itself, arising from the fact that Japanese words consist of various kinds of character types, including Chinese characters, hiragana and katakana, and so on. When Japanese choose names for their babies or select names for certain products, they tend to carefully choose and combine various Chinese characters to convey a particular impression or meaning. Given this tendency, we hypothesized that in the Japanese language, youth slang could be distinguished from other slang terms or from existing words by recognizing distinctive features of the characters used to represent them. Because youth slang words have various notations, many of them have a large number of synonyms. This can sometimes cause a failure in the emotion estimation of a sentence if one is using an emotion estimation model based on the existing emotion tagged corpus. To solve this problem, we assumed that it would be effective to add to the distinguishing feature expressions that were contextually similar to the words judged to be youth slang. Concretely, we believed that we could estimate the emotion in a youth slang term based on other expressions that have been used in a similar context with the youth slang.

2.2 Statistical Analysis

We began by attempting to identify characteristics of youth slang terms that might be used for detecting them in sentences. We randomly chose five examinees (subjects) in their twenties, together with 671 youth slang words. We then determined how many of the youth slang words in our list were known to the examinees. The recognition rate for the list of 671 words was 58.7 %, which showed that even young people do not know all of youth slang, In fact, there were many youth slang words that were unknown to our test group. Next, we had the examinees complete a questionnaire indicating their positive/negative impression of the youth slang terms.

Figure 1 displays the results, showing the calculations of positive and negative values applied to each youth slang word. These positive/negative impressions were divided into 50 levels and chosen by the examinees from this range. As indicated in

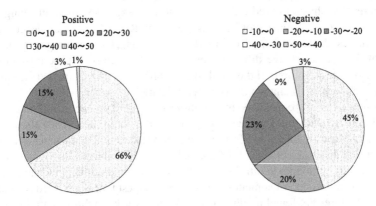

Fig. 1. Impressions of youth slang words.

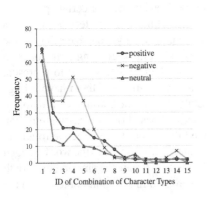

ID	combination of character types
1	[ZEN_KATA]
2	[ZEN_HIRA]
3	[ZEN_HIRA]+[ZEN_KATA]
4	[ZEN_KANJI]
5	[ZEN_KANJI]+[ZEN_KATA]
6	[ZEN_HIRA]+[ZEN_KANJI]
7	[ZEN_JSYMBOL]+[ZEN_KATA]
8	[ZEN_HIRA]+[ZEN_JSYMBOL]
9	[HAN_LOWER]
10	[HAN_UPPER]
11	[ZEN_HIRA]+[ZEN_JSYMBOL]+[ZEN_KATA]
12	[ZEN_HIRA]+[ZEN_LOWER]
13	[ZEN_HIRA]+[ZEN_JSYMBOL]+[ZEN_KANJI]+[ZEN_KATA]
14	[ZEN_HIRA]+[ZEN_KANJI]+[ZEN_KATA]
15	[ZEN_JSYMBOL]+[ZEN_KANJI]+[ZEN_KATA]

Fig. 2. Appearance frequency of positive/negative/neutral according to the combination pattern of character types.

the figure, the number of negative impressions was nearly double the number of positive impressions. This result dovetailed with the results of an existing study on youth slang [4]. Figure 2 shows youth slang word frequencies for various character type combination patterns. Calculations were made for positive, negative and neutral impressions, respectively. We used the combination patterns defined in the Moji module of Ruby [5]. Table 1 shows some of the character types defined in the Moji module. As indicated in Fig. 2, frequencies varied. However, there were not large differences among the positive, negative and neutral impressions. This result suggests that it would be difficult to classify youth slang into positive/negative/ neutral by using only combinations of character types.

Table 1. Examples of the character types defined in Moji module

Character type	Ellipsis	Examples
Two-byte Kanji Character	ZEN_KANJI	海, 誤, 漢, etc.
One-byte Capital Letter	HAN_UPPER	A, B, C, etc.
Two-byte Symbol Character	ZEN_ASYNBOL	?, !, @, {, etc.
One-byte Numeric Character	HAN_NUMBER	1, 2, 3, 4, 5, etc.

Table 2. The number of the registered entry words in the context similar word list

	stw	cfilt		
Window	–	7	3	
Size	10	100	100	200
# of Direction Word	257	447		
# of Registered Similar Word	25,700	44,700		

3 Proposed Method

3.1 Feature Extraction

As described, we are proposing a method to determine whether a word is youth slang by referring to certain characteristics (such as character types) of the characters used to write the word. We found that youth slang has at least two distinguishing features: it usually does not use complicated kanji characters not commonly used, and it tends to use short words such as abbreviations of existing words. Based on this observation, we used the following as our distinguishing features:

- Character type (18 types)
- Sum of the stroke counts for the characters
- Usage of kanji character (for personal name/for other)
- School grade where the students learn the kanji character (7 grades)
- Level of the kanji character in the kanji test (10 difficulty levels)
- Radical of the kanji character (226 kinds)
- Degree of word intimacy of all substrings (7 degrees)

Features 4 to 6 were extracted from only kanji characters. We believed that the same slang word can convey a different meaning depending on the character type used. Accordingly, we judged whether the word is expressed in kanji characters to be significant and concluded that the same word written with a different character type should be judged as a different word. The word intimacy degree in feature 7 is a quantified measure how friendly the examinee felt the word was as reported in the questionnaire experiment [6]. This we divided into six levels (level 1 to 6). All substrings were extracted from the word, and the number of matches with words belonging to each intimacy level was counted as a feature.

3.2   Classification Algorithm

For binary classification, machine learning methods such as Support Vector Machine (SVM) have been traditionally used. Because the character type of the slang word and the word intimacy degree would be used as features in our research, it seemed unlikely to result in high dimension as compared to a vector using words as the feature.

For cases in which the feature vector has a high dimension, various methods to reduce the dimension have been proposed. However, even if all the features (1 to 7) that we identified are used, the number of dimensions would be at most 271, which is lower than the size of the vocabulary in the general training corpus. Therefore, we did not attempt a compression of vector dimension. We chose to use three kinds of binary classification algorithms that were applicable in the machine learning tool Classias [7]: averaged_perceptron, pegasos.hinge, and truncated_gradient.logistic. The parameters were not tuned since there was no bias with respect to the number of slang words and the number of standard words. In the classification of slang and other words, it is preferable to judge slang words from words that are not registered in dictionaries. However, because the degree of difference between the slang word string and existing strings was considered to be important, existing words that were registered in dictionaries were used as standard words. In this paper, words whose word intimacy degrees were comparatively high (intimacy degree 6 and 7) were randomly chosen.

3.3   Emotion Estimation Algorithm

We considered the context feature to estimate the emotion of the slang term. Two types of contexts were used. In the first, we extracted N words around the target word to create the context feature. In the other, we extracted N words around the target word but limited the words to specific expressions, such as an emotion expression, an evaluation expression, or a slang expression. These context features were trained by word2vec [8], and the word vector representation database (WVRDB) was generated. We obtained a similar word list for each slang word by calculating the similarity between the input word and the other word based on WVRDB. We added similar words to the feature of the input sentence and estimated the emotion based on the word emotion dictionary which was trained with the Youth Slang Emotion Corpus (YSEC) [9, 10]. We used the Maximum Entropy method as the training algorithm. Figure 3 (A) shows the flow of the training of word vector representation from a tweet text [11]. We used the original noise filtering algorithm for the tweet sentences. Figure 3 (B) shows the flow of similar word expansion and emotion estimation using the word emotion dictionary.

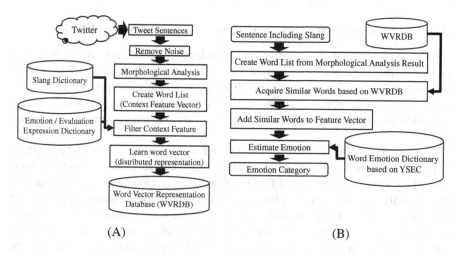

(A)   (B)

Fig. 3. (A): Training flow, (B): Estimation flow.

4   Experiment

4.1   Detecting Experiment of Slang Word or Standard Word

We conducted an evaluation experiment to confirm the effectiveness of the proposed method. As baseline methods to compare with the proposed method, we used three approaches: *crngram*, which extracts a character n-gram as a feature; *rngram*, which adds pronunciation of the character string then extracts the character n-gram from the pronunciation string; and *crngram*, which combines *cngram* and *rngram*. These methods were used to extract all the n-grams (n = 1, 2, 3). Experimental results are presented in Fig. 4. The graph indicates the recall rates of successfully detected slang words. It shows

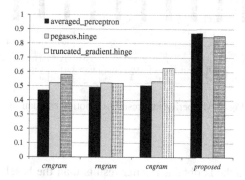

Fig. 4. Result of the experiment (slang words – existing words)

that the recall rate for the proposed method was higher than the recall rate for the baseline methods that used character n-grams or character pronunciation n-grams.

4.2 Emotion Estimation Experiment of the Sentence Including Slang Words

To evaluate our method of expanding slang words based on context, we compared our approach to the baseline method using words as the feature. The test data for the experiment were produced by randomly extracting uttered sentences without noise from the corpus of collected weblogs and Twitter tweets, and annotating tags manually. In all, 976 sentences were extracted from weblogs and 1,079 sentences were extracted from Twitter.

Window size was set to 3, 5, 7, 10 and vector dimension was set to 100 and 200 as the parameters of word2vec. Eight kinds of tags were annotated to the sentences in the youth slang emotion corpus (YSEC) to indicate the emotion category. Because it is difficult to distinguish anger from hate, we treated both of these emotions as belonging to the same category. The emotion categories, then, were joy, anxiety, hate, hope, love, sorrow, surprise, and neutral. After adding the weights of the word emotion categories, if the correct category was included in the top two obtained categories, we treated the result as successful. The experimental results are shown in Fig. 5. The 'cfilt' label indicates that context filtering was used; 'stw' indicates that the training of the word vector was done based on the tweet corpus. The corpus collected approximately 50 million tweets by using randomly selected standard words as search queries. Others are based on another tweet corpus that collected approximately 500 million tweets by using emotion expressions as search queries. These tweet corpora were collected over approximately four months by using tweepy, which is an API library for python.

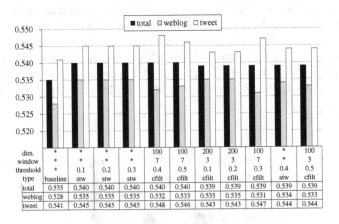

dim.	*	*	*	*	100	100	200	200	100	*	100
window	*	*	*	*	7	7	3	3	7	*	3
threshold	*	0.1	0.2	0.3	0.4	0.5	0.1	0.2	0.3	0.4	0.5
type	baseline	stw	stw	stw	cfilt	cfilt	cfilt	cfilt	cfilt	stw	cfilt
total	0.535	0.540	0.540	0.540	0.540	0.540	0.539	0.539	0.539	0.539	0.539
weblog	0.528	0.535	0.535	0.535	0.532	0.533	0.535	0.535	0.531	0.534	0.533
tweet	0.541	0.545	0.545	0.545	0.548	0.546	0.543	0.543	0.547	0.544	0.544

Fig. 5. Experimental result of emotion estimation.

4.3 Discussions

The two proposed methods obtained better results than the baseline methods. It was found that filtering the context feature was effective for the emotion estimation of tweets.

However, it was also found that context feature filtering did not result in high accuracy rates for weblog sentences. One of the reasons for this was that function words such as sentence ending expressions, which are considered important as a feature, were ignored, even though some function words can add a difference to the expression by connecting to the content word. By removing these differences with context feature filtering, the context was wrongly determined to indicate a similar emotion. We surmise that estimation accuracy decreased as a consequence. There were very few influences of window size and vector dimension size affecting accuracy. Many slang words seem to have similar expressions in the tweet corpus. For that reason, even without setting the threshold of context similarity high, there was no obvious effect on decreasing accuracy. Table 2 shows a partial listing of the number of registered entry words in the context similar word list. The data showed that there were more registered entry words without context feature filtering, and thus there would be a greater number for the expanded word. On the other hand, even though the number for the expanded word was smaller, the accuracy rate was high for tweet sentences. We believe the reason for this is that slang words with a tendency of co-occurring with emotion expressions may be registered preferentially as entry words when conducting context feature filtering.

One problem in our proposed method is that an existing word vector expression cannot be applied when it is difficult or impossible to obtain a sufficient context feature from the expression – for example, when the word being evaluated tends to be used by itself. To deal with this problem, it would be necessary to develop a method that can make use of the character information used for determining a slang word and can, as a supplement, acquire context information from a similar character string when there is a character string that has be judged as similar.

5   Conclusions

We have proposed a method for distinguishing slang words from standard dictionary words based on certain character features. In an evaluation experiment, our approach achieved an approximately 30 % improvement in the recall rate as compared to the performance of the baseline method using character n-grams. However, we found that some slang words cannot be distinguished from standard words by using only superficial character features. It would seem necessary to judge a word to be slang based on its distinctive appearance tendencies on the Web. Not all unregistered words are slang. If nonsense words are automatically created by combining certain character types, it would be difficult to distinguish these words from slang words. The semantic background of the word is thus important to the process of identifying the word as slang. For example, one useful clue would be context information such as who wrote the word or for what purpose he/she wrote the sentence that includes the word.

In our proposed method to expand slang words as a way of improving emotion estimation, we found a difference in accuracy rates for various context extraction methods. From this we surmise that context filtering is effective not only for distinguishing slang words but also for use under certain conditions, such as when a sentence includes unknown expressions (e.g., spoken words or tweets with substantial noise). In future work, we will need to conduct detailed experiments in different settings and under various conditions in order to address the validity of the context filtering method.

Acknowledgments. This research was partially supported by JSPS KAKENHI Grant Numbers 15K16077, 15K00425, 15K00309.

References

1. Matsumoto, K., Ren, F.: Construction of Wakamono Kotoba emotion dictionary and its application. In: Gelbukh, A.F. (ed.) CICLing 2011, Part I. LNCS, vol. 6608, pp. 405–416. Springer, Heidelberg (2011)
2. Matsumoto, K., Kita, K., Ren, F.: Emotional vector distance based sentiment analysis of Wakamono Kotoba. China Commun. **9**(3), 87–98 (2012)
3. Matsumoto, K., Akita, K., Keranmu, X., Yoshida, M., Kita, K.: Extraction Japanese slang from weblog data based on script type and stroke count. Procedia Comput. Sci. **35**(2014), 464–473 (2014)
4. Yonekawa, A.: Wakamonogo wo kagakusuru. (Meiji Shoin) (1998). (in Japanese)
5. Moji module. http://gimite.net/gimite/rubymess/moji.html
6. Amano, N., Kondo, K.: NTT Database Series Nihongo-no Goitokusei: Lexical Properties of Japanese, CD-ROM version, Sanseido (2008)
7. Classias. http://www.chokkan.org/software/classias/index.html.en
8. Mikolov, T., Chen, K., Corrado, G. and Dean, J.: Efficient estimation of word representations in vector space. In: Proceedings of the Workshop at ICLR (2013)
9. Matsumoto, K., Kita, K. Ren, F.: Emotion estimation from sentence using relation between Japanese slangs and emotion expressions. In: Proceeding of the 26th Pacific Asia Conference on Language, Information, and Computation (PACLIC 2012), pp. 343–350 (2012)
10. Ren, F., Matsumoto, K.: Semi-automatic creation of youth slang corpus and its application to affective computing. IEEE Trans. Affect. Comput. **7**(2), 176–189 (2016)
11. Twitter Website. https://twitter.com/

Factors Affecting Sentiment Prediction of Malay News Headlines Using Machine Learning Approaches

Rayner Alfred[1]([✉]), Wong Wei Yee[1], Yuto Lim[2], and Joe Henry Obit[1]

[1] Faculty of Computing and Informatics, Universiti Malaysia Sabah, Kota Kinabalu, Malaysia
{ralfred,joehenry}@ums.edu.my, huiyee0529@hotmail.com
[2] School of Information Science, Japan Advanced Institute of Science and Technology,
Nomi, Japan
ylim@jaist.ac.jp

Abstract. Most sentiment analysis researches are done with the help of supervised machine learning techniques. Analyzing sentiment for these English text reviews is a non-trivial task in order to gauge public perception and acceptance of a particular issue being addressed. Nevertheless, there are not many studies conducted on analyzing sentiment of Malay news headlines due to lack of resources and tools. The Malay news headlines normally consist of a few words and are often written with creativity to attract the readers' attention. This paper proposes a standard framework that investigates factors affecting sentiment prediction of Malay news headlines using machine learning approaches. It is important to investigate factors (e.g., types of classifiers, proximity measurements and number of Nearest Neighbors, k) that influence the prediction performance of the sentiment analysis as it helps to study and understand the parameters that can be tuned to optimize the prediction performance. Based on the results obtained, Support Vector Machine and Naïve Bayes classifiers were capable to obtain higher accuracy compared to the k-Nearest Neighbors (k-NN) classifier. In term of proximity measurement and number of Nearest Neighbors, k, the k-NN classifier achieved higher prediction performance when the Cosine similarity is applied with a small value of k (e.g., 3 and 5), compared to the Euclidean distance because it measures can be affected by the high dimensionality of the data.

Keywords: Sentiment analysis · Opinion mining · Naïve bayes · k-Nearest neighbors · Text classification

1 Introduction

Sentiment analysis, sometimes called opinion mining, is a natural language processing (NLP) technique which performs computational study of emotions, opinion, attitudes, appraisal, evaluation, subjectivity and sentiments in source of materials [1]. Sentiment analysis helps to study and understand the general mind-state of the communities at a particular time with regard to some issues. Nowadays, there are many reviews and opinions on news which have a great impact on social life of the communities. The perceptions are largely conditioned on how people see the world. Peoples often seek out

© Springer Nature Singapore Pte Ltd. 2016
M.W. Berry et al. (Eds.): SCDS 2016, CCIS 652, pp. 289–299, 2016.
DOI: 10.1007/978-981-10-2777-2_26

the opinions of others for a decision. Sentiment Analysis helps to determine the perspective of different sources of information, and yet another possible application would be the processing of answers to opinion questions [2]. The knowledge from the huge amount of unstructured information from news is the key of influencers of behaviors.

Nowadays, people are able to access to all sorts of information on the Web. The large number of online news as well as the explosive growth of social media sites have induced large amount of text articles available online. The news at every moment changes the perception or sentiment which may influence the decision making. IBM Research, New Intelligence for a Smarter Planet estimated that roughly 80 % to 85 % of the useful data stored on modern computer systems is in unstructured format. Peoples are very good at dealing with small amounts of unstructured data, but have difficulties in handling the vast quantities of data produced in even moderately sized legal cases. As a result, an efficient and effective tool is required to process and manage the vast amounts of unstructured data provided. Moreover, human cannot read and process big data visualization in a short time, but machines or computer can process big data visualization for humans more efficiently.

News can be obtained from different resources, either in the printed materials or the online materials which provide enormous amount of facts and opinions to people. Readers often read the news headlines to determine the news content. The news headlines are typically constrained within few words, constantly updated and made publicly available in order to encourage readers keen on reading the full text of the news. The elements of the predefined set in sentiment analysis are usually 'negative' and 'positive', but they can also be any other elements such as 'relevant' and 'irrelevant', 'in favor of' or 'against', or other more than two elements such as 'negative', 'positive', 'neutral' or a range of numbers such as from 1 to 10 [3]. Generally, sentiment analysis helps to collect information about the positive and negative aspects of a particular topic. Therefore, large intention has been given to sentiment analysis because of its wide range of possible applications in recent years.

There are several machine learning techniques that are widely used for sentiment analysis in additional to lexicon based method and linguistics method [4]. Machine learning techniques can be categorized into supervised learning approaches and unsupervised learning approaches that can be used to analyze sentiment. Supervised machine learning techniques have been widely used for document classification which achieved great success in sentiment analysis. Some of the supervised machine learning approaches are Support Vector Machine (SVM), Naïve Bayes (NB), k-Nearest neighbor (k-NN), Decision Tree and etc.

Although the news headlines always picture the important information provided in the content of the news, some people have difficulties in understanding and interpreting the news based on a short piece of text such as news headlines. Thus, this paper aims to investigate factors affecting sentiment prediction of Malay news headlines using supervised machine learning approaches. In this work, three supervised learning approaches will be designed and implemented that analyze sentiments of Malay news headlines which are Support Vector Machine (SVM), Naïve Bayes (NB) and k-Nearest Neighbor

(k-NN). In addition to that, this paper also investigates the effects of using various proximity measurements and number of nearest neighbors, k, on the performance of the k-NN in analyzing sentiment of the Malay news headlines.

The rest of this paper is organized as followed. Section 2 outlines some related works in sentiment analysis on Malay language. Section 3 discusses the methodology used to investigate factors affecting sentiment prediction of Malay news headlines using supervised machine learning approaches. Section 4 discusses the experimental design and the experimental results obtained. Section 5 concludes this paper.

2   Related Works

Many researches have been conducted on improving algorithms for sentiment analysis and opinion mining. Various types of datasets can be analyzed to extract the hidden sentiments in order to gauge public perception and acceptance on a particular issue such as movie reviews, microblogging sites and blogs. Sentiment analysis has become more important due to the growing amount of the unstructured documents created in digital form [5]. The machine learning paradigm has become one of the main classification approaches in this area [5]. It generates a classifier from the training set based on the characteristics of the documents that already classified. An experimental study has been conducted on automatic classification of Malay proverbs using Naïve Bayesian algorithm [5]. The Malay proverbs are classified into five categories; life, family, destiny, social and knowledge. Based on the findings obtained, the multinomial Naïve Bayes model performed better than multivariate Bernoulli Naïve Bayes model in classifying Malay proverbs. There is also study conducted on the analysis of differences of Opinion Mining research works based on language, writing style and feature selection parameters [6]. Various techniques are used such as Support Vector Machine (SVM), Naïve Bayes (NB), k-Nearest Neighbors (k-NN) and decision tree were used for classifying sentiments in both writing style of formal and informal text [6]. The performance accuracy of Malaysian language achieved was 66.28 % for the formal text and 69.09 % for the informal text in their research. They have concluded that a better understanding of features of language being used is required to increase the performance of opinion mining task. Another research has been conducted on the sentiment classification for Indonesian message published in social media [7]. SVM and Maximum Entropy (ME) are used as the classification algorithms with machine learning features. The informal words that exist in the sentence are transformed into formal words by using the deletion of punctuation mark, tokenization, conversion number to letter and the reduction of repetition letter in order to formalize abbreviation. The messages are classified into four classes; *neutral, question, positive sentiment* and *negative sentiment*. They concluded that the preprocessing of noisy text and using dictionaries (positive sentiment word, negative sentiment word, and question word) may increase the accuracy of sentiment classification in social media messages and the best classification method is SVM that yields 86.66 % accuracy [7].

The classification of Malay poems using SVM technique through Radial Basic Function (RBF) and linear kernel function has been implemented to classify Malay poems by theme, as well as poetry or non-poetry [8]. The results show the potential of using SVM technique in classifying poems into various classifications of which previous approaches only focused on classifying prose only. The type of feature selection method used to extract relevant words also may affect the performance of the sentiment classification models for Malay sentiment analysis [9]. The highest performance on Malay sentiment classification is obtained with the SVM classifier while the worst performance is obtained with the k-NN classifier [9]. However, there in not many efforts put in analyzing sentiments in Malay literary text such as poetry and Malay news headlines. Generally, there are limited studied on sentiment analysis and text classification on text written in Malay language. The SVM, NB and k-NN classifiers are selected to be investigated in this work since they are the most common and simplest classifiers among various techniques. SVM provides consistently high classification performance based on the findings obtained from other researches while Naïve Bayes has been successfully applied to document classification in many research efforts and has a good reputation for Natural language Processing (NLP) and it is highly sensitive to feature selection [10]. They have analyzed that NB is the optimal in probability sense as NB assigns the documents to the class with the highest probability. The k-NN classifier is chosen to be implemented in this research as it is based on the assumption that the classification of an instance is most similar to the classification of other instances that are nearby in the vector space. The k-NN algorithm does not rely on prior probabilities compared to other text classification techniques such as Naïve Bayes classifier and it is computationally efficient.

3   Methodology

This research focuses on analyzing sentiment of Malay news headlines. This section outlines the methodology used to investigate factors affecting sentiment prediction of Malay news headlines using supervised machine learning approaches.

3.1   Corpus Collection

All the selected Malay new headlines are extracted and used as training and testing data from *Berita Harian* (www.bharian.com.my), *Sinar Harian* (www.sinarharian.com.my) and *Utusan Malayia* (www.utusan.com.my). The average length of Malay news headline is usually between four to twelve words. All these news headlines are manually annotated and stored into a text file as database. There are 600 Malay news headlines are collected and annotated manually into positive sentiment and negative sentiment as dataset based on a listing of positive and negative words which is done by referring to the Multilingual Sentiment tool available in Data Science Labs licensed by General Sentiment, Inc. [11].

3.2 Preprocessing

Pre-processing the corpus is the process of cleaning and preparing the text from noise data for classification. Malay news headlines usually contain lots of noise such as punctuation, stop word and capital letter of news headlines. It helps to improve the performance of the classifier and speed up the classification process in sentiment analysis. The whole process of pre-processing involves five steps: case folding, symbol elimination, tokenization, handling of negation and removal of stop words. In the case folding process, all capital letters will be converted to small letters. Then, in symbol elimination, all non-alphanumeric symbols will be eliminated before the tokenization process is performed. In the negation handling process, all the negation words will be converted to a single token. For instance, given a phrase "*tak sempat*", it will be converted to a single word "*!sempat*". Finally, the removal of stop words from the index is performed based on the common stop words by Agus et al. [12]. All these processes are implemented by using the Java language.

3.3 Term Weighting

After the preprocessing, a bag of word is formed based on unigram technique as terms-documents vector, where the vector represents the weight of terms in documents. Given that Malay news headlines are short textual documents, un-stemmed single terms are used as features. A weight is assigned to each term by using the TF-IDF term weighting scheme. The weighting was done by applying the TF-IDF computation which is shown below.

$$TF - IDF = \left(1 + \log f_{i,j}\right) * \log \frac{N}{n_i} \tag{1}$$

where, $f_{i,j}$ refers to the frequency of the term t_i in the document d_j, N is the total number of documents and n_i refers to the number of documents in which term t_i exists in.

3.4 Sentiment Classification

Three classifiers will be used in classifying these Malay news headlines based on the sentiment extracted, namely, Support Vector Machine, Naïve Bayes and k-NN approaches, which are used because of the simplicity, effectiveness, and high accuracy of these approaches. The following are the brief descriptions of these approaches.

Support Vector Machines (SVM) is a rather recent and more successful machine learning method to be introduced into the field. This method is built upon the principle of Structural Risk Minimization which is to find a hypothesis that can guarantee the lowest true error [19]. True error is the probability of making an error in a randomly selected test sample that is unseen. The SVM method solves binary classification by constructing a vector space. Documents will be represented as points in the vector space. With the vector space, the next step is to find a hyperplane (decision surface) that can best separate the points into 2 distinct classes. The optimal hyperplane provides the maximum distance that is closest to both classes of documents. The maximum margins limits the area of obtaining the optimal hyperplane and any point that falls on them are known as support vectors. The hyperplane that is parallel to the margin hyperplanes is considered the best suited to be the optimal hyperplane. The optimal hyperplane is the result of the training of the algorithm. Classifying a new document involves calculating its relative position distance to the hyperplane. Mathematically, the hyperplane is a vector representation, the solution:

$$\vec{h} = \sum_{i=1}^{n} \alpha_i y_i \vec{d_i} \alpha_i \geq 0 \tag{2}$$

where, α_i = dual optimization problem, y_i = +1 or −1, representing Positive or Negative and d_i = document vector. Any of the $\vec{d_i}$ which has its α_i greater than zero are the support vectors since they contribute to the overall vector equation [19].

Naïve Bayes algorithm is chosen to be applied in this research for sentiment analysis. Naïve Bayes uses conditional probability as a platform. The Bayes formula is used to calculate the probability of document d belongs to a class C, and the equation is shown as below.

$$P(C|d) = \frac{P(d|C) \times P(C)}{P(d)} \tag{3}$$

where $P(C)$ denotes the probability of document belonging to a class C. The probability that a document is positive or negative is the estimation or the likelihood for an event to take place if it is pre-determined to be positive or negative. This learning process is called a supervised learning that requires the existence of pre-classification examples for training, as shown below for the Naïve Bayes [13],

$$P(x_i|class) = \frac{\sum tf(X_i, document \in class) + \alpha}{\sum N_{document \in class} + V} \tag{4}$$

where, X_i refers to a word obtained from the Malay news headline, $\sum tf(x_i, document \in class)$ refers to the sum of raw term frequencies of word x_i obtained from all documents in the training sample that belong to class, α refers to an additive smoothing parameter (α = 1 for Laplace smoothing), $\sum N_{document \in class}$ is the sum of all term frequencies in the training dataset for class and finally, V is the size of the vocabulary (number of different words in the training set). The class-conditional probability of encountering the headlines can be calculated as the product from the likelihoods of the individual word which is under the naïve assumption of conditional independence as derived the formula as below.

$$P(headlines|class) = P(X_1|class) \times P(X_2|class) \times \ldots \times P(X_3|class) \tag{5}$$

Therefore, the overall of Bayes theorem formulated in this work in order to calculate the probability of a news headline belongs to a class of negative or positive is derived as below:

$$P(class|headline) = \frac{P(class)P(headline|class)}{P(headlines)} \tag{6}$$

where

$$P(class) = \frac{number\ of\ document\ in\ class\ (positive\ or\ negative)}{total\ number\ of\ document} \tag{7}$$

k-NN algorithm works by inspecting the k closest neighbors in the dataset to a new occurrence that needs to be classified, and making a prediction based on what classes the majority of the k neighbors belong to [14]. The value of k is advised to be odd number to make tie less likely, for example, values of $k = \{1, 3, 5, 7, 9, 11\}$ are usually used to determine the classification of the document. There are two types of proximity measurement in k-NN classifier. Basically, standard Euclidean distance is commonly used in k-NN classifier to calculate the distance between two points in an n-dimensional space, where the n is the number of attributes in the dataset. A smaller distance measurement indicates that both documents are most similar compared to a larger value distance measurement. The Euclidean distance formula is as shown below,

$$D(x, y) = \sum_{k=1}^{n} (x_k - y_k)^2 \tag{8}$$

$$\cos(d1, d2) = \frac{d1 \times d2}{||d1|| \times ||d2||} \tag{9}$$

Another proximity measurement is to measure the degree of two documents (e.g., $d1$ and $d2$) in order to determine the similar characteristic between two documents is known as similarity measurement (e.g., Cosine Similarity as shown in Eq. 9). The more similar the two documents are, the more parallel they are in the feature space and the greater the cosine value which is nearer to value of 1. The Cosine value of 0 indicates that the two documents are dissimilarity.

4   Experimental Design and Results

This section describes the experiments that are carried out in this work and discusses the results obtained. The experiments are outlined in order to achieve the following objectives;

(a) The effects of using different classifiers on the Sentiment Prediction of Malay News Headlines.
(b) The effects of using various proximity measurements on the Sentiment Prediction of Malay News Headlines using the k-NN classifier.
(c) The effects of using different number of nearest neighbors, k, on the Sentiment Prediction of Malay News Headlines using the k-NN classifier.

There are a total of 600 Malay news headlines that are collected and annotated manually for the outlined experiments. These 600 Malay news headlines are then separated into 3 sets in which each set consists of 200 Malay news headlines respectively. Three folds cross validation will be conducted for the Support Vector Machine and Naïve Bayes classifiers. On each round of the experiment, two folds were used as the training data sets, and the remaining fold will be used as the testing data.

There are three experiments that will be carried out by using the Support Vector Machine and Naïve Bayes classifiers. However, the other experiments that involve the k-NN classifier will be carried out by using the Euclidean distance and Cosine similarity and also with various number of nearest neighbors, $k = \{1, 3, 5, 7, 9, 11\}$, in order to determine the best value of k that produces the best prediction performance. This is because if the value of k selected is too large, outliers from dominating class will be included in the estimation process and the prediction accuracy of the k-NN classifier is degraded. On the other hand, if the value of k chose is too small then ambiguous, noisy or miss-labelled documents come into local estimation boundary for class estimation of test document [15].

$$Accuracy = \frac{TP + TN}{TP + TN + FN + FP} \tag{10}$$

The performance of the Support Vector Machine (SVM), Naïve Bayes (NB) and k-NN classifiers are evaluated by calculating the prediction accuracy as shown in Eq. (10). The prediction accuracy is calculated based on the number of true positive (TP), false positive (FP), true negative (TN) and false negative (FN). True positive indicates that the documents are classified correctly into positive class as predefined class of positive whereas false negative indicates that the documents are classified incorrectly into negative class as predefined class of positive. On the other hand, true negative indicates that the documents are classified correctly into negative class as predefined class of negative and false positive indicates that the documents are classified incorrectly into positive class as predefined class of negative.

Based on the experimental results obtained as shown in Table 1, well trained supervised machine learning algorithms can perform very good classification on sentiment analysis of Malay news headlines. In terms of accuracy, SVM and NB classifiers can reach more than 80 % (up to 84.8 % and 82.2 % average accuracies) of the prediction classification correctly. For the k-NN classifier, using the Euclidean distance allows user to achieve the average accuracy in the range of 59.8 % to 67 % and using the Cosine similarity allows user to achieve the average accuracy in the range of 78.0 % to 79.8 % where $k=1$, 3, 5, 7, 9 and 11. SVM and NB classifiers performed better than k-NN classifier in this work. This is due to the fact that NB classifier assumes that attributes are conditionally independent to each other given the class. It is one of the oldest formal classification algorithms, and yet even in its simplest form it is often surprisingly effective [17, 18]. Based on the findings, it can be stated that the prediction performance of the k-NN classifier is affected by the number of nearest neighbour, k. When the value of k value is too large, the result can be sensitive to noise points. In other words, the k-NN classifier may include too many points from other classes if the value of k is too large.

Table 1. Prediction accuracies for each different experiment

Classifier			Accuracy (%)											
	SVM	NB					k-NN							
Experiments			Euclidean Distance						Cosine Similarity					
			k=1	k=3	k=5	k=7	k=9	k=11	k=1	k=3	k=5	k=7	k=9	k=11
1	81.5	79.5	66.0	61.0	58.5	59.0	59.0	58.0	80.0	78.5	80.0	78.5	78.5	78.5
2	87.3	85.0	70.0	65.0	62.5	62.0	61.0	61.0	80.5	77.0	80.5	80.0	77.0	78.5
3	85.7	82.0	65.0	59.0	58.5	58.0	59.0	59.0	79.0	78.5	79.0	78.5	78.5	79.0
Mean	84.8	82.2	67.0	61.7	59.8	59.8	59.7	59.3	79.8	78.0	79.8	79.0	78	78.7

When comparing the proximity measurements used in k-NN, using the Cosine similarity as the proximity measurement in k-NN has produced better prediction accuracy results compared to when using the Euclidean distance. Based on the results obtained, the prediction performance of the k-NN coupled with the Euclidean distance is decreasing as the number of nearest neighbours increases. A more consistent average of prediction accuracy is achieved with 59.8 % when k=5 and k=11 for the k-NN coupled with the Euclidean distance. The prediction performance of the k-NN coupled with the Cosine similarity decreases as the number of nearest neighbour increases. The prediction performance of the k-NN coupled with the Cosine similarity is high when k is lower where the average of the mean achieved at 79.8 %. The Cosine proximity measure is found to be more suitable to be used rather than the Euclidean distance for the sentiment prediction of the Malay news headlines because some distance measures can be affected by the high dimensionality of the data [18]. Euclidean distance tends to find a high-dimensional representation of the documents whereas cosine similarity tends to find a low-dimensional representation of documents. Euclidean distance measure becomes less discriminating as the number of attributes increases. In this paper, most of the Malay news headlines are classified into negative class in Euclidean distance as it is being dominated by negative class where it sorts the training documents in order to find the number of nearest neighbours for the test documents.

5 Conclusion

The process of pre-processing is very important to obtain a clean data from noise data. TF-IDF has been used as a weighting scheme in this paper in order to calculate the proximity measurement for the k-NN classifier. Support Vector Machine and Naïve Bayes algorithms only require a small amount of training data to estimate the parameters necessary for classification. Support Vector Machine (SVM), Naïve Bayes (NB) and k-NN classifiers are capable to automatically identify and categorize the Malay news headlines into positive and negative sentiment. Nevertheless, Support Vector Machine and Naïve Bayes classifiers are found to be able to produce higher prediction accuracies compared to the k-NN algorithm. Both the SVM and NB classifiers are simple and efficient to train and to test the data. However, for the k-NN algorithm, the prediction performance is affected by the type of proximity measurement used and the value of k used. The k-NN algorithm coupled with the Cosine similarity showed better result compared to the k-NN algorithm coupled with the Euclidean distance.

A large number of distinct terms are produced but some of the distinct terms are not relevant for sentiment analysis process such as the words of "MAS", "RM", and "PM". As a result, the dimension of the feature vector space is high and this affects the efficiency of the classifiers when these useless distinct terms are included in the documents' representation. Feature selection techniques can be applied in order to reduce these irrelevant features. A feature selection technique can be used to remove irrelevant features for sentiment analysis. However, different feature selection methods will give different prediction performance. The best option is to combine several feature selection techniques that can produce the best set of relevant features that can be used to improve the prediction performance.

References

1. Liu, B.: Sentiment Analysis and Opinion Mining. Morgan & Claypool Publishers, San Rafael (2012)
2. Cassinelli, A., Chen, C.-W.: CS 224 N Final Project Boost up! Sentiment Categorization with Machine Learning Techniques. Stanford University: The Stanford Natural Language Processing Group (2009)
3. Gebremeskel, G.: Sentiment Analysis of Twitter posts about news. University of Malta: Department of Computer Science and Artificial Intelligence (2011)
4. Thelwall, M., Buckley, K., Paltoglou, G.: Sentiment in twitter events. J. Am. Soc. Inform. Sci. Technol. **62**(2), 406–418 (2011)
5. Noah, S.A., Ismail, F.: Automatic classifications of Malay proverbs using naïve bayesian algorithm. Inf. Technol. J. **7**(7), 1016–1022 (2008)
6. Kaur, J., Saini, J.R.: An analysis of opinion mining research works based on language, writing style and feature selection parameters. Int. J. Adv. Netw. Appl. (2013)
7. Naradhipa, A.R., Purwarianti, A.: Sentiment classification for indonesian message in social media. In: International Conference on Electrical Engineering and Informatics 17–19 July, Bandung, Indonesia (2011)
8. Jamal, N.: Masnizah mohd and shahrul azman noah: poetry classification using support vector machines. J. Comput. Sci. **8**(9), 1441–1446 (2012)
9. Alsaffar, A., Omar, N.: Study on feature selection and machine learning algorithms for Malay sentiment classification. In: ICIMU2014, Putrajaya, Malaysia (2014)
10. Zhang, W., Gao, F.: An improvement to naive bayes for text classification. Proc. Eng. **15**, 2160–2164 (2011)
11. Multilingual sentiment-Data Science Labs. Accessed https://sites.google.com/site/datascienceslab/projects/multilingualsentiment
12. Kwee, A.T., Tsai, F.S., Tang, W.: Sentence-level novelty detection in English and Malay. In: Theeramunkong, T., Kijsirikul, B., Cercone, N., Ho, T.-B. (eds.) PAKDD 2009. LNCS, vol. 5476, pp. 40–51. Springer, Heidelberg (2009)
13. Raschka, S.: Naive Bayes and Text Classification: Introduction and Theory. Cornell university library, Ithaca (2014)
14. Kalaivai, P.: Sentiment classification of movie reviews by supervised machine learning approaches. Indian J. Comput. Sci. Eng. (IJCSE) **4**(4), 317–323 (2013)
15. Patel, F.N., Soni, N.R.: Increasing accuracy of k-NN classifier for text classification. Int. J. Comput. Sci. Inform., ISSN (PRINT) **3**(2), 2231–5292 (2013)
16. Khamar, K.: Short text classification using kNN based on distance function. Int. J. Adv. Res. Comput. Commun. Eng. **2**(4) (2013)

17. Ashari, A., Paryudi, I., Tjoa, A.M.: Performance comparison between naïve bayes, decision tree and k-nearest neighbor in searching alternative design in an energy simulation tool. Int. J. Adv. Comput. Sci. Appl. **4**(11) (2013)
18. Wu, X., Kumar, V., Quinlan, J.R., Ghosh, J. Yang, Q., Motoda, H.: Top 10 algorithms in data mining. © Springer-Verlag London Limited (2007)
19. Joachims, T.: Text categorization with support vector machines: learning with many relevant features. In: Nédellec, C., Rouveirol, C. (eds.) ECML 1998. LNCS, vol. 1398. Springer, Heidelberg (1998)

Assessing Factors that Influence the Performances of Automated Topic Selection for Malay Articles

Rayner Alfred[✉], Leow Jia Ren, and Joe Henry Obit

Faculty of Computing and Informatics, Universiti Malaysia Sabah, Sabah, Malaysia
{ralfred,joehenry}@ums.edu.my, jiaren23@gmail.com

Abstract. Malay language is a major language that is in used by citizens of Malaysia, Indonesia, Singapore and Brunei. As the language is widely used, there are abundant of text articles written in Malay language that are available on the internet. This has resulted in the increasing of the Malay articles published online and the number of articles has increased greatly over the years. Automatically labeling Malay text articles is crucial in managing these articles. Due to lack of resources and tools used to perform the topic selection automatically for Malay text articles, this paper studies the factors that influence the performances of the algorithms that can be applied to perform a topic selection automatically for Malay articles. This is done by comparing the contents of the articles with the corresponding topics and all Malay articles will be assigned to the appropriate topics depending on the results of the classification process. In this paper, all Malay articles will be classified by using the k-Nearest Neighbors (k-NN) and Naïve Bayes classifiers. Both classifiers are used to classify and assign a topic to these Malay articles according to a predefined set of topics. The effectiveness of classifying these Malay articles using the k-NN classifier is highly dependent on the distance methods used and the number of Nearest Neighbors, k. Thus, this paper also assesses the effects of using different distance methods (e.g., Cosine Similarity and the Euclidean Distance) and varying the number of clusters, k. Other than that, the effects of utilizing the stemming process on the performance of the classifiers are also studied. Based on the results obtained, the proposed approach shows that the k-NN classifier performs better than the Naïve Bayes classifier in classifying the Malay articles into their respective topics. In addition to that, the stemming process also improves the overall performances of both classifiers. Other findings include the application of Cosine Similarity as the distance measure has improved the performance of the k-NN classifier.

Keywords: Topic selection · Feature extraction · Classification · Clustering

1 Introduction

Malay language is one of the languages that is widely spoken by people in the Southeast Asia especially in Malaysia, Singapore, Brunei, Indonesia and South of Thailand [18]. In fact, Malay language is the major language that is used by the citizens of Malaysia, Singapore and Brunei. As the language is widely used, there are abundant of text articles written Malay language and they are available on the internet [12]. The number of Malay

© Springer Nature Singapore Pte Ltd. 2016
M.W. Berry et al. (Eds.): SCDS 2016, CCIS 652, pp. 300–309, 2016.
DOI: 10.1007/978-981-10-2777-2_27

articles published on the internet increases dramatically due to the advancement of internet-related technologies. As the number of these Malay articles or documents increases greatly, the efficiency of documents classification based on a predefined set of topics is becoming more important [19].

Topic selection is defined as the task of selecting the appropriate topic for an article based on the contents of the article. Topic selection has becoming as one of the important techniques in the studies as it enables documents to be assigned to a specific topic category from the set of predefined topics [5]. In other researches, topic selection is also known as topic identification [5], topic spotting [20], text categorization [9], or text classification [2]. A conventional topic selection framework consists of a pre-processing, feature extraction, feature selection and classification stages [5]. This paper studies the effectiveness of the topic selection methods using the k-NN and Naïve Bayes classifiers. The effectiveness of classifying these Malay articles using the k-NN classifier is highly dependent on the distance methods used and the number of Nearest Neighbors that is used, k. Thus, this paper also assesses the effects of using different distance methods (e.g., Cosine Similarity and the Euclidean Distance) and varying the values of k. Other than that, the effects of utilizing the stemming process on the performance of the classifiers are also studied.

The rest of this paper is organized as followed. Section 2 explains some of related works. Section 3 outlines the framework of the classification based topic selection for Malay articles. Section 4 describes the experimental setup and discusses the results obtained in investigating the effects of using different distance methods (e.g., Cosine Similarity and the Euclidean Distance) and varying the values of k on the performance of the k-NN classifier. The effects of utilizing the stemming process on the performance of the classifiers are also studied. Finally, Sect. 5 concludes this paper.

2   Related Works

There are researches conducted and models proposed that investigate methods to find the best and optimum way to perform the topic selection process for English articles [23]. However, there are not many works conducted that focus on the task of performing topic selection for Malay articles [3]. Thus, this has motivated this paper to propose a general framework of topic selection for Malay articles. Nevertheless, a study has been conducted to assess the performances of several machine learning methods coupled with several different features selection methods in performing the topic selection for Malay articles [3]. Based on the results obtained, the k-Nearest Neighbor classifier performance is found to be better compared to the others classifier. However, the text preprocessing process had being neglected in the experiments done by Alshalabi. Text pre-processing is a vital process in any topic selection or text classification as it is the main task of extracting the features from the text documents. It is also known as the feature extraction process in topic selection as generally the features exist in the text documents are all words [1]. The importance of this process can be seen as many similar researches with tasks involving text documents include this process [1, 11]. Text pre-processing is the first step in a topic extraction or topic selection process as it reduces the noise in the documents or texts by removing the unnecessary terms [13]. The following includes the three pre-processing tasks such as tokenization, stop words removal, punctuation removal, stemming and finally the construction of document-texts matrix that represents the Term-Frequency and Inverse Document Frequency (TF-IDF) [2, 24].

Tokenization is one of the lexical analysis techniques. Tokenization is a process of forming tokens from an input stream of characters where a token is a string of one or more characters. Tokenization is the process of breaking a stream of text into tokens that can be words, sentences, phrases, symbols or other meaningful elements or terms to help contribute to the task that is under study [5]. Stop words removal is a process of removing the words that have little lexical meaning in the documents [5]. These words are unlikely to contribute to the distinctiveness of the texts and examples of the stop words are the conjunctions, pronouns and prepositions words. Punctuation removal is a process of removing all the punctuation that exists in the text during the text pre-processing. Punctuation is the usage of the special characters such as "/", "?", "!", ";" and many more. The uses of these special characters are to enhance and disambiguate the meaning of a sentence and those characters should be removed as it does not bring any lexical meaning which contributes to the analysis of topic selection to identify the topic of the text. Stemming refers to the transformation of a word from its root words to its stem. The stemming algorithm used to stem Malay text was different with those used to stem English text [15]. In order to stem Malay text, the suffixes, prefixes and infixes must be removed in a proper order [15]. As the Malay words are formed with a combination containing one or more syllable, thus is possible to perform stemming by analyzing the pattern of the syllable than analyzing the word character by character [6]. Lee's proposed stemmer consists of two components where first they perform the process of syllabification where it separate a Malay word into its syllables and the second component is the Malay stemming rules where it is used to identify the morphological structures of the Malay word to be stemmed. Term Frequency– Inverse Document Frequency (TF-IDF) is one of the terms weighting model that is used in the text processing to represent a document. It includes the use of the bag-of-words model, which is widely used in Information Retrieval (IR) and text mining [2]. TF-IDF is often used and also is the most popular to be used as the weighting scheme. Term Frequency, TF is the frequency that the one term t in a document d, represents the count of the particular term exists in the document. Inverse Frequency Document, IDF is the inverse document frequency of that term t across all the documents in the document sets [16]. The IDF influences the weighting of the terms that exist in all the documents in the document sets as the IDF value for the terms that exists in all documents will be zero or less, resulting the TF-IDF value will become insignificant.

In determining the topic label of a single document, a text classification can be used to classify the documents into the appropriate topic. Text classification is a task of assigning a category to any single piece of text or document or articles. Thus, the text classification technique also is used to automate the topic selection as it allows the documents to be identified by a label, subject or topic. Text classification also allows the grouping of the documents by common topics. There are two most commonly used algorithms which is the k-Nearest Neighbors (k-NN) algorithm and the Naïve Bayes algorithm. The k-Nearest Neighbor algorithm, k-NN, is the most widely used in classification and clustering algorithm as it is simple, fast and effective [22]. It is also known as the lazy learning algorithm [10]. The k-NN algorithm is also applied as a classifier and k-NN classifier is also one of the favorite and widely used classifiers due to its ease in implementation yet powerful as it often shows good performance [21]. The k-NN classifier will classify the text documents accordingly the k number of neighbors. It classifies the text documents to the class or topic with the most votes which are determined based on the k nearest neighbors [4]. Naïve Bayes is a probabilistic classification that based on the Bayesian probability. It requires training data to learn the classes' probabilities. This classifier is called naïve as the Naïve Bayes classification is a classifier that assumes all the attributes are conditionally independent given the classes [17]. Despite that the naïve assumptions and its simplicity to be implemented, the Naïve Bayes classifier proved to be an effective classifier [7]. This classifier classifies a document, d in to a class, c by learning a probability value from the training data [14].

3 Topic Selection for Malay Articles

This section outlines the framework used to automate the topic selection for Malay articles.

3.1 Text Processing Tasks

Several text preprocessing tasks are described before classifying the articles into its respective topics. Three major tasks will be performed in this phase which is the tokenization, punctuation removal, Malay stop words removal and Malay words stemming. For the stop words removal, all the stop words found in the predefined list are removed. Example of stop words includes "*adalah*", "*ialah*", "*dan*" and many more. Malay words stemming is another text preprocessing task that will remove all the prefixes and suffixes from a Malay term or word so that the term will be transformed back into its root word. This paper will also investigate the effects of the stemming process on the topic selection for Malay articles. The following shows the list of prefix and suffix [6] (Table 1).

Table 1. List of Prefixes and Suffixes

Type	Substring
Prefix	'*ber*', '*per*', '*ter*', '*mem*', '*pem*', '*meng*', '*peng*', '*men*', '*pen*', '*pe*', '*me*', '*be*', '*ke*', '*se*', '*te*', '*di*'
Suffix	'*nya*', '*kan*', '*an*', '*i*', '*kah*', '*lah*', '*pun*', '*man*', '*ku*', '*mu*'

The following is the procedures of the Malay stemming algorithm that is used to perform the stemming process.

Step 1: Get the term from the term list.

Step 2: Check if the term is a root word by checking its number of letters where the threshold set is 5. If it is less than 5 letters, it is the root word, then exit the stemming process or else proceed.

Step 3:   Check the first few letters against the Prefix list, if a match found, remove the prefix and add the appropriate letter in front if necessary.

Step 4:   Check the term's letter threshold, if it is less than 5, exit or else proceed.

Step 5:   Check the last few letters against the Suffix list, if a match found, remove the suffix.

Step 6:   Check the term's letter threshold, if it is less than 5, exit or else proceed.

Step 7:   Repeat the whole process again from Step 2 to remove multiple Prefix and Suffix on the term.

There are several more rules that will be needed as some prefixes will modify the first letter of the root word when the prefix is added to it. Thus, there is a set of rules for the first letter modification when applying the stemming as to restore the word back to its root word.

Rule 1:   if the first letter is 'm', then change it to 'p' or 'f'.

Rule 2:   if the first letter is 'n', then change it to 't'.

Rule 3:   if the first letter is 'y', then change it to 's'.

Rule 4:   if the first letter is vowel ('a', 'e', 'i', 'o', 'u'), then add 'k' at the front.

3.2   Term Weighting

A weight is assigned to each term by using the TF-IDF term weighting scheme. The weighting was done by applying the TF-IDF computation which is shown below.

$$TF - IDF = \left(1 + \log f_{i,j}\right) * \log \frac{N}{n_i} \tag{1}$$

where, $f_{i,j}$ refers to the frequency of the term t_i in the document d_j, N is the total number of the documents and n_i refers to the number of n documents that the term t_i exist in.

3.3   Identification of Topics Based on Classification

The topic selection for each article will be done by using the k-Nearest Neighbor and Naïve Bayes classifier. The k-NN classifier will classify the document into its topics or classes based on the classes of the k nearest neighbors of the document. The k-NN classifier uses the distance measure to compute the distance between documents. In this paper, the distance measure that will be used for comparison in their effectiveness is the Cosine similarity and the Euclidean distance methods. The Cosine Similarity is one of the methods used in order to compute the distance between two documents by computing the degree of similarity of the two documents. The Cosine similarity will produce a value where the closer the value to 1 means that both of the documents are similar to each other. The formula of the cosine similarity is shown below.

$$S_{dj,cp} = \sum_{dt \in N_k(d_j)} similarity(d_j, d_t) \times T(d_t, c_p) \tag{2}$$

where, $N_k(d_j)$ refers to the set of the k nearest neighbors in d_j in the training set, *similarity(d_j,d_t)* refers to the Cosine formula for vector model and finally, $T(d_t,c_p)$ refers to the function that returns value of 1, if d_j is belongs to class c_p and 0 otherwise. The cosine formula is denoted as

$$sim\left(d_j, d_t\right) = \frac{d_j.d_t}{\left|d_j\right| \times \left|d_t\right|} \tag{3}$$

where, d_j is a vector of document j and d_t is a vector of training document t.

Euclidean distance is also another method which is commonly used to calculate the distance between two documents. The smaller value of Euclidean distance between two documents indicates that they are close to one and another. The Euclidean distance is computed as shown below.

$$Dist_{d_i,d_j} = \sqrt{\sum_{N=1}^{N} (x_{d_i} - x_{d_j})^2} \tag{4}$$

where, $Dist_{d_i,d_j}$ refers to the distance of d_i from d_j, N is number of terms or features in d_i and x_{d_i} and x_{d_j} are terms or features exists in the documents.

The Naïve Bayes will also be applied to classify the text documents by computing the probability of the document with respect to the class. It will assign the document to the class with the highest probability value computed [17].

The evaluation method used to evaluate the performance of each of the proposed approach is by applying the accuracy measure. It is a simply accuracy evaluation which computes the percentage of the documents whether it is correctly classified or not.

$$Accuracy = \frac{Number\ documents\ correctly\ assigned}{Total\ number\ of\ test\ documents} \times 100\% \tag{5}$$

4   Experimental Setup

This section describes the experiments that are carried out in this work. The experiments are outlined in order to achieve the following objectives,

(a) To investigate the effectiveness of the applied stemming algorithm in reducing the number of terms or features extracted.
(b) To investigate the best k value for the k-NN classifier.
(c) To identify which classifier performs better in the proposed approach.

There are 1000 Malay news articles that will be used in this study. These news articles are retrieved from *Bernama* archive and *theStar* website. The news articles are retrieved and annotated manually in order to make sure that these documents are categorized into the appropriate topic labels. These topic labels include Financial, Politics, Sports, Entertainments and General. The following describes how the experiments are carried out.

Experiment 1: The dataset will not undergo the stemming process and the distance method used to compute the distance is the Cosine Similarity and the experiment is repeated with different values of k for k-NN classifier where $k = 1, 3, 5, 7$.

Experiment 2: Experiment 1 is repeated but using Euclidean distance.

Experiment 3: Repeat Experiment 1 by adding the stemming process into the pre-processing phase. Then, compute the distance by using the Cosine Similarity and repeat the experiment with different values of k for k-NN classifier.

Experiment 4: Repeat Experiment 3 by changing the distance computation with Euclidean distance computation.

Experiment 5: Repeat Experiments 1 and 3 by changing the Naïve Bayes classifier.

For the experiments, each of the experiment is validated by using the 2/3 fold cross validation method. The data sets are firstly divided into 3 different sets so that a different set of documents will be used as the training and test data while performing the cross validation.

The results shows that k-NN classifier performs better when the stemming process is performed prior to the classification task, the Cosine similarity is used as the distance measure and when $k = 3$. The k-NN classifier has a higher accuracy with Cosine similarity as the distance measure compare to Euclidean distance as the distance measure. Thus, a conclusion can be drawn from the result in Table 2 is that the k-NN classifier performs a better accuracy in classifying the documents when it uses cosine similarity as its distance measure, when its $k = 3$ and with the stemming algorithm.

Table 2. Prediction accuracies obtained for the k-NN classifier with different approaches

	Without Stemming								With Stemming							
	Cosine Similarity				Euclidean Distance				Cosine Similarity				Euclidean Distance			
k, (k-NN)	1	3	5	7	1	3	5	7	1	3	5	7	1	3	5	7
CV1	97.3	**97.3**	96.7	97.3	95.3	96.0	**95.3**	95.3	97.3	**98.0**	96.7	96.7	95.3	96.0	**94.7**	94.7
CV2	92.0	**96.7**	94.7	94.0	84.7	90.0	**90.7**	88.7	92.7	**98.7**	96.7	94.7	83.3	93.3	**93.3**	92.0
CV3	96.0	**97.3**	96.0	95.3	94.0	94.7	**94.7**	89.3	96.7	**98.0**	97.3	96.7	93.3	94.7	**94.7**	89.3
Mean	95.1	**97.1**	95.8	95.5	91.3	93.6	**93.6**	91.1	95.6	**98.2**	96.9	96.0	90.6	94.7	**94.2**	92.0
Std. Dev	2.76	**0.35**	1.01	1.66	5.78	3.15	**2.50**	3.65	2.50	**0.40**	0.35	1.15	6.43	1.35	**0.81**	2.70

On the other hand, the performance of the Naïve Bayes classifier also performs better with the stemming algorithm. The accuracy of the classifier is higher when it applied the stemming algorithm, 88.7 % ± 3.46 compared to without stemming which is 83.1 % ± 5.37. However, the result also shows that the k-NN classifier outperforms the Naïve Bayes classifier. The k-NN classifier has the accuracy above 90 % for all the approaches while Naïve Bayes classifier only reaches 83.1 % and 88.7 % in both approaches, with and without performing the stemming process. This shows the k-NN classifier was indeed performed better than Naïve Bayes classifier in topic selection of Malay articles (Table 3).

Table 3. Prediction accuracies obtained for the k-NN and Naive Bayes classifiers

	k-Nearest Neighbour Classifier				Naive Bayes Classifier	
	Without Stemming		With Stemming		Without Stemming	With Stemming
	Cosine Similarity	Euclidean Distance	Cosine Similarity	Euclidean Distance		
CV1	97.2	95.5	97.2	95.2	80.0	86.7
CV2	94.4	88.5	95.7	90.5	89.3	92.7
CV3	96.2	93.2	97.2	93.0	80.0	86.7
Mean	**95.9**	92.4	**96.7**	92.9	83.1	**88.7**
Std. Dev	1.42	3.54	0.85	2.35	5.37	3.46

The results also show that the stemming does help improving the performances of the classifier. Performing the stemming process also improves the prediction classification by reducing the number of features or terms in each document.

5   Conclusion

In conclusion, the proposed approach for topic selection for Malay articles has been described in detailed in this paper. Based on the results obtained from the experiments carried out in this paper, the proposed approach to topic selection for Malay articles by performing only feature extraction does show promising results. The k-NN classifier is the best classifier to be used for topic selection of Malay articles. The k-NN classifier also performs better with Cosine Similarity as its distance method and the best k value for the k-NN classifier was also identified in the experiment where the best value for k was 3. The obtained prediction performance is found to be better when the stemming process is performed prior to the classification process as redundant features are eliminated that could decrease the prediction performance. This can be seen in the results obtained from the experiments where both k-NN classifier and Naïve Bayes classifiers have improvement in the accuracy when stemming algorithm was included. However, the stemming algorithm still required refinement and it can be further improve. As mentioned before, a suggestion on improving the stemming algorithm will be adding the use of root word dictionary which is a list of root words so that the modification of the words after removal of prefix can be done correctly and effectively.

References

1. Khan, A., Baharudin, B., Lee, L.H., Khan, K.: A review of machine learning algorithms for text-documents classification. J. Adv. Inf. Technol. **1**(1), 4–20 (2010)
2. Baeza-Yates, R.A., Ribeiro-Neto, B.A.: Modern Information Retrieval, edn. 2. ACM Press Books, Addison-Wesley Professional ISBN-10: 0321416910 (2011)
3. Salim, J., Ismail, M., Suwarno, I., Alshalabi, H., Tiun, S., Omar, N., Albared, M.: Experiments on the use of feature selection and machine learning methods in automatic malay text categorization. Procedia Technol. **11**, 748–754 (2013). ISSN 2212-0173

4. Uguz, Harun: A two-stage feature selection method for text categorization by using information gain, principal component analysis and genetic algorithm. Knowl.-Based Syst. **24**(7), 1024–1032 (2011)

5. Echeverry-Correa, J.D., Ferreiros-López, J., Coucheiro-Limeres, A., Córdoba, R., Montero, J.M.: Topic identification techniques applied to dynamic language model adaptation for automatic speech recognition. Expert System with Applications **42**(1), 101–112 (2015). ISSN: 0957-4174

6. Lee, J., Othman, R.M., Mohamad, N.Z.: Syllable-based Malay word stemmer, computers & informatics (ISCI). In: 2013 IEEE Symposium on, Langkawi, pp. 7–11 (2013). doi:10.1109/ISCI.2013.6612366

7. Jiang, L., Zhang, H.: Learning instance greedily cloning naive Bayes for ranking. In: Fifth IEEE International Conference on Data Mining (ICDM 2005), pp. 8–12 (2005). doi:10.1109/ICDM.2005.87

8. Sankupellay, M., Subbu, V.: Malay-language stemmer. Sunway Academic J. **3**, 147–153 (2006)

9. Manning, C.D., Raghavan, P., Schütze, H.: Introduction to Information Retrieval. Cambridge University Press, Cambridge (2008). ISBN-10: 0521865719

10. Meenakshi, Singla, S.: Review paper on text categorization techniques. Int. J. Innovative Res. Comput. Commun. Eng. **3**(11), 809–813 (2015). ISSN: 2320-9801

11. Samat, N.A., Murad, M.A.A., Abdullah, M.T., Atan, R.: Malay documents clustering algorithm based on singular value decomposition. Int. J. Comput. Sci. Netw. Sec. (IJCSNS) **8**(10), 357–361 (2008)

12. Ismail, N.K., Saad, N.H.M., Omar, S.B.S., Sembok, T.M.T.: 2D visualization of terms and documents in Malay language. In: 5th International Conference on Information and Communication Technology for the Muslim World (ICT4 M), Rabat, pp. 1–6 (2013). doi:10.1109/ICT4M.2013.6518919

13. Koulali, R., El-Haj, M., Meziane, A.: Arabic topic detection using automatic text summarisation. In: 2013 ACS International Conference on Computer Systems and Applications (AICCSA), Ifrane, pp. 1–4 (2013). doi:10.1109/AICCSA.2013.6616460

14. Thakur, S.K., Singh, V.K.: A lexicon pool augmented naive bayes classifier for nepali text. In: 2014 Seventh International Conference on Contemporary Computing (IC3), Noida, pp. 542–546 (2014). doi:10.1109/IC3.2014.6897231

15. Sembok, T.M.T., Bakar, Z.A., Ahmad, F.: Experiments in Malay information retrieval. In: 2011 International Conference on Electrical Engineering and Informatics (ICEEI), Bandung, pp. 1–5 (2011). doi:10.1109/ICEEI.2011.6021578

16. Yong-qing, W., Pei-yu, L., Zhen-fang, Z.: A feature selection method based on improved TFIDF. In: Third International Conference on Pervasive Computing and Applications, 2008, ICPCA 2008, Alexandria, pp. 94–97 (2008). doi:10.1109/ICPCA.2008.4783657

17. Qin, Z.: Naive bayes classification given probability estimation trees. In: 2006 5th International Conference on Machine Learning and Applications (ICMLA 2006), Orlando, FL, pp. 34–42 (2006). doi:10.1109/ICMLA.2006.36

18. Sharum, M.Y., Abdullah, M.T., Sulaiman, M.N., Murad, M.A.A., Hamzah, Z.A.Z.: MALIM — A new computational approach of malay morphology. In: 2010 International Symposium on Information Technology, Kuala Lumpur, pp. 837–843 (2010). doi:10.1109/ITSIM.2010.5561561

19. Ko, Y., Seo, J.: Automatic text categorization by unsupervised learning. In: Proceedings of the 18th conference on Computational linguistics - Volume 1 (COLING 2000). Association for Computational Linguistics, Stroudsburg, PA, USA, pp. 453–459 (2000). doi:http://dx.doi.org/10.3115/990820.990886

20. Wiener, E., Pedersen, J.O., Weigend, A.S.: A neural network approach to topic spotting. In: Fourth Annual Symposium on Document Analysis and Information Retrieval (SDAIR 1995) (1995)
21. Viswanath, P., Hitendra Sarma, T.: An improvement to k-nearest neighbor classifier. In: IEEE Recent Advances in Intelligent Computational Systems (RAICS), Trivandrum, 2011, pp. 227–231 (2011). doi:10.1109/RAICS.2011.6069307
22. Qu, C., Yuan, R., Wei, X.: KNNCC: an algorithm for k-nearest neighbor clique clustering. In: 2013 International Conference on Machine Learning and Cybernetics, Tianjin, pp. 1763–1766 (2013). doi:10.1109/ICMLC.2013.6890883
23. Tanha, J., de Does, J., Depuydt, K.: An LDA-based topic selection approach to language model. In: Adaptation for Handwritten Text Recognition, Proceedings of Recent Advances in Natural Language Processing, pp. 646–653, Hissar, Bulgaria, September 7–9 (2015)
24. Leong, L.C., Basri, S., Alfred, R.: Enhancing Malay Stemming Algorithm with Background Knowledge. In: Anthony, P., Ishizuka, M., Lukose, D. (eds.) PRICAI 2012. LNCS, vol. 7458, pp. 753–758. Springer, Heidelberg (2012)

Author Index

Printed in the United States
By Bookmasters